Springer Proceedings in Physics

Volume 244

Indexed by Scopus

The series Springer Proceedings in Physics, founded in 1984, is devoted to timely reports of state-of-the-art developments in physics and related sciences. Typically based on material presented at conferences, workshops and similar scientific meetings, volumes published in this series will constitute a comprehensive up-to-date source of reference on a field or subfield of relevance in contemporary physics. Proposals must include the following:

- name, place and date of the scientific meeting
- a link to the committees (local organization, international advisors etc.)
- scientific description of the meeting
- list of invited/plenary speakers
- an estimate of the planned proceedings book parameters (number of pages/articles, requested number of bulk copies, submission deadline).

More information about this series at http://www.springer.com/series/361

Alexander D. Pogrebnjak · Maksym Pogorielov ·
Roman Viter
Editors

Nanomaterials in Biomedical Application and Biosensors (NAP-2019)

Springer

Editors
Alexander D. Pogrebnjak
Department of Nanoelectronics
Sumy State University
Sumy, Ukraine

Maksym Pogorielov
Center of Collective Use
of Scientific Equipment
Sumy State University
Sumy, Ukraine

Roman Viter
Optical Biosensors and Functional
Nanomaterials Laboratory
University of Latvia
Riga, Latvia

ISSN 0930-8989 ISSN 1867-4941 (electronic)
Springer Proceedings in Physics
ISBN 978-981-15-3995-4 ISBN 978-981-15-3996-1 (eBook)
https://doi.org/10.1007/978-981-15-3996-1

This Springer imprint is published by the registered company Springer Nature Singapore Pte Ltd.
The registered company address is: 152 Beach Road, #21-01/04 Gateway East, Singapore 189721, Singapore

Preface

This book covers novel and innovative technologies used for development, modeling, chemical and physical investigation and biomedical (in-vitro and in-vivo) trials of nanomaterials and nanocomposites for medical applications and sensors. Novel method for nanoparticle development and manufacturing highlighted as well as their safety and promising application are under consideration. This book opens a new frontier in metal, metal oxide nanoparticle, hierarchical nanostructures and organic coatings as a sensor for gases, inorganic and organic materials, including biosensors for bacteria and cancer detection. Organic nanoparticle composites for medical application (tissue engineering, tissue replacement, regeneration, etc.) including hydroxyapatite-NPs are under the special focus, including in-vitro and preclinical investigation. Nanoparticle and nanocomposites for antibacterial application are discussed in the present book with a detailed focus on NPs–bacteria interaction and cell toxicity study. Orthopedic and dental implant coatings discussed and detailed described their biological effect and safety.

Sumy, Ukraine — Alexander D. Pogrebnjak
Sumy, Ukraine — Maksym Pogorielov
Riga, Latvia — Roman Viter

Contents

Contributors

M. A. Abdulzhanova Al-Farabi Kazakh National University, Almaty, Kazakhstan

K. Akatan East Kazakhstan State University named after S. Amanzholov, Oskemen, Kazakhstan

V. V. Andryushchenko Department of Surgery, Traumatology, Orthopedics and Phthisiatry, Sumy State University, Sumy, Ukraine

M. Azarenkov V. N. Karazin Kharkiv National University, Kharkiv, Ukraine

Emine Bagci Faculty of Science, Department of Chemistry, University of Selcuk, Konya, Turkey

A. N. Belousov Laboratory of Applied Nanotechnology of Belousov, Belousov, Ukraine;
Kharkov Medical Academy of Postgraduate Education, Kharkiv, Ukraine;
Kharkov Regional Center of Blood Service, Kharkiv, Ukraine

E. Yu. Belousova Laboratory of Applied Nanotechnology of Belousov, Belousov, Ukraine;
Kharkov Medical Academy of Postgraduate Education, Kharkiv, Ukraine

A. V. Bilduykevich Institute of Physical Organic Chemistry, National Academy of Science of Belarus, Minsk, Belarus

I. A. Bolshakova Magnetic Sensor Laboratory, Lviv Polytechnic National University, Lviv, Ukraine

S. Bolshanina Sumy State University, Sumy, Ukraine

V. V. Boyko National University of Life and Environmental Sciences of Ukraine, Kiev, Ukraine

U. Bubniene Department of Physical Chemistry, Faculty of Chemistry and Geosciences, Vilnius University, Vilnius, Lithuania;
Laboratory of Nanotechnology, State Research Institute Center for Physical Sciences and Technology, Vilnius, Lithuania

M. V. Bulavin Frank Laboratory of Neutron Physics, Joint Institute for Nuclear Research, Dubna, Russia

V. Chishkala V. N. Karazin Kharkiv National University, Kharkiv, Ukraine

V. P. Chornii National University of Life and Environmental Sciences of Ukraine, Kiev, Ukraine;
Taras Shevchenko National University of Kyiv, Kiev, Ukraine

V. Deineka Sumy State University, Medical Institute, Sumy, Ukraine

I. Demchenko Faculty of Chemistry, University of Warsaw, Warsaw, Poland

M. A. Deshmukh Department of Physical Chemistry, Faculty of Chemistry and Geosciences, Vilnius University, Vilnius, Lithuania;
Faculty of Chemistry and Geosciences, NanoTechnas - Centre of Nanotechnology and Material Science, Vilnius University, Vilnius, Lithuania;
Department of Physics, RUSA - Center for Advanced Sensor Technology, Dr. Babasaheb Ambedkar Marathwada University, Aurangabad, Maharashtra, India

B. Dryhval Sumy State University, Sumy, Ukraine

T. Dubrava Institute for Problems of Cryobiology and Cryomedicine NASU, Kharkiv, Ukraine

S. Dudin V. N. Karazin Kharkiv National University, Kharkiv, Ukraine

J. Dudko Sumy State University, Sumy, Ukraine

L. Dulebova Technical University of Kosice, Kosice, Slovak Republic

T. Dychenko Sumy State University, Sumy, Ukraine

A. Goltcev Institute for Problems of Cryobiology and Cryomedicine NASU, Kharkiv, Ukraine

K. Grundsteins Institute of Atomic Physics and Spectroscopy, University of Latvia, Riga, Latvia

O. N. Grytsenko Lviv National Polytechnic University, Lviv, Ukraine

Ersin Guler Faculty of Science, Department of Chemistry, University of Selcuk, Konya, Turkey

V. Holubnycha Sumy State University, Medical Institute, Sumy, Ukraine

Ye. Husak Sumy State University, Medical Institute, Sumy, Ukraine

L. V. Ignatova Al-Farabi Kazakh National University, Almaty, Kazakhstan

V. D. Ivchenko Sumy National Agrarian University, Sumy, Ukraine

Ł. Janus Faculty of Chemical Engineering and Technology, Cracow University of Technology, Kraków, Poland

O. Kalantaryan V. N. Karazin Kharkiv National University, Kharkiv, Ukraine

A. Kistaubayeva Biotechnology department, Al-Farabi Kazakh National University, Almaty, Kazakhstan

S. Kononenko V. N. Karazin Kharkiv National University, Kharkiv, Ukraine

V. Korniienko Sumy State University, Medical Institute, Sumy, Ukraine

Ya. Ya. Kost Magnetic Sensor Laboratory, Lviv Polytechnic National University, Lviv, Ukraine

A. N. Kucherenko Lviv National Polytechnic University, Lviv, Ukraine

T. Kuech College of Engineering, University of Wisconsin-Madison, Madison, USA

S. A. Kulikov Frank Laboratory of Neutron Physics, Joint Institute for Nuclear Research, Dubna, Russia

M. A. Kumeda Sumy State University, Sumy, Ukraine

Ahmed Nuri Kursunlu Faculty of Science, Department of Chemistry, University of Selcuk, Konya, Turkey

M. Ya. Kuznetsova Lviv National Polytechnic University, Lviv, Ukraine

I. M. Lukavenko Department of Surgery, Traumatology, Orthopedics and Phthisiatry, Sumy State University, Sumy, Ukraine

N. Magazov Veritas, East Kazakhstan State Technical University named after D. Serikbayev, Oskemen, Kazakhstan

M. Makarova The Filatov Institute of Eye Diseases and Tissue Therapy, National Academy of Medical Sciences of Ukraine Odessa, Odessa, Ukraine

N. Malushin Research Institute of Physics, Odessa I. I. Mechnikov National University Odessa, Odessa, Ukraine

E. I. Malygon Kharkov Medical Academy of Postgraduate Education, Kharkiv, Ukraine;
Kharkov Regional Center of Blood Service, Kharkiv, Ukraine

A. S. Masyuk Lviv National Polytechnic University, Lviv, Ukraine

D. Matysek Faculty of Mining and Geology, Technical University of Ostrava, Ostrava, Czechia

J. Michalska Silesian University of Technology, Gliwice, Poland

O. Mishchenko Osteoplant Research and Development, Dębica, Poland

N. Molchaniuk The Filatov Institute of Eye Diseases and Tissue Therapy, National Academy of Medical Sciences of Ukraine Odessa, Odessa, Ukraine

V. S. Moravskyi Lviv National Polytechnic University, Lviv, Ukraine

S. Mulenko G. Kurdyumov Institute of Metal Physics NAS of Ukraine, Kiev, Ukraine

P. Myronov Sumy State University, Sumy, Ukraine

I. Mysiura V. N. Karazin Kharkiv National University, Kharkiv, Ukraine

S. G. Nedilko Taras Shevchenko National University of Kyiv, Kiev, Ukraine

D. Neumaier Advanced Microelectronic Center Aachen, AMO GmbH, Aachen, Germany

O. Oleshko Sumy State University, Medical Institute, Sumy, Ukraine

Y. Otarov Veritas, East Kazakhstan State Technical University named after D. Serikbayev, Oskemen, Kazakhstan

M. Otto Advanced Microelectronic Center Aachen, AMO GmbH, Aachen, Germany

Ya. Oztekin Department of Physical Chemistry, Faculty of Chemistry and Geosciences, Vilnius University, Vilnius, Lithuania;
Faculty of Chemistry and Geosciences, NanoTechnas - Centre of Nanotechnology and Material Science, Vilnius University, Vilnius, Lithuania;
Faculty of Science, Department of Chemistry, Selcuk University, Konya, Turkey

M. Piątkowski Faculty of Chemical Engineering and Technology, Cracow University of Technology, Kraków, Poland

I. Plikusiene Department of Physical Chemistry, Faculty of Chemistry and Geosciences, Vilnius University, Vilnius, Lithuania;
Laboratory of Nanotechnology, State Research Institute Center for Physical Sciences and Technology, Vilnius, Lithuania

T. V. Plisko Institute of Physical Organic Chemistry, National Academy of Science of Belarus, Minsk, Belarus

M. Pogorielov Sumy State University, Medical Institute, Sumy, Ukraine;
Osteoplant Research and Development, Dębica, Poland

A. Pogrebnjak Department of Nanoelectronics, Sumy State University, Sumy, Ukraine

L. N. Ponomarova Sumy State University, Sumy, Ukraine

Yu. I. Prylutskyy Taras Shevchenko National University of Kyiv, Kiev, Ukraine

R. Pshenychnyi Sumy State University, Sumy, Ukraine

M. I. Radishevskyi Magnetic Sensor Laboratory, Lviv Polytechnic National University, Lviv, Ukraine

J. Radwan-Pragłowska Faculty of Chemical Engineering and Technology, Cracow University of Technology, Kraków, Poland

A. Ramanaviciene Faculty of Chemistry and Geosciences, NanoTechnas - Centre of Nanotechnology and Material Science, Vilnius University, Vilnius, Lithuania

A. Ramanavicius Department of Physical Chemistry, Faculty of Chemistry and Geosciences, Vilnius University, Vilnius, Lithuania;
Laboratory of Nanotechnology, State Research Institute Center for Physical Sciences and Technology, Vilnius, Lithuania

I. Rassokha Institute for Problems of Cryobiology and Cryomedicine NASU, Kharkiv, Ukraine

V. Ratautaite Department of Physical Chemistry, Faculty of Chemistry and Geosciences, Vilnius University, Vilnius, Lithuania;
Laboratory of Nanotechnology, State Research Institute Center for Physical Sciences and Technology, Vilnius, Lithuania

U. Ritter Technical University of Ilmenau, Ilmenau, Germany

L. M. Rozhdestvenska Vernadsky Institute of General and Inorganic Chemistry, Ukrainian National Academy of Sciences, Kiev, Ukraine

V. Safonov V. N. Karazin Kharkiv National University, Kharkiv, Ukraine;
National Science Center "Kharkov Institute of Physics and Technology", Kharkiv, Ukraine

A. Sagidugumar Veritas, East Kazakhstan State Technical University named after D. Serikbayev, Oskemen, Kazakhstan

S. Saidildina Al-Farabi Kazakh National University, Almaty, Kazakhstan

A. Savchenko Sumy State University, Sumy, Ukraine

I. Savitskaya Biotechnology department, Al-Farabi Kazakh National University, Almaty, Kazakhstan

R. Savkina V. Lashkaryov Institute of Semiconductor Physics NAS of Ukraine, Kiev, Ukraine

V. P. Scherbatskyi Taras Shevchenko National University of Kyiv, Kiev, Ukraine

D. H. Shokatayeva Al-Farabi Kazakh National University, Almaty, Kazakhstan

F. M. Shurygin Magnetic Sensor Laboratory, Lviv Polytechnic National University, Lviv, Ukraine

W. Simka Silesian University of Technology, Gliwice, Poland

R. Skiba V. N. Karazin Kharkiv National University, Kharkiv, Ukraine

V. Skobeeva Research Institute of Physics, Odessa I. I. Mechnikov National University Odessa, Odessa, Ukraine;
Odessa I. I. Mechnikov National University, Odessa, Ukraine

M. S. Slobodyanyk Taras Shevchenko National University of Kyiv, Kiev, Ukraine

A. Smirnov V. Lashkaryov Institute of Semiconductor Physics NAS of Ukraine, Kiev, Ukraine

V. Smyntyna Department of Experimental Physics, Faculty of Mathematics, Physics and Information Technologies, Odessa I.I. Mechnikov National University, Odesa, Ukraine

O. Solodovnyk Sumy State University, Sumy, Ukraine

L. B. Sukhodub Sumy State University, Sumy, Ukraine

L. F. Sukhodub Sumy State University, Sumy, Ukraine

A. Talipova Biotechnology department, Al-Farabi Kazakh National University, Almaty, Kazakhstan

K. V. Terebilenko Taras Shevchenko National University of Kyiv, Kiev, Ukraine

A. Tereshchenko Department of Experimental Physics, Faculty of Mathematics, Physics and Information Technologies, Odesa National I.I. Mechnikov University, Odesa, Ukraine;
Department of Physical Chemistry, Faculty of Chemistry and Geosciences, Vilnius University, Vilnius, Lithuania;
Faculty of Chemistry and Geosciences, NanoTechnas - Centre of Nanotechnology and Material Science, Vilnius University, Vilnius, Lithuania

V. Tkachenko Research Institute of Physics, Odessa I. I. Mechnikov National University Odessa, Odessa, Ukraine

A. Turlybekuly Veritas, East Kazakhstan State Technical University named after D. Serikbayev, Oskemen, Kazakhstan

V. Ulyanov Odessa National Medical University, Odessa, Ukraine

O. I. V'yunov Vernadsky Institute of General and Inorganic Chemistry, Ukrainian National Academy of Sciences, Kiev, Ukraine

I. S. Vasil'evskii Institute of Functional Nuclear Electronics, National Research Nuclear University MEPhI, Moscow, Russia

O. V. Vasyliev Magnetic Sensor Laboratory, Lviv Polytechnic National University, Lviv, Ukraine

N. Voloshin National Technical University of Ukraine "Igor Sikorsky Kyiv Polytechnic Institute", Kiev, Ukraine

M. Walczak School of Materials, University of Manchester, Manchester, UK

Z. Wang Advanced Microelectronic Center Aachen, AMO GmbH, Aachen, Germany

S. Yakovin V. N. Karazin Kharkiv National University, Kharkiv, Ukraine

A. Yanovska Sumy State University, Sumy, Ukraine

V. V. Yavorskiy Kharkov Medical Academy of Postgraduate Education, Kharkiv, Ukraine;
Kharkov Regional Center of Blood Service, Kharkiv, Ukraine

O. V. Yazykov Department of Surgery, Traumatology, Orthopedics and Phthisiatry, Sumy State University, Sumy, Ukraine

A. Yusupova Sumy State University, Sumy, Ukraine

V. Zhurenko V. N. Karazin Kharkiv National University, Kharkiv, Ukraine

Y. G. Zmievskii Ministry of Education and Science of Ukraine, National University of Food Technologies, Kiev, Ukraine

A. Zykova V. N. Karazin Kharkiv National University, Kharkiv, Ukraine;
National Science Center "Kharkov Institute of Physics and Technology", Kharkiv, Ukraine

Chapter 1
Bacterial Cellulose/Hydroxyapatite Printed Scaffolds for Bone Engineering

A. Turlybekuly, A. Sagidugumar, Y. Otarov, N. Magazov, A. Pogrebnjak, I. Savitskaya, K. Akatan, A. Kistaubayeva, and A. Talipova

Abstract The bone tissue recovery after serious injuries has been a goal of regenerative medicine for more than a decades. So the issue of recovering the acetabular bone deficit has not been resolved. The typical practice of "deficient areas" filling is metal porous augments using, which is a short-term solution to this issue. It should be noted that the use of augments is not entirely justified, since for 5 years and more, bone tissue lysis also occurs due to the inevitable macrophage reaction of the body to friction products (debridement). One of the solutions, competing with auto- and/or alloplastic, is the use of composite biomimetic structures obtained by 3D printing. The paper describes the method of inkjet printing of a composite material based on bacterial cellulose(BC)/hydroxyapatite (HA). Printing inks was obtained by mixing the BC nanoparticles in an aqueous solution of $CaCl_2$ and Na_2HPO_4 with different proportions. Drying the resulting structures was carried out by freeze drying. The samples with different architecture were obtained. The samples' microstructure, functional composition were studied. It was found that obtained composite have homogeneous structure of HA paricle's spreading among BC matrix.

A. Turlybekuly (✉) · A. Sagidugumar · Y. Otarov · N. Magazov
Veritas, East Kazakhstan State Technical University named after D. Serikbayev, Oskemen, Kazakhstan
e-mail: aturlybekuly@gmail.com

A. Pogrebnjak
Department of Nanoelectronics, Sumy State University, Sumy, Ukraine

I. Savitskaya · A. Kistaubayeva · A. Talipova
Biotechnology department, Al-Farabi Kazakh National University, Almaty, Kazakhstan

K. Akatan
East Kazakhstan State University named after S. Amanzholov, Oskemen, Kazakhstan

A. D. Pogrebnjak et al. (eds.), *Nanomaterials in Biomedical Application and Biosensors (NAP-2019)*, Springer Proceedings in Physics 244,
https://doi.org/10.1007/978-981-15-3996-1_1

1.1 Introduction

Hydroxyapatite (HA) opens broad possibilities in bone engineering application due to being an inorganic component of natural bone. Despite to its high biocompatibility and osteoconductivity, low range of mechanical properties makes restricts its spreading in bone engineering. The permissible way to solve this problem is a developing HA based composite materials. The authors of [1–4] suggested to incorporate HA nanocrystals into organic matrix, which has high biocompatibility and mechanical strength.

Bacterial cellulose (BC) has high level of biocompatibility along with high mechanical properties comparable with aramids. It makes actual producing hybrid composite materials based on BC/HA. Such composites were investigated in [5–11], where was showed that the BC/HA nanocomposites are promising for applications in various biomedical fields: tissue engineering and drug delivery.

In case of bone tissue recovery after serious injuries it becomes important, a providing appropriate conditions for fast osteoregeneration. Inspite to the great progress in that area, there is lack of techniques for the acetabular bone deficit solving. The typical practice is a filling "deficient areas" with metal porous augments, which is a short-term solution to this issue. It should be noted that the use of augments is not entirely justified, since for 5 years and more, bone tissue lysis also occurs due to the inevitable macrophage reaction of the body to friction products (debridement) [12–16]. One of the solutions, competing auto- and /or alloplastic, is the use of composite biomimetic structures obtained by 3D printing [14, 15, 17]. The paper describes the method of preparation composite based on BC/HA inks for inkjet printing.

1.2 Method and Materials

The printing ink of composite material was obtained by mixing in an aqueous solution two different proportion of BC powder, $CaCl_2$ and Na_2HPO_4. The managing of BC content explains by its viscous properties. The molar ratio of Ca/P was 1.67 as in stoichiometric HA. The two BC/HA biocomposites were prepared by follows:

> 2.94 g of $CaCl_2 \cdot 2H_2O$ and 4.29 g of $Na_2HPO_4 \cdot 12H_2O$ were dissolve in 40 ml of bi-distilled water and mixed. After 0.15 g of BC was put into the solution
> 1.47 g of $CaCl_2 \cdot 2H_2O$ and 2.15 g of $Na_2HPO_4 \cdot 12H_2O$ were dissolve in 40 ml of bi-distilled water and mixed. After 0.3 g of BC was put into the solution.

The resulting materials' suspensions were milled by planetary milling machine for 3 h. The milled suspensions were washed up by distilled water and were filtered. The composite ink was put into 3D printer's cylinder. The work principle of 3D printing setup is schematically illustrated in Fig. 1.1 The biocomposite ink presses

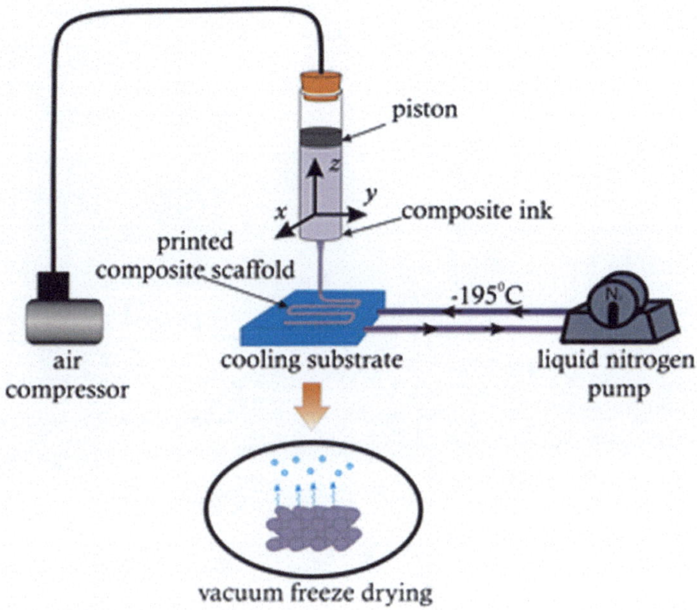

Fig. 1.1 The BC/HA scaffolds printing process

through needle filler by piston under pressure of air. The 3D moving noodles of biocomposite put onto cooled substrate and immediately freeze.

The resulting structures dried by freeze drying method. The microstructure of the samples was studied on optical microscope Olympus BX-51, a digital images were processed by SIAMS program. The functional composition was determined by FTIR spectrometry. It was found that obtained composite have homogeneous structure of HA paricle spreading among BC matrix.

1.3 Results and Discussion

The microstructure of BC/HA composites with different proportions are shown in Fig. 1.2. The suspension is homogenous. The particles are evenly spaced, Fig. 1.2a. At suspension that was obtained by second method (where BC content increased for 4 times) the HA microparticles are incorporated into BC matrix, Fig. 1.2c. It also makes some problems at printing, due to high viscosity of ink. We solved it by increasing fluidity of water solution. It is notable that increasing of BC content enlarges average microparticles size from 2.98 to 3.36 μm, according to Table 1.1.

Also, the maximum square of particles increased from 9.51 to 12.25 μm^2. We suppose that particles' sizes increasing caused by viscosity of BC material, which blocks suspension fluidity at milling process (Fig. 1.3).

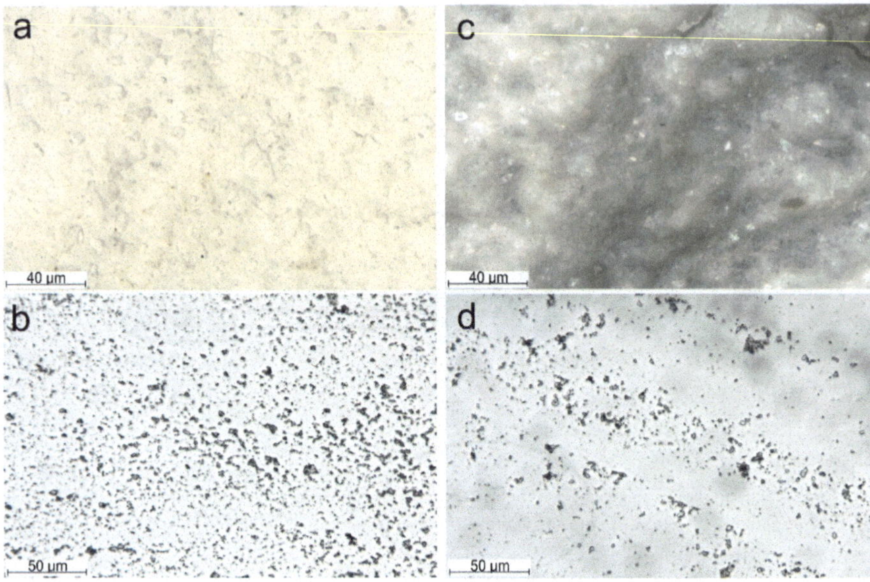

Fig. 1.2 The BC/HA biocomposites microstructure. **a, b**—for lower content of BC; **c, d**—for BC increased content (4 times); **a, c**—light reflected images; **b, d**—light transmitted images

Table 1.1 Statistical values of biocomposite paticles' dimensions

Parameters	Statistical values					
	BC/HA—1			BC/HA—2		
	Min.	Max.	Ave.	Min.	Max.	Ave.
S μm^2	1.01	132.51	9.51	1.01	344.47	12.25
D μm	1.13	13.58	2.98	1.13	22.04	3.36

The Fourier transform infrared spectroscopy was used to evaluate the HA/BC interactions. All obtained IR spectra were referenced on HA spectra, (Fig. 1.4c). The 562 and 602 cm^{-1} bands in HA spectra belong to the O-P-O groups (ν4 bending mode) [18, 19]. The highest band at 1026 cm^{-1} refers to asymmetric stretching (ν3) of P-O vibrations. The carbonate groups enhancing at 1456 to 1716 cm^{-1} caused by BC addition and indicates the bonding of BC with HA.

1.4 Conclusion

The BC/HA biocomposite materials with different BC content for use in 3D printing were obtained. The microstructure, functional groups composition were investigated.

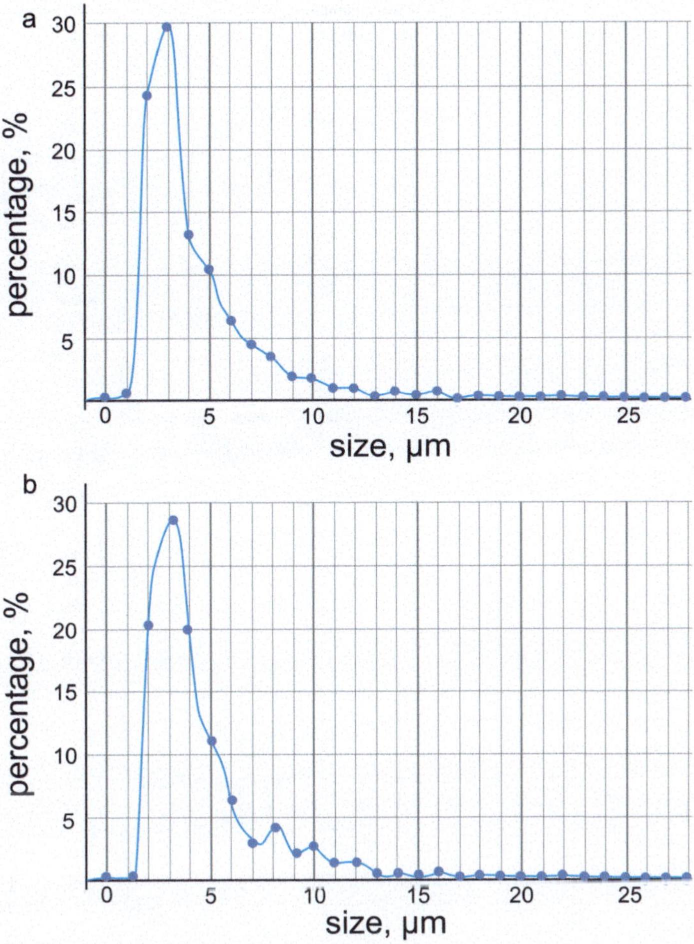

Fig. 1.3 Measurement of the size and shape of particles. Differential numerical particle size distribution

It was shown, that increasing of BC content has direct influence on particles' size at suspension. The method of BC/HA biocomposite 3D printing is suggested.

References

1. A. Turlybekuly, B.H. Shaimardanova, S.V. Plotnikov, A.D. Pogrebnjak, N.K. Erdybaeva, G.K. Uazyrkhanova, Q. Aqatan, Bull. Univ. Karaganda-Phys. **3**, 74 (2018)

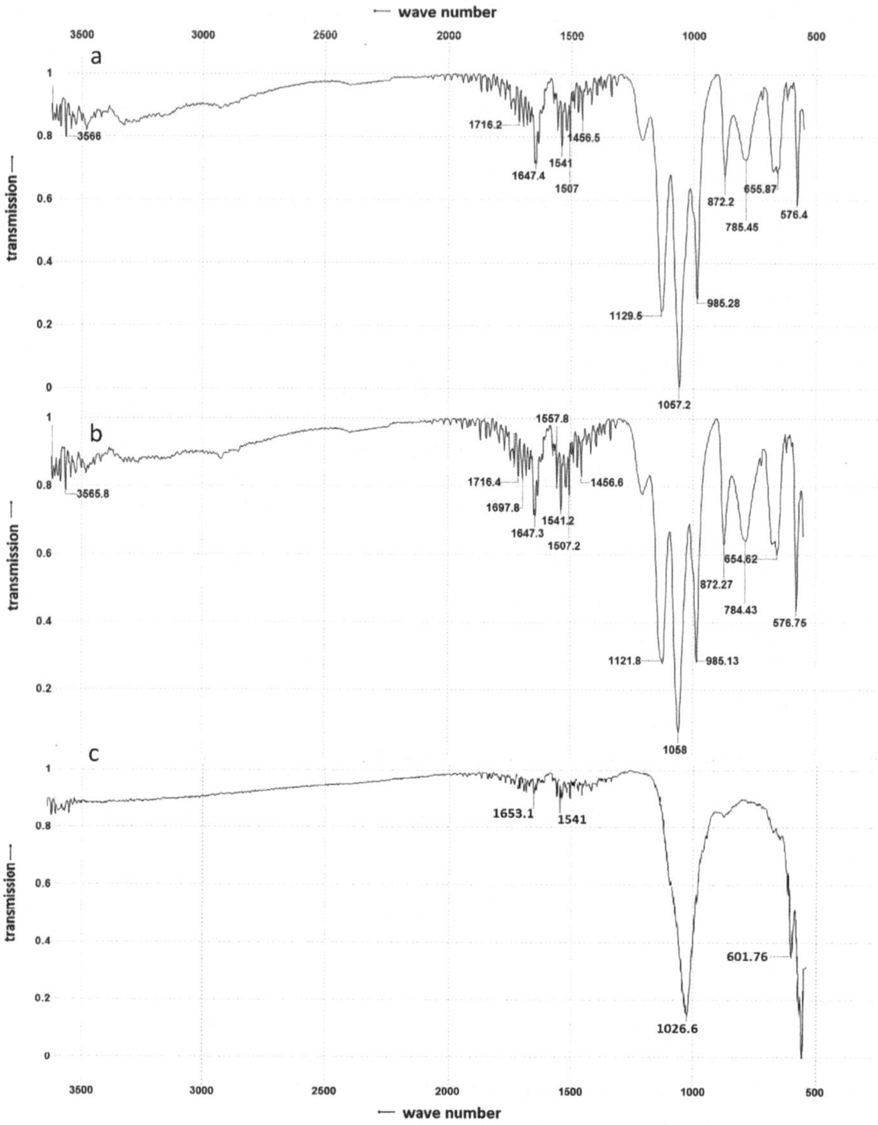

Fig. 1.4 FTIR spectra of the synthesized samples: BC/HA—1 (curve a), BC/HA—2 (curve b) and HA (curve c)

2. A. Turlybekuly, B.H. Shaimardanova, S.V. Plotnikov, A.D. Pogrebnjak, L.N. Yerokhina, G.K. Uazyrkhanova, A. Kasymbayev, E. Shaymardan, D.S. Dogadkin, Bull. Univ. Karaganda-Phys. **4**, 51 (2018)
3. A. D. Pogrebnjak, L. F. Sukhodub, L. Sukhodub, O. V. Bondar, A. Turlybekuly, *Advances in Thin Films, Nanostructured Materials, and Coatings* (2019) pp. 361–368
4. A. Pogrebnjak, L. Sukhodub, L. Sukhodub, O. Bondar, M. Kumeda, B. Shaimardanova, Z. Shaimardanov, A. Turlybekuly, Ceram. Int. **45**, 7504 (2019)

5. S.-J. Ahn, Y.M. Shin, S.E. Kim, S.I. Jeong, J.-O. Jeong, J.-S. Park, H.-J. Gwon, D.E. Seo, Y.-C. Nho, S.S. Kang, C.-Y. Kim, J.-B. Huh, Y.-M. Lim, Biotechnol. Bioprocess. Eng. **20**, 948 (2015)
6. N. Tazi, Z. Zhang, Y. Messaddeq, L. Almeida-Lopes, L.M. Zanardi, D. Levinson, M. Rouabhia, AMB Express **2**, 61 (2012)
7. S. Saska, H.S. Barud, A.M.M. Gaspar, R. Marchetto, S.J.L. Ribeiro, Y. Messaddeq, Int. J. Biomater. **2011**, 1 (2011)
8. N. Yin, S. Chen, Y. Ouyang, L. Tang, J. Yang, H. Wang, Prog. Nat. Sci. Mater. Int. **21**, 472 (2011)
9. L.-H. Fu, C. Qi, Y.-J. Liu, W.-T. Cao, M.-G. Ma, Sci. Rep. **8**, 8292 (2018)
10. P. Qi, S. Ohba, Y. Hara, M. Fuke, T. Ogawa, S. Ohta, T. Ito, Carbohydr. Polym. **189**, 322 (2018)
11. K. V. Massari, G. O. Marinho, J. L. Silva, L. A. Holgado, A. L. Leão, M. M. Chaves, A. Kinoshita, Tissue reaction after subcutaneous implantation of a membrane composed of bacterial cellulose embedded with hydroxyapatite. Dent. Oral Craniofacial Res. **1**, 25–30 (2015)
12. H. Mei, S. Zhao, W. Chen, Q. Wang, H. Liang, Trans. Nonferrous Met. Soc. China **28**, 1368 (2018)
13. D. Feldman, J. Macromol. Sci. Part A **52**, 322 (2015)
14. P.A. Rühs, F. Storz, Y.A. López Gómez, M. Haug, P. Fischer, Npj Biofilms Microbiomes **4**, 21 (2018)
15. M. Sayed, H.F. El-Maghraby, F. Bondioli, S.M. Naga, J. Appl. Pharm. Sci. **8**, 23 (2018)
16. M. Moniri, A. Boroumand Moghaddam, S. Azizi, R. Abdul Rahim, A. Bin Ariff, W. Zuhainis Saad, M. Navaderi, R. Mohamad, Nanomaterials **7**, 257 (2017)
17. Q. Wang, J. Sun, Q. Yao, C. Ji, J. Liu, Q. Zhu, Cellulose **25**, 4275 (2018)
18. R. Hu, C.-J. Lin, H.-Y. Shi, J. Biomed. Mater. Res. Part A **80A**, 687 (2007)
19. I.S. Neira, Y.V. Kolen'ko, O.I. Lebedev, G. Van Tendeloo, H.S. Gupta, F. Guitián, M. Yoshimura, Cryst. Growth Des. **9**, 466 (2009)

Chapter 2
Delivery of Probiotic to Microbiome by Layer-by-Layer Encapsulation

M. A. Abdulzhanova, I. Savitskaya, A. Kistaubayeva, D. H. Shokatayeva, A. Pogrebnjak, and L. V. Ignatova

Abstract The probiotic alginate microcapsules of a matrix type were obtained by extrusion method. The coating with an additional layer (layer-by-layer) was carried out by placing the alginate microcapsules with bacteria included in them in a 0.4% solution of chitosan or pullulan. Microcapsules look like spherical particles, white, with a smooth surface. The size of microcapsules is 102–145 microns. The efficiency of *Lactobacillus acidophilus* AA-1 cells immobilization for uncoated (alginate) and coated with chitosan and pullulan microcapsules was 96.35 ± 1.65; 95.28 ± 2.31 and 94.43 ± 2.31, respectively. The survival rate of strains encapsulated in alginate-chitosan and alginate-pullan under model gastrointestinal conditions in vitro increased on average by 50–70% compared with free cells.

2.1 Introduction

Technologies for the functional nutrition products production are developing at a rapid pace. A great part of this market consists of nutrition products containing probiotics.

The viability of probiotic microorganisms, i.e. the number of live and active cells in a certain volume (g or ml) of probiotic food products at the time of consumption is a crucial characteristic of these products quality, which determines their effectiveness [1]. However, a significant part of probiotic cells lose their activity due to damage and death of microorganisms during the production of starter culture, storage of products, and especially in the process of passing through the gastrointestinal tract [2]. Therefore, it is important to ensure high survival of probiotic bacteria during the production and storage of the product and during its consumption.

M. A. Abdulzhanova · I. Savitskaya (✉) · A. Kistaubayeva · D. H. Shokatayeva · L. V. Ignatova
Al-Farabi Kazakh National University, Almaty, Kazakhstan
e-mail: irasava_2006@mail.ru

A. Pogrebnjak
Department of Nanoelectronics, Sumy State University, Sumy, Ukraine

A. D. Pogrebnjak et al. (eds.), *Nanomaterials in Biomedical Application and Biosensors (NAP-2019)*, Springer Proceedings in Physics 244,
https://doi.org/10.1007/978-981-15-3996-1_2

One of the ways to solve the problem of increasing the viability of probiotics is the use of adsorption and spatial immobilization of bacteria under mild conditions [3]. Among them, increasing attention of scientists is attracted to the process of encapsulation as the most promising method for protecting and adapting cells and active substances in human body [4]. Biocapsulation refers to the creation of various polymer systems in form of hydrogel nano- and microparticles, nano- and microcapsules or polymer films with immobilized biomaterial, which can be represented by various BAS (proteins, including enzymes, DNA, peptides, low molecular weight hormones, antibiotics, etc.), as well as microbial, plant and animal cells [5]. Encapsulation provides a significant increase in cell survival under conditions of gastrointestinal tract [6, 7]. In this regard, encapsulated probiotics are used in various fermented dairy products, such as yogurt, cheese, sour cream, frozen dairy desserts, to obtain biomass, as well as in dry preparations [8].

One of the popular strategies for encapsulating probiotic cells is extrusion process, since it does not require high temperatures, use of expensive equipment and organic solvents [9]. Therefore, in order to obtain microcapsules, an extrusion method was chosen in this work. This method consists of mixing a probiotic inoculum in a hydrocolloid solution, followed by extrusion through a nozzle, and then the resulting droplets are collected in a solution of gelling agent such as sodium alginate. It is a natural polymer, widely used as an encapsulating material, since it forms a universal, biocompatible and non-toxic matrix to protect probiotic strains from harmful factors [10]. However, there are more and more reports that alginate microcapsules have insufficient strength and can break down under adverse conditions [11]. In particular, the alginate gel formed is highly sensitive to pH values, which can affect both the release and protection of encapsulated probiotic [12].

The encapsulation material can be modified by physical or chemical means to strengthen the matrix. One of such methods is coating of microcapsules with an additional layer (by-layer) of other natural polymers, for example, chitosan, which is widely used to obtain microcapsules by this method [6, 9, 13, 14]. Alginate microcapsules coated with chitosan were used as a "positive control" in this work. Another microbial "food" polysaccharide, investigated in the work for the possibility of creating an additional coating from it, is pullulan. It is capable of forming elastic, durable films, which are used as packaging material for food products, used for edible decoration in confectionery and bakery products [15–17]. This strategy was used in the work, the aim of which is—Microencapsulation of probiotics into polysaccharide matrices to improve their stability, viability and effective gastrointestinal delivery for functional products.

2.2 Materials and Methods

2.2.1 Preparation of Bacteria for Microencapsulation

In this study Lactobacillus acidophilus AA-1 isolated by our group from traditional yogurt was used. The strain was stored in glycerol (50%) at −20° C. The stored cells were activated twice on MRS agar plates before use. After 48 h of growth, one colony was selected, inoculated into 20 ml MRS broth and incubated for 24 h at 37 °C. After this, the culture was transferred to fresh MRS broth and incubated at 37 °C for 18 h under anaerobic conditions. The cells were collected in the stationary growth phase by centrifugation at 6000 g for 15 min at 4 °C. The supernatant was discarded and cell pellet was washed twice with sterilized distilled water. The precipitate was resuspended in 3 ml of sterilized 0.85% sodium chloride solution. The concentration of cells after this treatment was around 2×10^{10} CFU/ml. Freshly prepared concentrated cell suspension was quantitatively evaluated by plating on MRS agar and used immediately.

2.2.2 Preparation of Encapsulated Probiotics

The following coating materials were used: sodium alginate (A) from Hi Media, chitosan (Ch) and pullulan (P) from Serva. A suspension of probiotic cells was suspended in a solution of an encapsulating substance (2% A) in a 1:5 ratio to obtain a suspension containing approximately 10^{10} CFU/ml of cells. Then this mixture was extruded through a needle with a 0.6 mm diameter into a sterile 1% calcium chloride solution. For extrusion a syringe dispenser "Armed MP-2003" was used. The distance between the needle and calcium chloride solution was 25 cm. The drops immediately formed gel spheres. These spheres were allowed to settle for 30 min for complete solidification.

Alginate microcapsules with included bacteria were placed in a 0.4% solution of Ch or P, incubated for 40 min on a shaker at 130 rpm. This speed of mixing avoided the aggregation of microcapsules. Then the capsules were removed from solution using a sterilized sieve (50 μm) and washed with sterile distilled water. The whole procedure was performed using autoclaved (121 °C, 15 min) materials and under sterile conditions in a laminar air flow box. Capsules were stored in sterile vials with lids at 4 °C and used in further experiments.

The average diameter of microcapsules was measured using a laser particle size analyzer (Winner 2000ZD, Jinan Winner Particle Instruments Stock Co., Ltd., China). Optical microscopy of wet microcapsules was performed using a microscope (MDL-150-TPI) and a digital camera (Samsung 14.2). The morphology of frozen dried microcapsules was also evaluated using a scanning electron microscope (JEOL, JM6360).

2.2.3 Quantification of Viable Bacteria in Capsules

1.0 g of microcapsules was added to a mixed solution of 0.06 mol/l sodium citrate and 0.2 mol/l sodium bicarbonate and stirred for 1 h at 37 °C to loosen the coating. The viability of released cells was determined by plating serial dilutions of resulting suspension on MRS agar. Colony-forming units were counted after 48 h of anaerobic incubation at 37 °C. Free cells were not subjected to homogenization, since comparison of CFU of free cells before and after homogenization did not show a significant difference.

The encapsulation efficiency (EI) was determined by the formula: EI % = (N × M)/N0 × 100, where N is the number of viable cells released from 1.0 g of microcapsules, M is the total mass of the collected microcapsules, and N0 is the number of free cells before microencapsulation.

2.2.4 Determination of Resistance of Encapsulated Bacteria to Conditions that Mimic Human Gastrointestinal Tract

To determine the stress resistance of encapsulated cells in environments that mimic gastrointestinal tract, test samples were cultured in media with different pH levels, the content of digestive enzymes and bile salts [18]. Artificial gastric juice (AGJ) was made using a 0.2% solution of sodium chloride, the pH of which was raised to 2.0 with 1 M HCl. Pepsin with a concentration of 3.2 g/l and 1 g of lipase were added to the solution, then the mixture was filtered through a 0.22 μm membrane for sterilization. Then, 0.5 g of fresh microcapsules and a suspension of free cells (2×10^{10} CFU/ml) were suspended in 9.5 ml of AGJ and incubated at 37 °C for 120 min at 150 rpm. AGJ samples were neutralized with phosphate buffer (0.2 M, pH 7.0, 4 °C) at the end of incubation.

Duodenal conditions were simulated by dissolving bile salts in physiological saline to a 0.6% final concentration, 10 g/l of each trypsin and pancreatin were added, the pH of solution was adjusted to 6.5 with 0.1 M NaOH and sterilized by membrane filtration. 0.5 g of fresh microcapsules were added to 9.5 ml of bile solution and incubated at 37 °C for 1 h at 150 rpm.

To create a colon model, artificial intestinal juice (AIJ) was prepared by dissolving KH_2PO_4 in distilled water to a final concentration of 6.8 g/l and at pH 7.4. 10 g/l of trypsin was added to the solution, filtered through a membrane with 0.22 μl for sterilization. 0.5 g of microcapsules were placed into 9.5 ml of AIJ and incubated at 37 °C for 4 h at 150 rpm. At the end of incubation free ones were washed with physiological saline, microcapsules were destroyed by homogenization in phosphate buffer, and cell survival was quantified by the method described above using standard plating on MRS agar.

2.2.5 Statistical Analysis

Statistical comparison was performed using unpaired t-test and one-way analysis of variance (ANOVA) followed by Dunnett's test for multiple comparisons. All statistical analyses were performed using SPSS 16.0 software package (SPSS Inc., USA).

2.3 Results and Discussion

Live probiotic bacteria of *Lactobacillus* genus are most often used in probiotic fermented milk products. The strain *Lactobacillus acidophilus* AA-1 has all the necessary probiotic characteristics: a high level of acid production, adhesive and antagonistic activity. It does not have pathogenicity and toxicity, i.e. has the required level of biosafety.

As a result of extrusion in a hydrocolloid alginate solution, probiotics formed a matrix-type microcapsule. Due to the fact that probiotics are present in the entire structure of particles, they can be exposed to environmental conditions, which can reduce their viability during storage and use. To coat alginate microcapsules with an additional layer, they were kept for 30–40 min in a 0.4% solution of chitosan or pullulan. Figure 2.1 shows the 3 types of microcapsules obtained by the method described above: containing alginate (A) or a mixture of alginate-chitosan (A–Ch) and alginate-pullulan (A–P) as a material-sealant.

Microcapsules are spherical particles of white color with a smooth surface. In appearance, all 3 types of capsules did not differ from each other. The absence of external differences was confirmed by analysis on a scanning electron microscope. Alginate microcapsules and coated microcapsules had a similar oval shape. Moreover, the combination of polymers as an additional wall does not change the typical spherical shape. A compact, penetration-resistant surface can serve as a strong physical barrier that protects probiotic cells as they pass through the gastrointestinal tract. The SEM study showed bacteria in the capsule matrix. They are present in all 3 types of microcapsules.

Table 2.1 shows data obtained in experiments on determination the encapsulation efficiency (%). For uncoated (A) and coated with Ch and P microcapsules, it was 96.35 ± 1.65; 95.28 ± 2.31 and 94.43 ± 2.31, respectively. The difference was not statistically significant ($P > 0.5$). The interactions between bacteria and materials are crucial for determining the efficiency of encapsulation. The high encapsulation efficiency obtained in our work showed that encapsulation process was soft and the materials were compatible with the probiotic strain.

Microcapsule size has an important effect on viability of probiotics and sensory effects on food. As a rule, larger microcapsules provide better protection against probiotics [19], but the product may have undesirable sensory properties [20]. It was shown that microcapsules in the size ranging of 100–200 μm provide optimal

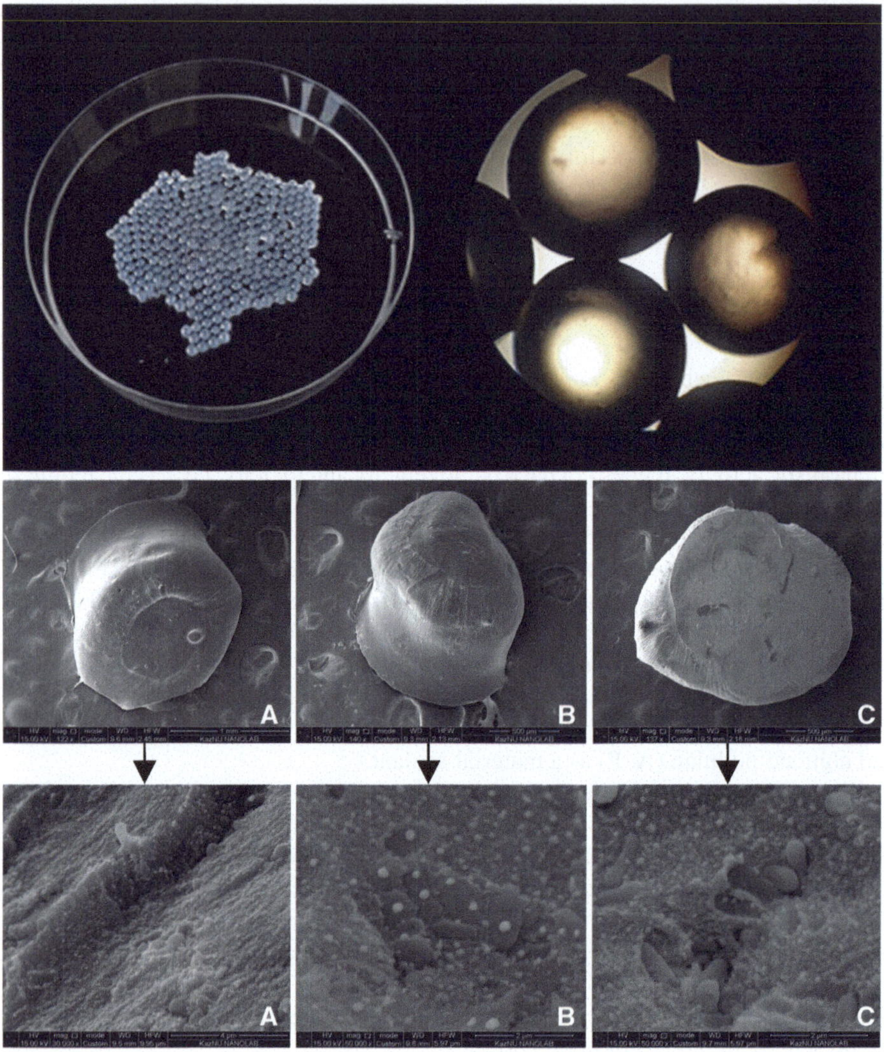

Fig. 2.1 The morphology of A (A), A-Ch (B) and A-P (C) microcapsules loaded with *Lactobacillus acidophilus* AA-1

Table 2.1 Encapsulation efficiency and size of probiotic microcapsule containing *L. acidophilus* AA-1 cells

Type of microcapsule	Encapsulation efficiency (%)	Size of microcapsules (mcm)
Alginate	96.35 ± 1.65	102.35 ± 4.31
Alginate-chitosan	95.28 ± 2.31	136.42 ± 5.73
Alginate-pullulan	94.43 ± 2.31	145.37 ± 6.25

balance between these two conflicting requirements [21]. In our study, the size of all microcapsules falls within this optimal range. After coating with chitosan and pullulan, the average microcapsule size increased by 34 and 43 μm, respectively. This is partly due to the thickness of coating layer and partly with the aggregation of microcapsules.

One of the most important biopharmaceutical indicators of encapsulated probiotics is their ability to maintain integrity in acidic environment of stomach, alkaline environment of duodenum and disintegrate in neutral environment of large intestine. For this, studies were conducted on comparative survival of encapsulated cells of *L. acidophilus* AA-1 in the model stomach, small and large intestine. Samples were cultivated in media with different pH levels, content of hydrolytic enzymes and bile salts.

The greatest inhibitory effects showed low pH levels. Within 2 h under the action of AGJ (pH 2.0) there was a significant death of free cells: the number of viable cells decreased by 6.4 lg (Fig. 2.2).

The smallest degree of protection of probiotic culture cells from the action of low pH levels was provided by alginate microcapsules. After 2 h of exposure in a model stomach, the number of viable cells was only 6.4 lg CFU/g, i.e. their number decreased by 3.3 lg. Alginate capsules coated with Ch and P provided greater cell resistance to low pH values: the titer of living cells was reduced only by 1.6 lg and 1.4 lg, respectively. Consequently, the number of cells dying from exposure to low pH values in A–Ch and A–P pullan microcapsules was reduced almost 2 times in comparison with alginate ones.

In both tests for tolerance to gastric juice and bile, the mortality of probiotic cells was significantly reduced when they were protected by microencapsulation compared with free cells. The reduction in probiotic population was 5.4 lg CFU/g in simulated duodenum (solution of bile salts). This result indicates that probiotic cells were highly susceptible to simulated gastrointestinal fluids, which justifies the importance of encapsulating this microorganism. Indeed, probiotics loaded into capsules of A–Ch and A–P and coated with Ch, as well as those contained in microcapsules of A–P, showed a decrease in viability of approximately 2 lg CFU/g and 0.9 lg CFU/g, respectively. In the work of M. Chávarri, similar results were shown for chitosan-coated alginate microparticles loaded with *Lactobacillus gasseri*. These

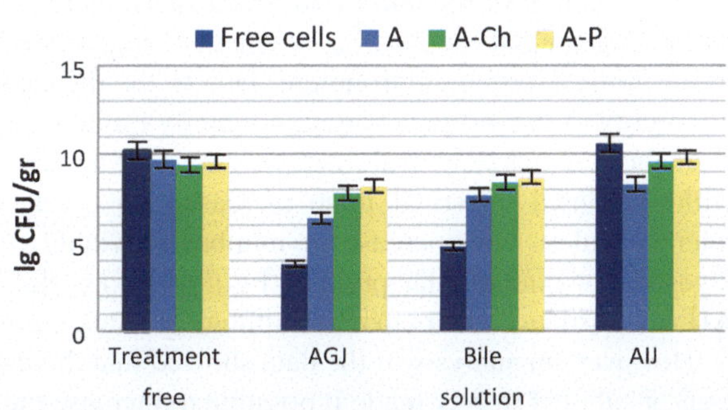

Fig. 2.2 Survival of free and encapsulated *L. acidophilus* AA-1 in the model gastrointestinal tract

authors reported that encapsulation protected probiotics compared with free cells, reaching more than 7 lg CFU/g, whereas non-encapsulated cells could not withstand this treatment [22].

The pH of the model duodenum was 6.8, which is close to the optimum pH for this bacterial strain. But bile salts can penetrate membranes of bacterial cells, damage DNA and proteins, cause leakage of intracellular components and ultimately lead to the bacterial death [23]. The formation of an additional barrier between bile salts and cells played main protective role of these microcapsules in a bile tolerance analysis. Our results show that the barrier properties of Ch and especially P make it a promising encapsulating material for protecting probiotics from both gastric juice and bile. This can be explained by appearance of a polyelectrolyte complex shell on the structure surface and possible further propagation of reaction between macromolecules of polyelectrolytes into the core. Thus, based on the data obtained, the maximum protection of probiotic against stress effects occurring in upper parts of digestive tract was provided by A–P microcapsules.

A necessary condition for implementation of beneficial properties of a probiotic is not only the prevention of damage from hostile gastric juice and bile, but also targeted release into large intestine. The revealed dependence of the probiotic cells survival on the type of encapsulating material indicates a different permeability of substances used, which determines the diffusion capabilities of capsule membranes.

The main problem in production of microcapsules based on natural polymers is formation of a membrane that provides, on the one hand, the mechanical strength of a capsule, and on the other hand, permeability to nutrients needed by cells. The increase in strength of capsule membranes occurs both due to the formation of mixed adsorption layers, and by increasing the amount of film-forming substance on the particles of encapsulated product. At the same time, an increase in density due to the deposition of secondary membranes can lead to a decrease in permeability of microcapsule membranes [24].

It was shown that the obtained A–Ch and A–P probiotic microcapsules had significant permeability, due to which the cells of microorganisms diffused out of the capsules into the media, and there was a withdrawal of waste products. Moreover, the membranes of these two microcapsules provided almost the same diffusion rate of *L. acidophilus* AA-1 cells of them (Fig. 2.3).

Since cells enclosed in capsules were capable of reproduction, at a certain stage of population development there was a sharp increase in the risk of microcapsules membrane breakthrough, as a result of which the cells could almost completely occur in the external environment. Due to the diffusion of cells from capsules, the possibility of this process was significantly limited.

In fact, compared with alginate capsules, the number of viable cells after treatment with simulated intestinal juice increased by 0.2 lg CFU/g for the extra-coated microcapsules. The increase in number of viable cells was associated with the digestion of intercellular proteins by trypsin, i.e. the digestive effect of trypsin in AIJ can introduce errors in determining the exact number of cells.

However, an analysis of the data showed that the inclusion of microorganisms in a encapsulating matrix made it possible to increase the stereostability of cells under

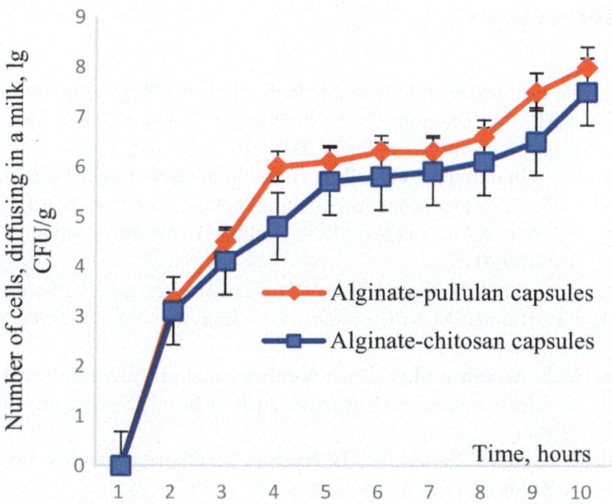

Fig. 2.3 The diffusion rate of immobilized probiotic cultures from microcapsules

adverse conditions, and the type of material that forms the capsules had a significant effect on survival of probiotic microorganisms. Under various adverse conditions, encapsulating materials behaved differently, providing different degrees of survival for probiotic cultures.

The experiments carried out in work and processing of obtained data allow us to conclude that immobilization of probiotic cultures in gel-forming substances of natural origin will contribute to increase of stress resistance of microorganisms in human body when passing through gastrointestinal tract.

2.4 Conclusion

The smallest degree of protection of probiotic cells from the gastric juice and of bile salts solutions low pH levels action of was provided by alginate microcapsules. A–Ch and A–P microcapsules do not undergo changes and remain without damage after exposure for 2 h under stressful conditions, and provide a high degree of survival of probiotic included in them under model conditions of gastrointestinal tract. In addition to in vitro assessments, there is still a need for in vivo studies on viability of microencapsulated probiotics and their beneficial effects on human health.

Nevertheless, the use of microencapsulation in biotechnology ensures efficient delivery of viable probiotic cells to intestine, thereby increasing therapeutic and prophylactic efficacy of healthy foods and probiotic preparations.

Acknowledgements This work was supported by grant AP05134040 «Microencapsulation of probiotics into polysaccharide matrices for the development of functional food products based on mare's milk» from the Science Committee of the Ministry of Education and Science of the Republic of Kazakhstan.

References

1. M. Fernandez, J.A. Hudson, R. Korpela, G. Reyes-Gavilan, BioMed Res. Int., pp. 1–13 (2015)
2. A.M. Mortazavian, R. Mohammadi, S. Sohrabvandi, *New Advances in the Basic Clinical Gastroenterology: InTech* (2012), pp. 121–146
3. C.P. Champagne, Crit. Rev. Biotechnol. **109–134**, 14 (2014)
4. C. Tomaro-Duchesneau, S. Saha, M. Malhotra, I. Kahouli, S. Prakash, J. Pharm. **5**, 1–19 (2013)
5. S.J. Sri, A. Seethadevi, K.S. Prabha, P. Muthuprasanna, P. Pavitra, Int. J. Pharm. Bio Sci., **3**,1–23 (2012)
6. A.K. Anal, H. Singh, *Trends in Food Science and Technology* (2017), pp. 240–25
7. P.E. Ramos, M.A. Cerquera, J.A. Teixeira, A.A. Vecente, Crit. Rev. Food Sci. Nutr. (2018), pp. 1864–1877
8. M.E. Morales, M.A. Ruiz, Nutraceuticals **1**, 627–668 (2016)
9. A. Mortazavian, S.H. Razavi, M.R. Ehsani, S. Sohrabvandi, Iranian J. Biotechnol. **5**, 1–18 (2017)
10. F. Pavli, C. Tassou, G.J.E. Nychas, N. Chorianopoulos, Int. J. Mol. Sci. (2018), p. 150
11. A.B. Shori, HAYATI J. Biosci. **24**, 1–5 (2017)
12. B.L. Pasin, C.G. Azón, A.M. Garriga, Técnicas yaplicaciones. Revista Venezolana de Ciencia y Tecnología de Alimentos **3**(1), 130–151 (2012)
13. W. Tong, X. Song, Ch. Gao, Chem Soc Rev **41**, 6103–6124 (2012)
14. A. Bepeyeva, J.M.S. de Barros, H. Albadran, A.K., J. Food Sci., pp. 1–6 (2017)
15. P. Oguzhan, F. Yangilar, African J. Food Sci. Technol. **4**(3), 57–63 (2013)
16. R.S. Singh, N. Kaur, V. Rana, J.F. Kennedy, Carbohyd. Polym. **171**, 102–121 (2017)
17. K.R. Sugurman, V. Ponnusami, Carbohyd. Polym. **173**, 573–591 (2017)
18. H. Chen, X. Li, J. Funct. Foods **29**, 248–255 (2017)
19. K. Lee, T. Heo, Appl. Environ. Microbiol. **66**, 869–873 (2010)
20. R. Rajam, S. Kumar, J. Food Sci. Technol. **52**, 4029–4041 (2015)
21. K.S. Nag, H. Han, Singh. Int. Dairy J. **21**, 247–253 (2011)
22. M. Chávarri, I. Marañón, R. Ares, F.C. Ibáñez, Int. J. Food Microbiol. **142**, 185–189 (2010)
23. J. Silva, Appl. Environ. Microbiol. **66**, 2605–2612 (2013)
24. A.S. Carvalho, J. Food Sci. **68**, 2538–2541 (2013)

Chapter 3
Modernization of the Preservative Solution for Red Blood Cells by Magnetite Nanoparticles (ICNB)

A. N. Belousov, E. I. Malygon, V. V. Yavorskiy, and E. Yu. Belousova

Abstract This study was devoted to the learning of the use of nanotechnology to correct the functional activity of red blood cells (RBCs) at the storage stages at a positive temperature. It was established that saline NaCl, which had previously been processed by magnetite nanoparticles (ICNB) had a marked membrane-stabilizing effect, inhibits hemolysis and increasing the sedimentation stability of preserved RBCs. The complex analysis of the obtained data allowed to determine the primary mechanisms effect of the saline NaCl, which had previously been processed by ICNB on the preserved RBCs. The proposed method of additive modernization of preserved RBCs was adapted to the production process. The optimisation results were obtained in creating a simple and practical method of additive modernization of preservation solutions that does not violate the compliance requirements, improves the quality, efficiency and safety transfusion of RBCs.

3.1 Introduction

Red blood cells (RBCs) transfusion is a critical, life-saving treatment for severe anemia caused by disease or chemotherapy, or by blood loss due to trauma or major surgery. For several decades RBCs components have been prepared as concentrates suspended in nutrient additive solution, which preserves and extends the shelf-life of the RBCs component, allowing up to 6–7 weeks of refrigerated storage [1]. Nevertheless, during storage RBCs undergo a complex and progressive accumulation of physicochemical changes, collectively referred to as the RBCs storage lesion [2, 3].

A. N. Belousov (✉) · E. Yu. Belousova
Laboratory of Applied Nanotechnology of Belousov, Belousov, Ukraine
e-mail: an.belousov2012@ukr.net

A. N. Belousov · E. I. Malygon · V. V. Yavorskiy · E. Yu. Belousova
Kharkov Medical Academy of Postgraduate Education, Kharkiv, Ukraine

A. N. Belousov · E. I. Malygon · V. V. Yavorskiy
Kharkov Regional Center of Blood Service, Kharkiv, Ukraine

A. D. Pogrebnjak et al. (eds.), *Nanomaterials in Biomedical Application and Biosensors (NAP-2019)*, Springer Proceedings in Physics 244,
https://doi.org/10.1007/978-981-15-3996-1_3

Recent clinical studies have identified RBCs transfusion as an independent risk factor for increased morbidities and mortalities in certain groups of patients, including trauma, cardiac surgery and the critically-ill (reviewed in [4–6]). Additionally, some of these studies have identified that older stored RBCs are more strongly implicated in poorer outcomes compared to fresher RBCs [6]. In order to address these concerns, there is renewed interest to better understand the RBCs storage lesion and to find ways to ameliorate the deleterious effects of storage, thereby improving the quality, efficacy and safety of RBCs components for all transfusion recipients.

While increased research effort is being directed to better understand the effects of storage on RBCs and the potential impact on transfusion outcomes [7], slower progress is being made in finding ways to deter the detrimental effects of the RBCs storage lesion.

Over the past 15–20 years, research into the development of new additive solutions has focussed on ways to maintain higher intracellular levels of ATP and 2.3-DPG during storage of RBC components [1].

Despite the RBCs having been a favourite experimental model for cellular biologists and biochemists, RBCs storage research has repeatedly demonstrated that a lot of fundamental biology about RBCs is still not well understood. The complexity of the inter-relationship between RBCs biochemistry, cytoskeletal structure and membrane properties have made it difficult to predict how RBCs will respond to different storage conditions. Exposure of RBCs to non-physiological storage environments has pointed to the existence of previously unknown biochemical mechanisms in RBCs, including apoptotic-like processes, ion and osmotic channels that behave differently than expected, exposure of new or altered receptors possibly due to oxidative and/or protease/glycosidase activities or altered senescence [8–11].

The benefits gained by improved RBCs component quality should more than justify any real or perceived inconvenience to the blood services in implementing adjustments to their processing procedures or additional processing costs of the introduction of new generation RBCs additive solutions. The bigger challenge that has hindered the advancement of this field is the significant financial burden and risk for manufacturers of blood collection systems to obtain licensure and to bring a new RBCs storage system to a market that is inherently based on very low profit margins, such as the blood services sector.

The financial burden to technology developers of new RBC storage systems is largely due to regulatory requirements, particularly those mandated by the FDA. In addition to in vitro data, the FDA requires in vivo data on the 24 h post transfusion recovery of transfused autologous RBCs. Recently the FDA has tightened and increased the assessment and acceptance criteria making it potentially more difficult and expensive to bring new RBCs storage systems to market. Although the regulatory agencies are to be commended for focussing on the safety of new therapies and devices for patients, there are concerns that the regulatory requirements for RBC storage systems have become excessive and are hindering progress [12].

Another significant challenge for obtaining licensure of new RBCs storage systems is the inherent donor-related variability in stored RBCs quality. It has long been recognised that RBCs from some donors do not store well, as evidenced by higher levels of haemolysis at RBC component expiry 14 and poorer in vivo 24 h recovery data [13]. The relationship of specific donors and poorer quality of some stored RBCs components was confirmed in a recent paired cross-over study designed to compare manual and automated whole blood processing methods [14, 15]. Technology developers are unwilling to take on the risk that a random poor quality RBCs component could jeopardise the success of licensure tests and clinical trials of their new blood storage systems and their significant financial investment.

In Ukraine, the first standardized and biocompatible magnetite nanoparticles for medical use were manufactured and patented in 1998. These are intracorporal nanobiocorrector of brand ICNB, magnet-controlled sorbent of brand MCS-B, and biologically active nanodevice of brand Micromage-B [16].

It is well established that the magnetite nanoparticles effectively modulate the metabolic processes in leukocytes, regulate activity of the enzyme link of the antioxidant system in erythrocytes in healthy and sick patients [17–19]. Previously the complex investigations that were performed in the study of the influence on metabolism of cells by preparations of nanotechnology show that in whole standardized biocompatibility of magnetite nanoparticles have nonspecific and modulated effect on metabolic processes. Research of ultrastructure investigations of the reticuloendothelial system (liver, lungs and kidneys) it was proved that after injection of biocompatibility magnetite nanoparticles into a vein caused nonspecific activation of the metabolic processes, increase adaptive mechanisms and potential of organelle cells, acceleration of reparative processes a level of membranes and macromolecules [18, 20, 21]. Existing sorption and indirect (magnetic) effects not only allow selectively absorb the protein of surface membrane cells by magnetite nanoparticles (according to the principle of magnetophoreses), but also to prevent the oxidative modification of proteins by way of stabilizing the active groups, normalizing a state of receptors that are located on the surface membrane of cells, increasing activity of enzymes' membrane-bound [22–24].

Recent scientific work related to use of magnetite nanoparticles (ICNB) in contrast means in an MRI investigation of cancer reliably was shown that nanoparticles cause reversible changes associated with a temporary increase in the mobility of hydrogen protons in the pericellular fluid that inevitably modifies the metabolism in malignant cells [25]. The results of these investigations have not only widened the understanding of the mechanisms of action of nanoparticles on condition outside and intracellular spaces but also have revealed new aspects of the cellular(cells) metabolism, determined the membrane role of cellular enzymes in the regulation processes of metabolism [23, 26–29].

Also, it was established that extracorporally processing the blood by nanoparticles of MCS-B reliably reduces activity of Ca, Mg—ATPHese of erythrocytes.

Currently, studies have shown that magnetite nanoparticles are able to inhibit hemolysis of heparinized blood, increase the activity of ATP and 2.3 DPH in red blood cells, regulate transmembrane metabolism and inhibit eryptosis [23, 30, 31].

The above was the basis for the choice of the theme of this study, devoted to the learning of the use of nanotechnology to correct the functional activity of red blood cells at the storage stages at a positive temperature.

The main purpose of the first stage of the study is to develop a simple and practical method of additive modernization of preservation solutions that does not violate the compliance requirements, improves the quality, efficiency and safety transfusion of red blood cells.

3.2 Materials and Methods

Materials:

1. Standardized intracorporeal nanobiocorrector of ICNB was taken as nanoparticles. Magnetite nanoparticles synthesized by co-precipitation method. The main physics and chemical properties of ICNB the following data and also in Tables 3.1, 3.2, 3.3 and 3.4; Figs. 3.1 and 3.2 were presented:

   - Concentration of the colloidal solution of magnetite nanoparticles in physiology solution of NaCl is 0.0225%.
   - Theoretical osmolality of colloid solution is 500 mosmol/l.
   - Size of magnetite nanoparticles is 6–12 nm.
   - Total area of surface magnetite of nanoparticles $S_s = 800–1200\ m^2/g$.
   - Magnetization of saturation $I_s = 2.15$ кA/m.
   - ζ—potential = −19 mV.

Table 3.1 The calculated lattice parameters of the phases

Phase name	a (Å)	b (Å)	c (Å)	Alpha (град)	Beta (град)	Gamma (град)
Magnetite low	8.38	8.38	8.38	90	90	90
Magnetite low, syn	5.93	5.93	14.7	90	90	120
Johannsenite	9.89	9.05	5.28	90	105.5	90

Table 3.2 Determination of percent composition of the ICNB by X-ray spectrometer ARL OPTIM'X (semi-quantitative analysis)

Compound	wt%	Std. Err	El	Weight%/O_2	Std. Err	El	wt%	Std. Err
Fe_3O_4	97.37	0.09	Fe	68.40	0.07	Fe	97.62	0.09
CaO	2.26	0.07	Ca	1.71	0.05	Ca	2.3	0.07
P_2O_5	0.280	0.027	Px	0.122	0.012	Px	0.157	0.015
MnO	0.255	0.013	Mn	0.198	0.010	Mn	0.278	0.014
SiO_2	0.098	0.027	Si	0.046	0.013	Si	0.059	0.016
SO_3	0.032	0.013	Sx	0.0126	0.0051	Sx	0.0164	0.0066
Cl	0.0280	0.0090	Cl	0.0280	0.0090	Cl	0.0380	0.012

Table 3.3 X-ray analysis of ICNB in X-ray diffractometer Rigaku Ultima IV (CuKα, Kβ filter—Ni), one-coordinate DTeX semiconductor detector

Phase	Formula	Space group	No. card database ICDD
Magnetite low	$Fe_{2.886}\ O_4$	227: Fd-3m, choice-2	10861339 (ICDD)
Magnetite low, syn	Fe_3O_4	166: R-3m, hexagonal	10716766 (ICDD)
Johannsenite	Ca Mn + $2Si_2O_6$	15: C12/c1, unique-b, cell-1	380413 (ICDD)

Table 3.4 The phases of magnetite of nanoparticles (RIR—method; error 8 ± 3%)

Phases (method of corundum numbers)	Content (%)
Magnetite low	71
Magnetite low, syn (hexagonal)	29

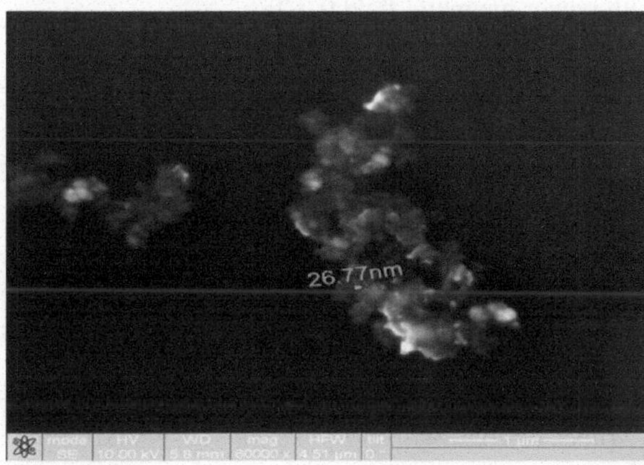

Fig. 3.1 Study of magnetite nanoparticles with use microscope ion-electronic raster-type Quanta 200 3 D

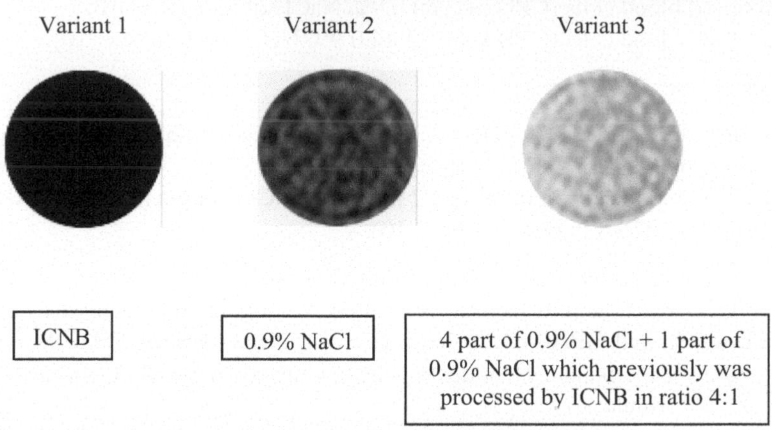

Fig. 3.2 Images of fluids that were studied in the research at the MRI

2. 0.9% NaCl solution
3. 0.9% NaCl solution which previously was processed by ICNB in ratio 4:1

Object of research:

- red blood cells (RBCs) into bags containing anticoagulant citrate, nutrient phosphate and dextrose (CPD);
- red blood cells (RBCs) into bags containing anticoagulant citrate, nutrient phosphate, dextrose and adenine (CPDA-1).

Methods: Before starting the experience with red blood cells (RBCs) we used visual assessment by comparison of the brightness of the imagines different variants of solutions. These were: ICNB, 0.9% NaCl solution and 0.9% NaCl solution that had been treated by magnetite nanoparticles (ICNB).

The tests were performed on the Siemens MR-tomagraphy Magneton Concerto with power magnetic-field 0.2 T.

The axial tomograms were received:

1. T1—the self-weighted sequences of Echo Spin of TR 50 ms, TE 17 ms the field of review the 250 mm, the thickness cut 2 mm.
2. T2—the self-weighted sequences Echo Gradient of TR 500 ms, TE 17 ms the field of review a 180 mm, the thickness cut 4 mm.

Of each bag of 3 ml amounts of red blood cells was distributed into 20 sterile glass tubes. Then, into the first 10 tubes of control were added of 2 ml amounts 0.9% NaCl solution. Into the next 10 tubes of test were added of 2 ml amounts 0.9% NaCl solution, which previously was processed by ICNB.

Thus, the distribution of tubes was as the follows:

Tubes of control:

- 3 ml of red blood cells (CPD) +2 ml 0.9% NaCl solution ($n = 10$);
- 3 ml of red blood cells (CPDA-1) +2 ml 0.9% NaCl solution ($n = 10$).

Tubes of test:

- 3 ml of red blood cells (CPD) +2 ml 0.9% NaCl solution that previously was processed by ICNB in ratio 4:1 ($n = 10$);
- 3 ml of red blood cells (CPDA-1) +2 ml 0.9% NaCl solution that previously was processed by ICNB in ratio 4:1 ($n = 10$).

The state of red blood cells was determined visually by the registration of signs of hemolysis. Also, hemolysis was controlled by photometric method by means Plasma/Low Hb and GPHP-01 devices. The centrifuge mark of SM-70M-07 was used to obtain supernatant. Hematocrit was calculated by means hematocrit ruler and using the formula:

$$\mathrm{HCT} = \frac{\text{Free Hb} \times (1 - \mathrm{HCT})}{\text{Common Hb}} \times 100\%$$

Morphology of the red blood cells was studied by direct microscopic method. Sedimentation stability of red blood cells was studied by Panchenkov's method. Change in the acidity of the red blood cells was performed by means of pH metric.

Tests were carried out in six stages: day 1—I, day 7—II, day 14—III, day 21—IV, day 28—V, day 35—VI.

The blood after performance of the biochemical investigation was stored in the refrigerating chamber at temperature +4 °C.

Statistically processing the obtained results was carried out by parametrical method of variation statistics by Student criterion. Processing the obtained data was carried out by means of Excel.

3.3 Results and Discussion

Results of visual assessment of the images brightness used in experiment liquids at the MRI are represented in Fig. 3.2.

Figure 3.3 illustrates the difference of image brightness in liquids compared at the MRI. The order of the brightness increase is the following: ICNB, 0.9% NaCl solution, 0.9% NaCl solution that was treated by nanoparticles of ICNB. The difference in brightness of images is explained in the following way:

Variant 1. Magnetite nanoparticles of ICNB reduce the mobility of hydrogen protons in the liquid medium (0.9% NaCl solution). Therefore, the image brightness was very low in the MRI.

Variant 2. Rising mobility of hydrogen protons in the intact 0.9% NaCl solution increased brightness in comparison with Variant 1.

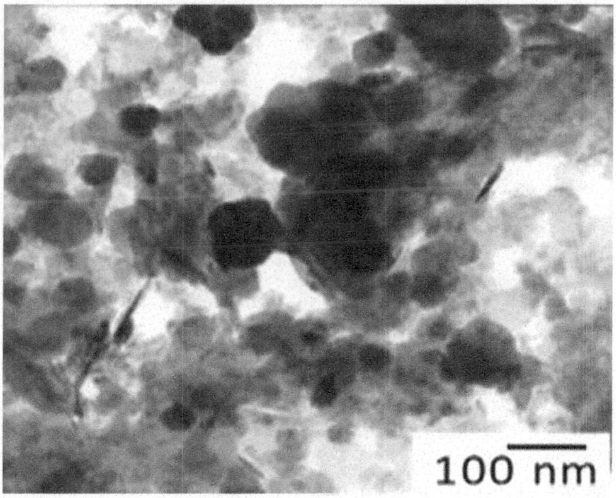

Fig. 3.3 Study of magnetite nanoparticles with use microscope electronic translucent JEM-2100

Variant 3. Mobility of hydrogen protons in 0.9% NaCl solution that has previously been processed ICNB nanoparticles is maximized. Therefore, the image brightness is much higher than before (in the previous variants).

Thus, previously conducted research clearly shows that the nanoparticles of ICNB change the mobility and the orientation of the hydrogen atoms in liquids that are registered in the visual evaluation of MRI.

The next set of studies was essential and aimed at studying of functional activity of red blood cells at the storage stages at a positive temperature after by modifying the mobility and spatial orientation of hydrogen protons in the pericellular fluid using magnetite nanopatricles of ICNB.

A study of the sedimentation stability of RBCs showed a highly significant difference between control and test data. Data of sedimentation stability of the RBCs at the stages of a study were presented in Fig. 3.4.

Figure 3.4 shows that sedimentation stability of RBC in test tubes are reliably more highly ($p < 0.001$) then in control tubes at stages of research. It should be said that negative surface charge of human RBCs's results primarily from the presence of ionogenic carboxyl groups of sialic acids on the cell surface [32–34]. The value of the charge is determined by the amount of adenosine threephosphate (ATP). ATP is macroergic compounds, the product of glycolysis. The sedimentation stability of RBC is determined by the amount of ATP.

In this case, the change of mobility and spatial orientation of the hydrogen protons in the extracellular liquid significantly increased the sedimentation stability of RBCs in the test compared to the control. For greater clarity results of RBCs sedimentation is shown in Fig. 3.5.

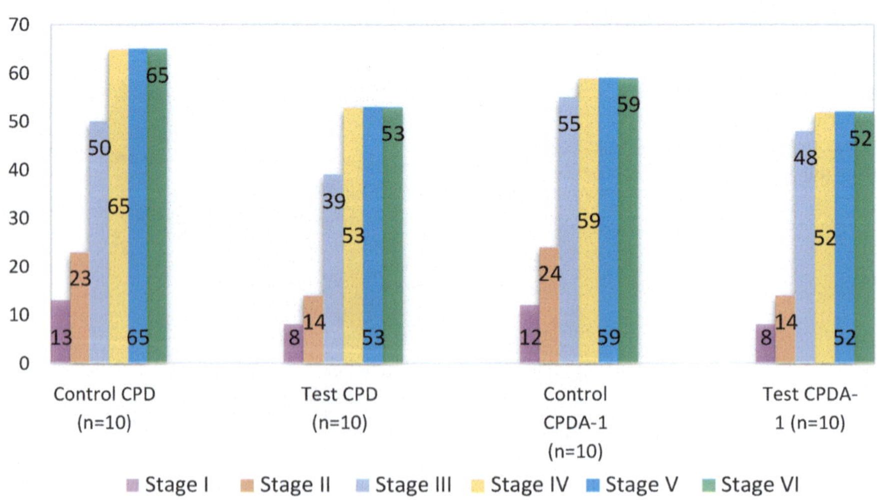

Fig. 3.4 Study of the sedimentation stability of RBCs at the stages (M ± m; $p < 0.001$)

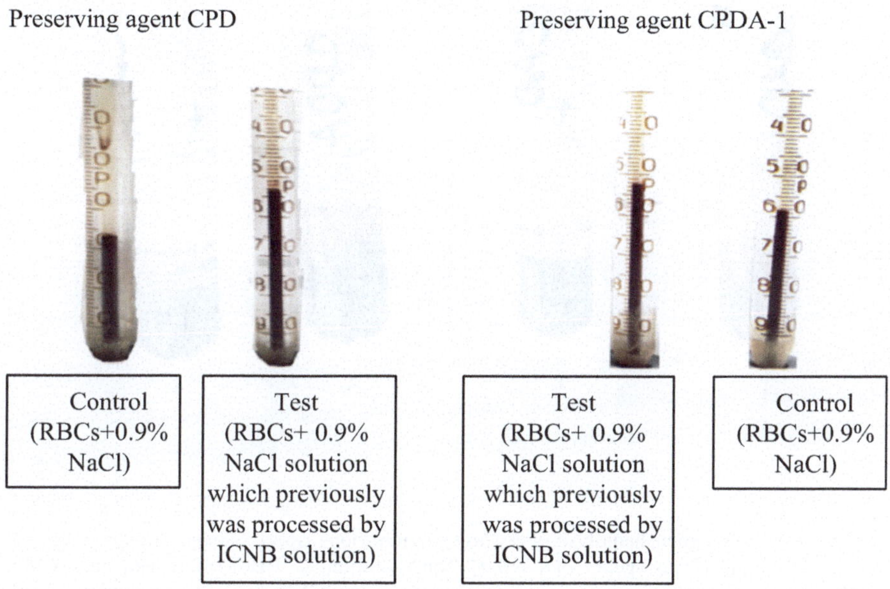

Fig. 3.5 Sedimentation stability of erythrocytes at the stage VI of the study

Figure 3.5 is showing that with preserved in the anticoagulant CPD the sedimentation stability in the control was 62 mm; in the test—53 mm. With preserved in the anticoagulant CPDA-1: control—58 mm; test—52 mm.

Thus, following the logic of the above reasoning, if improving sedimentation stability of RBC is associate with an increase in ATP, then the isotonic solution which previously was processed by ICNB should actively stabilize the membranes of RBCs and inhibit hemolysis. Therefore, the next investigation was to study the hemolysis processes preserved of the RBCs at various stages. Results of the visual assessment hemolysis of erythrocytes in various aspects of exposure are presenting in Fig. 3.6.

Figure 3.7 shows that in the control tubes at the stage VI of the study there are pronounced signs of hemolysis. In contrast, hemolysis is not recorded in test tubes. Visual analysis was supplemented by objective data of the photometric method, as well as the method of calculation of hematocrit.

Objective data of free Hb and calculated HCT at the stage VI of the study are presented in Table 3.5.

The data in Table 3.5 indicate that in test tubes the average content of free Hb in the VI stage of the study 6.25 ± 0.1 g/l were less than controls; and the average calculation HCT was lower by 1.42%

Thus, the objective indicators of Table 3.5 are highly reliable ($p < 0.001$) and confirm the obtained visual effect of inhibiting hemolysis of preserved RBCs by physiological solution, which was pretreated by ICNB. Microscopic observation of changes in the morphology of RBCs was a logical continuation of the study of the

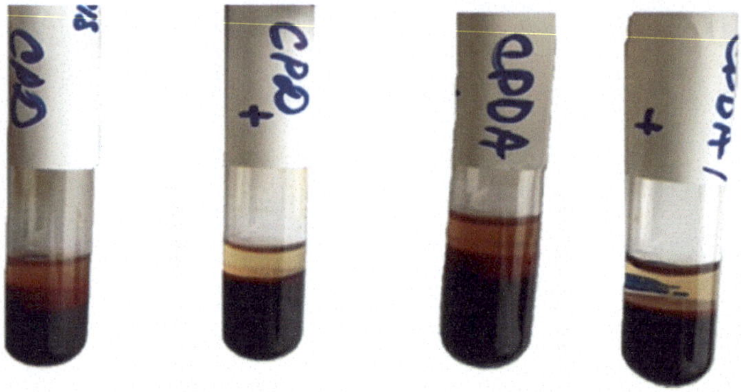

1 2 3 4

Fig. 3.6 Visual assessment hemolysis of erythrocytes in various aspects of exposure at the stage VI. *Notes* 1—RBC (preserving agent CPD) + 0.9% NaCl solution; 2—RBCs (preserving agent CPD) + 0.9% NaCl solution which previously was processed by ICNB; 3—RBCs (preserving agent CPDA-1) + 0.9% NaCl solution; 4—RBCs (preserving agent CPDA-1) + 0.9% NaCl solution which previously was processed by ICNB

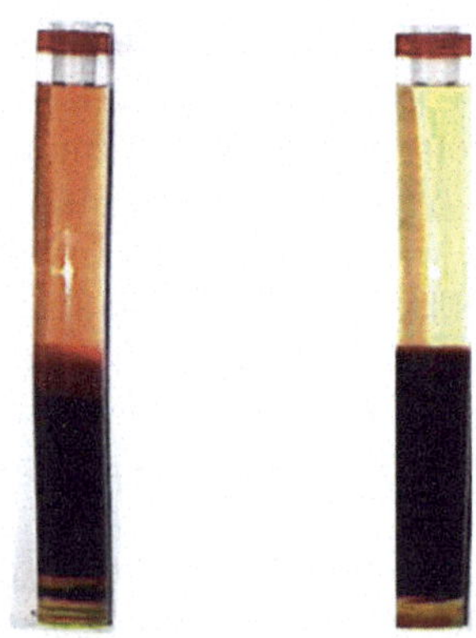

Control (RBCs+ 0.9% NaCl) solution	Test (RBCs+ 0.9% NaCl solution which previously was processed by ICNB)

Fig. 3.7 The presence of free Hb in the supernatant (preserving agent CPD)

Table 3.5 The indicators of free Hb and calculated HCT at the stage VI of the study at baseline values: Hb = 154 g/l; HCT = 0.46

Preservatives	Variants	Stage VI		*p*
		Free Hb (g/l)	Calculation HCT (%)	
CPD	Control (n = 10)	7.8 ± 0.1	1.84 ± 0.1	<0.001
	Test (n = 10)	1.7 ± 0.1	0.4 ± 0.1	<0.001
CPDA-1	Control (n = 10)	8.0 ± 0.1	1.9 ± 0.1	<0.001
	Test (n = 10)	1.6 ± 0.1	0.5 ± 0.1	<0.001

erythrocyte membranes stabilization effect. The results of microscopic examination of the morphology of erythrocytes in different variants of preservation and treatment are shown in Fig. 3.8.

Figure 3.8 clearly demonstrates that microscopically in the control variant at the VI stage of the study, widespread appearance of spheroechinocytes is observed. On the contrary, in the test variants, the shape of red blood cells at the stage VI was unchanged. Pathological changes in the RBC's shape and size in the control variants are most likely associated with inhibition of glycolysis processes [35]. Consequently, the number of ATP and 2.3 DPG decreases, the permeability of the erythrocyte membranes is disturbed, the state of the hemoglobin buffer changes. As a result, the pH of intracellular and extracellular media are changes.

The decrease in the formation of 2.3 DPG leads to the acidulation of intracellular environment of the RBCs. Deoxygenated hemoglobin which was previously formed actively binds the $[H^+]$, that comes from the extracellular environment and alkalizes the extracellular environment. The effect of RBC reduction, the appearance of widespread spheroechinocytes is observed in microscopy. Subsequently, oxyhemoglobin moves to the extracellular environment as a result of processes intensification destruction of the membranes of RBCs. The accumulation of oxygenated hemoglobin in the extracellular environment causes by shifting towards the acid of the pH.

The above mechanisms have been confirmed in the study of the dynamics of pH changes in the extracellular medium of preserved of the RBCs. The dynamics of pH changes in the extracellular medium of RBCs storage at key stages of the study on the example of preserving agent CPD is shown in Fig. 3.9.

Figure 3.9 demonstrates that despite the initial acidic environment of the preservative (pH CPD = 5–6) in the control and test in the extracellular medium at the first stage of the study alkaline pH is registered. The appearance of differences in the dynamics of change in the color of the pH indicator between the control and the test is clearly observed in the subsequent stages of the study. So, against the background of the appearance of hemolysis signs significant decrease of the pH to 7.1–7.2 in the control at the VI stage of the study is registered. On the contrary, the pH of the extracellular medium remains relatively stable and corresponds to the parameters 7.4–7.5 in the test at the VI stage of the study.

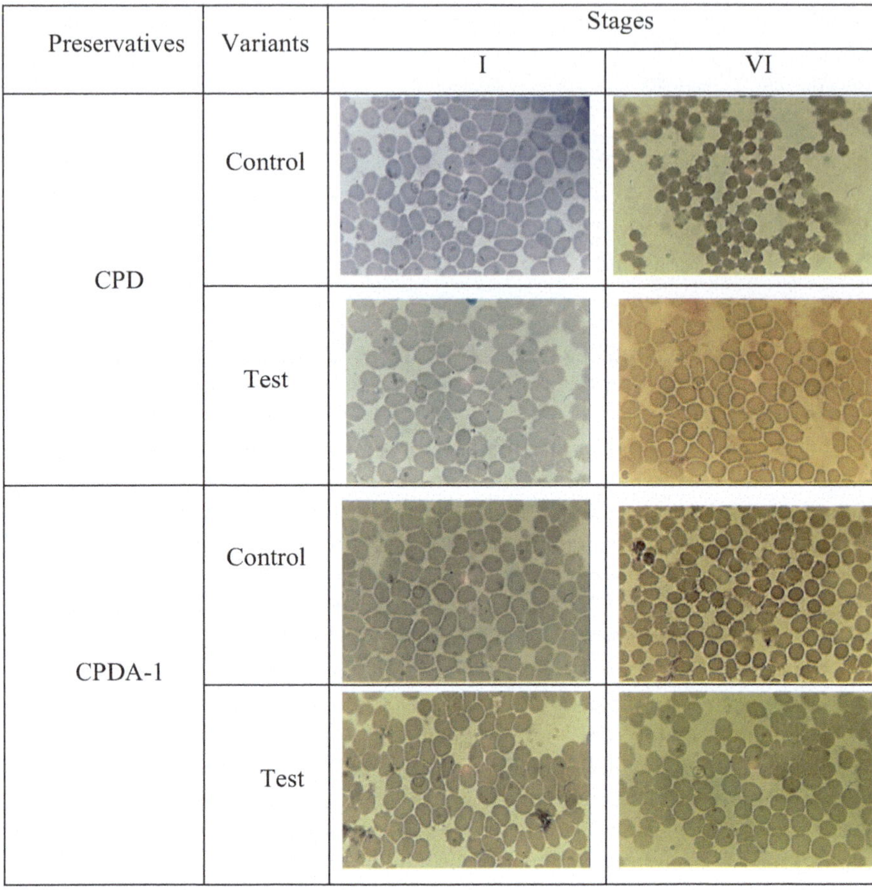

Fig. 3.8 The microscopic examination of the morphology of erythrocytes in different variants of preservation and treatment

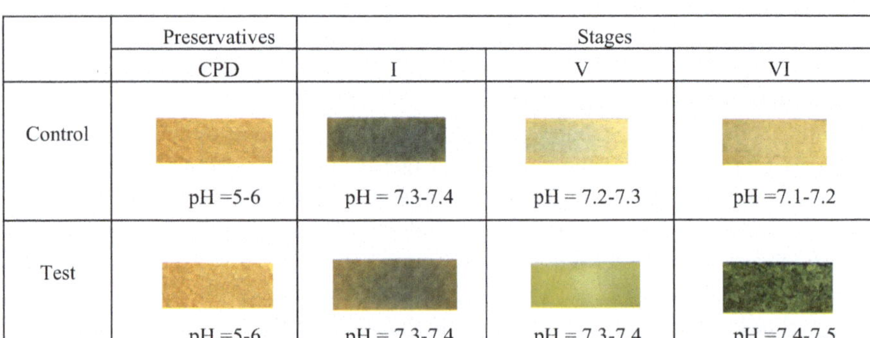

Fig. 3.9 Dynamics of pH changes in the extracellular medium of RBCs storage at key stages of the study on the example of preserving agent CPD

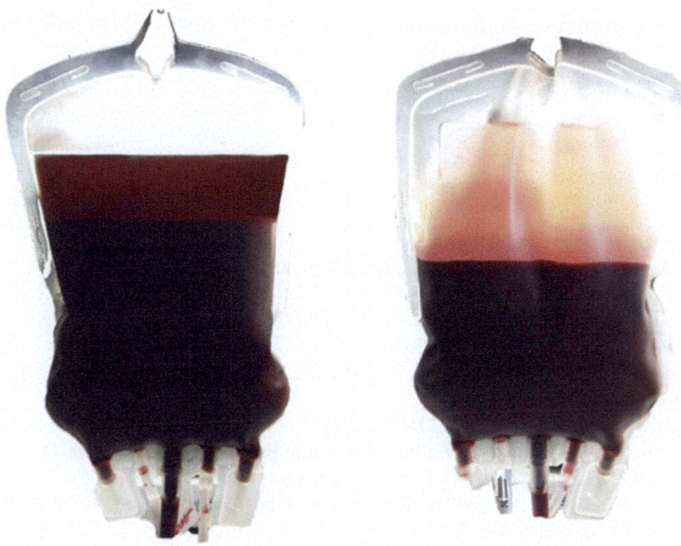

Fig. 3.10 The hemolysis inhibition effect with using method of additive modernization of preservative solutions adapted to the manufacture process at stage VI of the study

Thus, obtained result, show that change in cytoplasmic pH is both necessary and sufficient for the shape changes of human erythrocytes [36].

The effect of hemolysis inhibition by the method of additive modernization of preservation solutions, that adapted to the manufacture process at the VI stage of the study is shown in Fig. 3.10.

3.4 Conclusions

As a result of the studies it was found that physiologic solution NaCl which previously was processed by ICNB and added to the preserved of the red blood cells actively inhibits of hemolysis processes of RBCs at the storage stages at a positive temperature. A comprehensive analysis of data revealed the primary mechanisms of the effect modernized of the saline solution on the preserved RBCs. It was established that saline NaCl, which had previously been processed by magnetite nanoparticles (ICNB) had a marked membrane-stabilizing effect, inhibits hemolysis and increasing the sedimentation stability of preserved RBCs. In General, these effects provide the sustainability of the functional activity of preserved RBCs in during storage.

Thus, the first optimistic results were obtained on the way of creation a simple and practical method of additive modernization of preservation solutions that does not violate the compliance requirements, improves the quality, efficiency and safety transfusion of red blood cells.

References

1. J.R. Hess, Vox Sang. **91**, 13–19 (2006)
2. C.F. Högman, H.T. Meryman, Transfus. Med. Rev. **13**, 275–296 (1999)
3. J.R. Hess, Transfus Apher Sci. **43**, 51–59 (2010)
4. A.B. Zimrin, J.R. Hess, Vox Sang. **96**, 93–103 (2009)
5. L. Van de Watering, Vox Sang. **100**, 36–45 (2011)
6. D. Wang, J. Sun, S.B. Solomon et al., Transfusion **52**(6), 1184–1195 (2012). https://doi.org/10.1111/j.1537-2995.2011.03466.x
7. S.A. Glynn, Transfusion **50**, 1164–1169 (2010)
8. G.M. D'Amici, S. Rinalducci, L. Zolla, J. Proteome Res. **6**, 3242–3255 (2007)
9. A. Verhoeven, P. Hilarius, D. Dekkers et al., Vox Sang. **91**, 244–251 (2006)
10. R.L. Sparrow, M.F. Veale, G. Healey, K.A. Payne, Transfusion **47**, 966–968 (2007)
11. G.J. Bosman, E. Lasonder, Y.A. Groenen-Döpp et al., J. Proteomics. **73**, 396–402 (2010)
12. J.R. Hess, Transfusion **52**(8), 1827–1835 (2012). https://doi.org/10.1111/j.1537-2995.2011.03511.x
13. J.R. Hess, R.L. Sparrow, P.F. van der Meer et al., Transfusion. **49**, 2599–2603 (2009)
14. L.J. Dumont, J.P. AuBuchon, Transfusion **48**, 1053–1060 (2008)
15. M.J. McAteer, L.J. Dumont, J. Cancelas et al., Vox Sang. **99**, 34–43 (2010)
16. The use of magnetite nanoparticles in applied medicine. Int. J. Nano Dimension. **2,** 1(5), 25–28 (2011)
17. A.N. Belousov, E.Yu. Belousova, The first steps in discovery new mechanisms of cellular regulation in means by nanotechnology preparations in *X International Conference New Information Technologies in Medicine and Ecology* (Yalta, Gursuf, 2002), pp. 420–425
18. A.N. Belousov, E.Yu. Belousova, Mechanisms cell regulation by nanotechnology preparations (MCS-B), in *BioNanotech Conference & Expo 2012*. June 18–21 (Santa Clura, CA, USA, 2012)
19. A.N. Belousov Effect of magnet controlled sorbent on parameters of acid-base balance of the blood and the processes of glycolysis in erythrocytes. Pain Anesth. Intensive Care. Kiev. **1**, 263–265 (2000)
20. A.N. Belousov, V.P. Nevzorov ultrastructure of cells in the kidneys and lungs of rabbits after administration of magnetite, in *International Collection of Scientific Papers IV Scientific and Practical Conference on the Creation and Testing of New Drugs*, vol. 4 (Moscow, 1997), pp. 77–87
21. A.N. Belousov, V.P. Nevzorov, Ultrastructure of liver cells after administration of magnetite, in *International Collection of Scientific Papers IV Scientific and Practical Conference on the Creation and Testing of New Drugs*, vol. 4 (Moscow, 1997), pp. 71–77
22. A.N. Belousov Spectrum of application magnetite nanopaticles in medicine. Nanotech **2** Chapter 3, 154–157 (2009). ISBN: 978-1-4398-1783-4
23. A.N. Belousov Effect on hemolysis and transport ATPase activity of erythrocytes by means nanoparticles of magnetite controlled sorbent (MCS-B). Pain Anesth. Intensive Care. Kiev 1 (ad), pp. 26–28 (2012)
24. A.N. Belousov J. Cell Mol. Biol. **2**(1) 2017
25. A.N. Belousov, Application magnetite of nanoparticles (ICNB Preparation) as magnetically-resonant contrasting means during visualization of tumours. Clean Technol. Sustain. Ind. Organ. **10**, 379–381 (2013)
26. A.N. Belousov Prospects Med. Biol.. LSMU. V. IY **1**, 94–97 (2012)
27. A.N. Belousov, J. Biomed. Eng. Res. **2**(3), 147–152 (2014)

28. A.N. Belousov, J. Nanomater. Molec. Nanotechnol. Manuscript number: JNMN-14-25, 17-21 (2014)
29. A.N. Belousov, J. Clean Technol. Sustain. Ind. Org. **10**, 379–381 (2013). ISBN: 978-1-4822-5819-6
30. A.N. Belousov, J. Mater. Sci. Eng. A **4**(11), 367–372 (2014)
31. A.N. Belousov, J. Cell Mol. Biol. **2**(005), 1–5 (2017)
32. E.H. Eylar, M.A. Madoff, O.V. Brody et al., J. Biol. Chem. **237** (1992)
33. G.M.W. Cook, D.H. Heard, G.V.F. Seaman, Nature (Lond.) **191**, 44 (1961)
34. K.-M. Jan, S. Chien, J. Gen. Physiol. **6**(i), 638–654 (1973)
35. D. Sakota, R. Sakamoto, N. Yokoyama et al., Artif. Organs. **33**(9), 733–739 (2009). https://doi.org/10.1111/j.1525-1594.2009.00898.x
36. M.M. Gedde, D.K. Davis, W.H. Huestis, Biophys J. **72**(3), 1234–1246 (1997). doi: 10.1016/S0006-3495(97)78770-8)

Chapter 4
Application of Infrared Spectroscopy to Study the Effect of Magnetite Nanoparticles (ICNB) on Molecular Structure of the Membranes of Preserved RBCs

A. N. Belousov, E. I. Malygon, V. V. Yavorskiy, and E. Yu. Belousova

Abstract This study was devoted to the learning changes in the structure of erythrocyte membranes at the level of molecular bonds during their storage at a positive temperature by means method of infrared spectroscopy (IR). Objects of research were red blood cells (RBCs) into bags containing preservative CPD and RBCs into bags containing preservative CPDA-1. As membrane protective used saline which had previously been treated with magnetite nanoparticles (ICNB) by the Belousov's method. The physiological solution that was treated with nanoparticles was added to the preserved RBCs according to the developed method. Sample of control was the addition of intact saline. Analysis of changes occurring in the IR spectra of samples of control and test in the CPD medium was showed that during the first 28 days storage of: (1) In the suspension of control of the RBCs, noticeable destructive changes in the molecular structure membrane of erythrocytes at the 14th day storage begins. After three weeks, the molecular structure membrane of erythrocytes is completely destroyed; (2) In the sample of test there was a weakening and rupture of molecular bonds only at the 28th day storage of RBCs. Complete destruction of the structure of membrane of erythrocytes occurs at the 35th day storage. Analysis of changes occurring in the IR spectra of samples of control and test in the CPDA-1 medium was showed that during 49 days storage of: (1) In the suspension of control of RBCs noticeable destructive changes in the molecular structure begins in four weeks, and after six weeks storage the molecular structure of erythrocytes membrane are completely destroyed; (2) In the sample of test, a significant weakening of intra- and intermolecular bonds in the structure of erythrocytes membrane occurs after

A. N. Belousov (✉) · E. Yu. Belousova
Laboratory of Applied Nanotechnology of Belousov, Kharkiv, Ukraine
e-mail: an.belousov2012@ukr.net; an.belousov2012@yandex.ua

A. N. Belousov · E. I. Malygon · V. V. Yavorskiy
Kharkov Medical Academy of Postgraduate Education, Kharkiv, Ukraine

A. N. Belousov · E. I. Malygon · V. V. Yavorskiy · E. Yu. Belousova
Kharkov Regional Center of Blood Service, Kharkiv, Ukraine

A. D. Pogrebnjak et al. (eds.), *Nanomaterials in Biomedical Application and Biosensors (NAP-2019)*, Springer Proceedings in Physics 244,
https://doi.org/10.1007/978-981-15-3996-1_4

six weeks. However, the complete destruction of the structure is not observed. After seven weeks storage of erythrocytes obvious violations of the molecular structure of lipids and proteins that make up the RBCs are visible but some of the strongest compounds still remain. In general, the results clearly showed that the presented method of application of nanotechnology significantly increases the storage time of RBCs in different versions of preservatives due to mechanisms to reduce violations of the molecular structure of proteins and lipids in the erythrocyte membranes. Presented method of application of nanotechnology is not only safe for use in practice in the Blood Service, Transfusiology and Hematology, but also is the most promising innovation project.

4.1 Introduction

For several decades RBCs components have been prepared as concentrates suspended in nutrient additive solution, which preserves and extends the shelf-life of the RBCs component, allowing up to 6–7 weeks of refrigerated storage [1]. Nevertheless, during storage RBCs undergo a complex and progressive accumulation of physicochemical changes, collectively referred to as the RBCs storage lesion [2, 3]. Recent clinical studies have identified RBCs transfusion as an independent risk factor for increased morbidities and mortalities in certain groups of patients, including trauma, cardiac surgery and the critically-ill (reviewed in [4–6]). Additionally, some of these studies have identified that older stored RBCs are more strongly implicated in poorer outcomes compared to fresher RBCs [6]. In order to address these concerns, there is renewed interest to better understand the RBCs storage lesion and to find ways to ameliorate the deleterious effects of storage, thereby improving the quality, efficacy and safety of RBCs components for all transfusion recipients.

While increased research effort is being directed to better understand the effects of storage on RBCs and the potential impact on transfusion outcomes [7], slower progress is being made in finding ways to deter the detrimental effects of the RBCs storage lesion.

Despite the RBCs having been a favorite experimental model for cellular biologists and biochemists, RBCs storage research has repeatedly demonstrated that a lot of fundamental biology about RBCs is still not well understood. The complexity of the inter-relationship between RBCs biochemistry, cytoskeletal structure and membrane properties have made it difficult to predict how RBCs will respond to different storage conditions. Exposure of RBCs to non-physiological storage environments has pointed to the existence of previously unknown biochemical mechanisms in RBCs, including apoptotic-like processes, ion and osmotic channels that behave differently than expected, exposure of new or altered receptors possibly due to oxidative and/or protease/glycosidase activities or altered senescence [8–11].

The benefits gained by improved RBCs component quality should more than justify any real or perceived inconvenience to the blood services in implementing adjustments to their processing procedures or additional processing costs of the introduction of new generation RBCs additive solutions.

Currently, nanotechnology opens up new prospects for the study of not only the mechanisms of cell damage, but also the development of effective and safe methods of cells storage outside the body.

In Ukraine, the first standardized and biocompatible magnetite nanoparticles for medical use were manufactured and patented in 1998. These are intracorporeal nanobiocorrector of brand ICNB, magnet-controlled sorbent of brand MCS-B, and biologically active nanodevice of brand Micromage-B [12]. It is well established that the magnetite nanoparticles effectively modulate the metabolic processes in leukocytes, regulate activity of the enzyme link of the antioxidant system in erythrocytes in healthy and sick patients [13–15]. Previously the complex investigations that were performed in the study of the influence on metabolism of cells by preparations of nanotechnology show that in whole standardized biocompatibility of magnetite nanoparticles have nonspecific and modulated effect on metabolic processes. Research of ultrastructure investigations of the reticuloendothelial system (liver, lungs and kidneys) it was proved that after injection of biocompatibility magnetite nanoparticles into a vein caused nonspecific activation of the metabolic processes, increase adaptive mechanisms and potential of organelle cells, acceleration of reparative processes a level of membranes and macromolecules [14, 16, 17]. Existing sorption and indirect (magnetic) effects not only allow selectively absorb the protein of surface membrane cells by magnetite nanoparticles (according to the principle of magnetospheres), but also to prevent the oxidative modification of proteins by way of stabilizing the active groups, normalizing a state of receptors that are located on the surface membrane of cells, increasing activity of enzymes' membrane-bound [18–20].

Recent scientific work related to use of magnetite nanoparticles (ICNB) in contrast means in an MRI investigation of cancer reliably was shown that nanoparticles cause reversible changes associated with a temporary increase in the mobility of hydrogen protons in the pericellular fluid that inevitably modifies the metabolism in malignant cells [21]. The results of these investigations have not only widened the understanding of the mechanisms of action of nanoparticles on condition outside and intracellular spaces but also have revealed new aspects of the cellular(cells) metabolism, determined the membrane role of cellular enzymes in the regulation processes of metabolism [19, 22–24].

Also, it was established that extracorporeally processing the blood by nanoparticles of MCS-B reliably reduces activity of Ca, Mg–ATPHese of erythrocytes.

Currently, studies have shown that magnetite nanoparticles are able to inhibit hemolysis of heparinized blood, increase the activity of ATP and 2.3 DPH in red blood cells, regulate transmembrane metabolism and inhibit eryptosis [19, 21, 25].

In our recent studies, it has been found that physiologic solution NaCl which previously was processed by magnetite of nanoparticles (ICNB) and added to the preserved of the red blood cells actively inhibits of hemolysis processes of RBCs at the storage stages at a positive temperature (see Fig. 4.1). A comprehensive

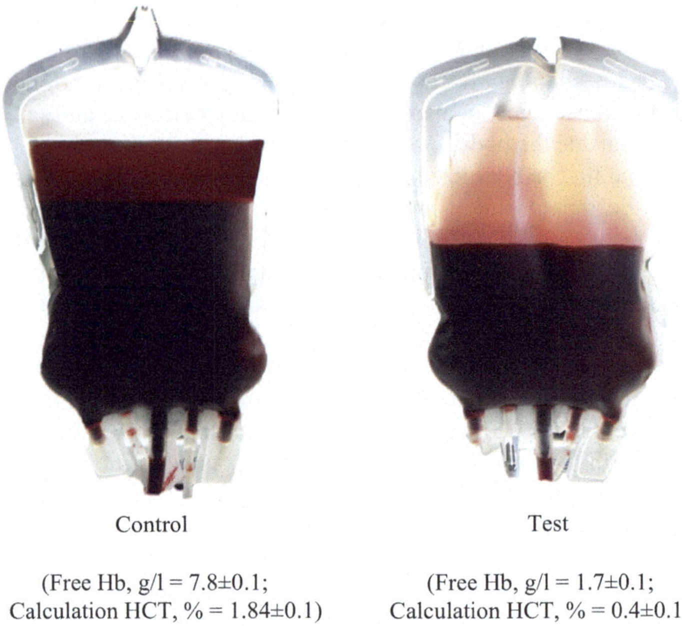

Fig. 4.1 The hemolysis inhibition effect with using method of additive modernization of preservative solutions adapted to the manufacture process on the 35th day of the study

analysis of data revealed the primary mechanisms of the effect modernized of the saline solution on the preserved RBCs. It was established that saline NaCl, which had previously been processed by magnetite nanoparticles (ICNB) had a marked membrane-stabilizing effect, inhibits hemolysis and increasing the sedimentation stability of preserved RBCs. In General, these effects provide the sustainability of the functional activity of preserved RBCs in during storage. The optimization results were obtained in creating a simple and practical method of additive modernization of preservation solutions that does not violate the compliance requirements, improves the quality, efficiency and safety transfusion of RBCs [26–28].

4.2 Materials and Methods

4.2.1 Materials

1. Standardized intracorporeal nanobiocorrector of ICNB was taken as magnetite nanoparticles. Magnetite nanoparticles synthesized by co-precipitation method. The main physics and chemical properties of ICNB the following data and also in Tables 4.1, 4.2, 4.3 and 4.4; Figs. 4.2 and 4.3 were presented:

Table 4.1 The calculated lattice parameters of the phases

Phase name	a (Å)	b (Å)	c (Å)	Alpha (град)	Beta (град)	Gamma (град)
Magnetite low	8.38	8.38	8.38	90	90	90
Magnetite low, syn	5.93	5.93	14.7	90	90	120
Johannsenite	9.89	9.05	5.28	90	105.5	90

Table 4.2 Determination of percent composition of the ICNB by X-ray spectrometer ARL OPTIM'X (semi-quantitative analysis)

Compound	wt%	Std. Err.	El	wt%/O_2	Std. Err.	El	wt%	Std. Err
Fe_3O_4	97.37	0.09	Fe	68.40	0.07	Fe	97.62	0.09
CaO	2.26	0.07	Ca	1.71	0.05	Ca	2.3	0.07
P_2O_5	0.280	0.027	Px	0.122	0.012	Px	0.157	0.015
MnO	0.255	0.013	Mn	0.198	0.010	Mn	0.278	0.014
SiO_2	0.098	0.027	Si	0.046	0.013	Si	0.059	0.016
SO_3	0.032	0.013	Sx	0.0126	0.0051	Sx	0.0164	0.0066
Cl	0.0280	0.0090	Cl	0.0280	0.0090	Cl	0.0380	0.012

Table 4.3 X-ray analysis of ICNB in X-ray diffractometer Rigaku Ultima IV (CuKα, Kβ filter–Ni), one-coordinate DTeX semiconductor detector

Phase	Formula	Space group	№ card database ICDD
Magnetite low	$Fe_{2.886}\ O_4$	227: Fd-3m, choice-2	10861339 (ICDD)
Magnetite low, syn	Fe_3O_4	166: R-3m, hexagonal	10716766 (ICDD)
Johannsenite	Ca Mn +2 Si_2O_6	15: C12/c1, unique-b, cell-1	380413 (ICDD)

Table 4.4 The phases of magnetite of nanoparticles (RIR-method; error 8 ± 3%)

Phases (method of corundum numbers)	Content (%)
Magnetite low	71
Magnetite low, syn (hexagonal)	29

- Concentration of the colloidal solution of magnetite nanoparticles in physiology solution of NaCl is 0.0225%;
- Theoretical osmolality of colloid solution is 500 mosmol/l;
- Size of magnetite nanoparticles is 6–12 nm;
- Total area of surface magnetite of nanoparticles Ss = 800–1200 m^2/g;
- Magnetization of saturation Is = 2.15 кA/m;
- ζ-potential = −19 mV.

2. 0.9% NaCl solution.

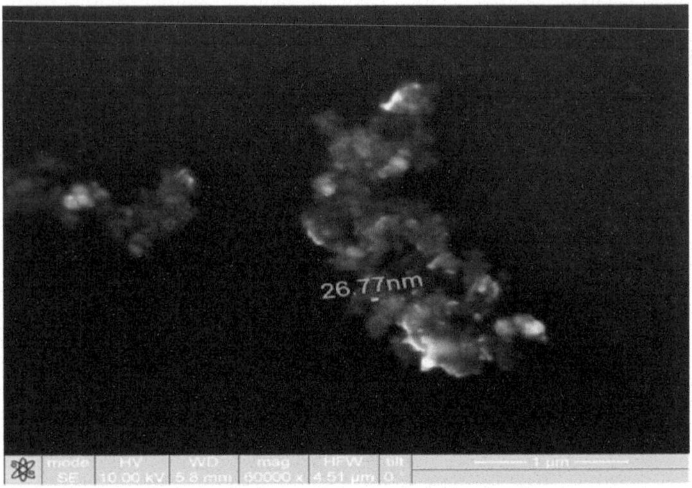

Fig. 4.2 Study of magnetite nanoparticles with use microscope ion-electronic raster-type Quanta 200 3 D

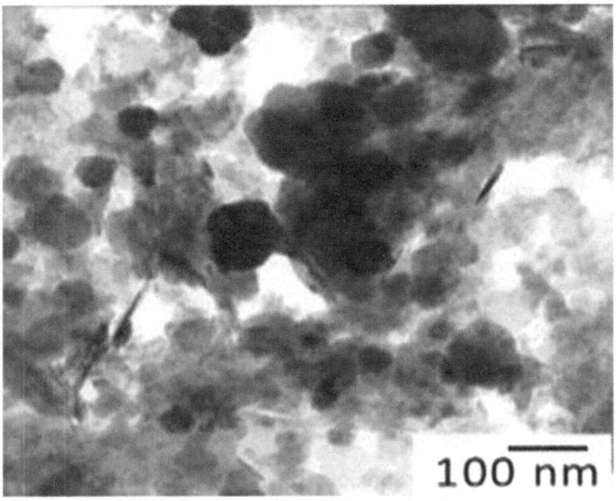

Fig. 4.3 Study of magnetite nano-particles with use microscope electronic translucent JEM-2100

3. 0.9% NaCl solution which previously was processed by ICNB in ratio 4:1.

4.2.2 Objects of Research

- Red blood cells (RBCs) into bags containing anticoagulant citrate, nutrient phosphate and dextrose (CPD);
- Red blood cells (RBCs) into bags containing anticoagulant citrate, nutrient phosphate, dextrose and adenine (CPDA-1).

Of each bag of 3 ml amounts of red blood cells was distributed into 20 sterile glass tubes. Then, into the first 10 tubes of control were added of 2 ml amounts 0.9% NaCl solution. Into the next 10 tubes of test were added of 2 ml amounts 0.9% NaCl solution, which previously was processed by ICNB.

Thus, the distribution of tubes was as the follows:

Tubes of control:

- 3 ml of red blood cells (CPD) + 2 ml 0.9% NaCl solution (n = 10);
- 3 ml of red blood cells (CPDA-1) + 2 ml 0.9% NaCl solution (n = 10).

Tubes of test:

- 3 ml of red blood cells (CPD) + 2 ml 0.9% NaCl solution that previously was processed by ICNB in ratio 4:1 (n = 10);
- 3 ml of red blood cells (CPDA-1) + 2 ml 0.9% NaCl solution that previously was processed by ICNB in ratio 4:1 (n = 10).

IR Spectrophotometer-29 (LOMO), working in NSC Kharkov Institute of Physics and Technology of the National Academy of Sciences of Ukraine, was used for registration of absorption spectra of an aqueous solution of erythrocytes in the IR range. The spectra were recorded in the range of 4000–400 cm^{-1} (from 2 to 25 mkm—middle infrared region).

Calibration was carried out on the spectrum of polystyrene with known frequencies of absorption maxima. The amendment amounted to an average of 10–5 cm^{-1}.

Liquid water is characterized by strong absorption in the study range, so to record the spectra of aqueous solutions is necessary to obtain a thin layer of the substance. For this purpose, two thin plates of CaF_2 material transparent in the mid-IR range, that does not have its own characteristic absorption bands, which could affect the interpretation of the IR spectrum of the aquatic solution, were used. A drop of the test liquid was crushed between two round plates CaF_2 and installed in the measuring channel of the device. The same plates were installed in the comparison channel, but without liquid. Spectral recording was carried out immediately after placing the sample in the device and lasted 10 min. The temperature in the chamber was 25–30 °C.

Tests were carried out in eight stages: day 1—I, day 7—II, day 14—III, day 21—IV, day 28—V, day 35—VI, day 42—VII, day 49—VIII.

The blood after performance of the biochemical investigation was stored in the refrigerating chamber at temperature +4 °C.

Statistically processing the obtained results was carried out by parametrical method of variation statistics by Student criterion. Processing the obtained data was carried out by means of Excel.

4.3 Results and Discussion

IR absorption spectra at the initial stage of the study of RBCs suspensions in the control and test in ranges 400–1300 cm^{-1} and 1200–4000 cm^{-1} are shown in Fig. 4.4.

Since the basis of the studied suspensions of RBCs is liquid water, the most intense and broad bands of the spectrum correspond to different types of vibrations of molecules H_2O:

1. The strong absorption in the range of 800–600 cm^{-1} corresponds to the librational modes of H_2O (Fig. 4.4).
2. Deformation vibrations of N–O–N, due to changes in the angles of valence bonds, look like a narrow intense band 1630 cm^{-1} (Fig. 4.3).
3. Composite vibration mode of deformation and libration vibrations of H_2O molecules lies in the range of 2150–2100 cm^{-1} (Fig. 4.4).
4. The main band of valence vibrations caused by changes in the bond lengths in the water molecule covers the spectral range of 3600–3000 cm^{-1} (Fig. 4.4).

In addition, the spectrum has a number of bands corresponding to the preservative for RBCs, which is contained in saline:

- Series of small range peaks: 415, 435, 480, 510, 530 and 580 cm^{-1} (Fig. 4.4).
- Wide band with highs in the area of 1100 cm^{-1} (Fig. 4.5).
- Weak intensity peaks in the region of 1260 and 1400 cm^{-1} (Fig. 4.7).

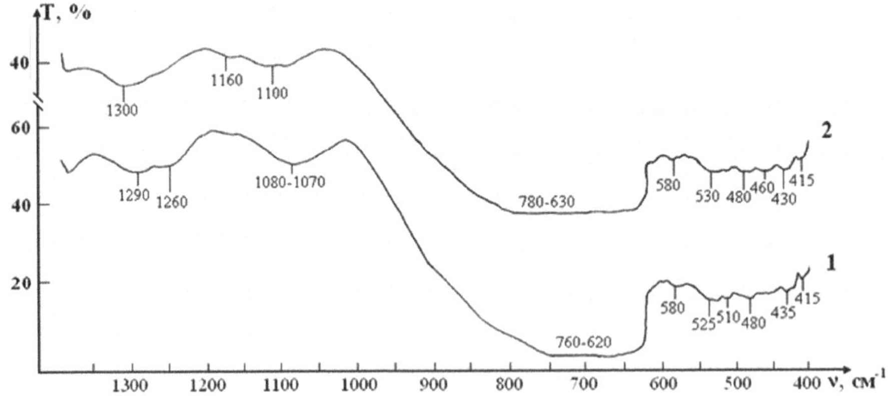

Fig. 4.4 IR absorption spectra of initial erythrocyte suspensions in the range 400–1300 cm^{-1}: curve 1—control; curve 2—test

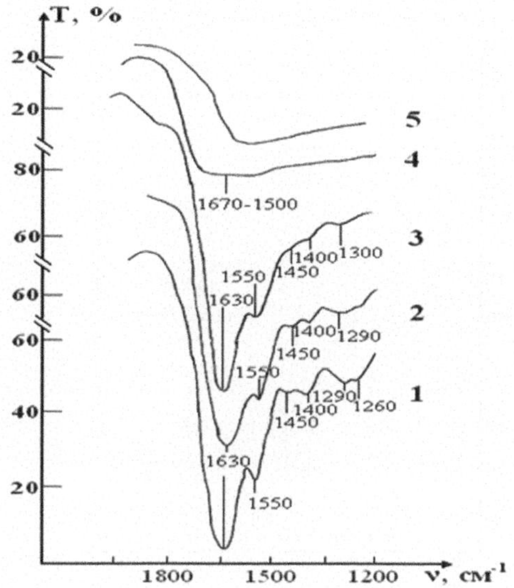

Fig. 4.5 IR absorption spectra of the sample of control consisting of RBCs suspension in native saline solution in CPD medium: curve 1—stage I (day 1); curve 2—stage II (day 7); curve 3—stage III (day 14); curve 4—stage IV (day 21); curve 5—stage V (day 28)

Thus, three characteristic peaks can be distinguished in the spectrum, which are directly related to intramolecular fluctuations in the structure of erythrocytes: 1290, 1450 and 1550 cm^{-1} (Fig. 4.6). Identification of absorption bands in the IR spectra of the initial RBCs suspension is presented in the Table 4.5.

The spectral range in which the characteristic bands corresponding to the main fragments of the structure of RBCs are presented in Figs. 4.5 and 4.6. These bands can be identified as follows:

Experiment 1 (preservative CPD)

A sample of control of the RBCs suspension in preservative CPD with intact saline solution on five stages of the study is presented in the Fig. 4.6. The peak of weak intensity of 1290 cm^{-1} can be attributed to fluctuations in the ring structure of the protein, as well as Amide III [29]. The weak peak of 1450 cm^{-1} refers to the deformation vibrations $\delta(CH_3)$ in the structure of lipids and proteins [30, 31]. Sharp band of average intensity at 1550 cm^{-1} caused by valence vibrations of ν(N–H) + ν(C–N) Amide II [32]. Figure 4.3 shows that after a week (stage II) of sample exposure the intensity of all these peaks associated with C-N valence vibrations and N-H deformation vibrations (1290, 1450 and 1550 cm^{-1}) is decreases. This may indicate the beginning of changes in the secondary structure of the protein. After two weeks (stage III), the band 1290 cm^{-1} is shifted towards high frequencies ($\rightarrow$1300 cm^{-1}) and turns into a small bend, the band 1450 cm^{-1} becomes barely

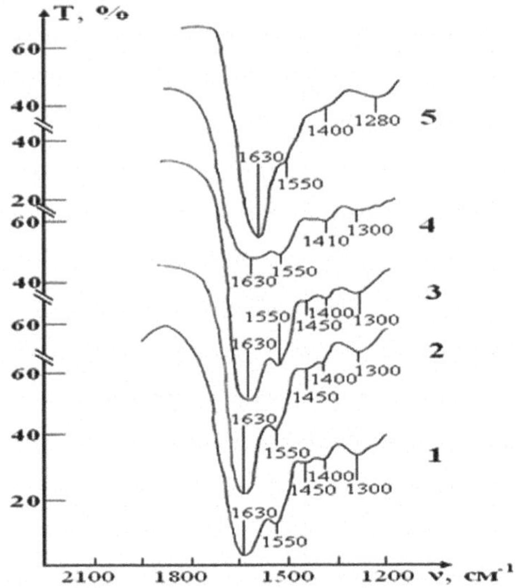

Fig. 4.6 IR absorption spectra of a sample consisting of a suspension of erythrocytes in a saline solution treated with magnetite nanoparticles in CPD medium: curve 1—stage I (day 1); curve 2—stage II (day 7); curve 3—stage III (day 14); curve 4—stage IV (day 21); curve 5—stage V (day 28)

noticeable shoulder, and the most intense peak of Amide II in the region of 1550 cm^{-1} is significantly weakened.

These changes indicate the degradation of the molecular structure of the substance and the weakening of intra- and intermolecular bonds. After three weeks of storage (stage IV) in the control IR spectrum completely smoothed, disappear all the bands of red blood cells, there is a wide band of blurred vibrations in the range of O–H (1650–1500 cm^{-1}). After four weeks of storage (stage V) changes in the spectrum are not observed. Given the latter, study of IR absorption spectra were impractical in the next planned stages (stages VI, VII).

A sample of test of the RBCs suspension in preservative CPD with saline solution that previously was processed by ICNB is presented in the Fig. 4.7.

The IR spectrum of erythrocyte suspension in the sample of test at the stage I is almost identical to the spectrum in the control, except that instead of the doublet 1290–1260 cm^{-1} there is a band 1300 cm^{-1} with a barely noticeable shoulder 1240 cm^{-1} (see Fig. 4.7). Over the next two weeks, no noticeable changes in the spectrum that would indicate disturbance of the molecular structure of RBCs were observed.

The intensity of the bands in zone 1300 and 1550 cm^{-1} are decreases, and the band 1450 cm^{-1} completely disappears on 21st day (stage IV) storage of RBCs. These changes indicate that the molecular structure of proteins and lipids of erythrocyte membranes begins to deteriorate. Hardly noticeable excesses associated with intramolecular vibrational against the background of water and bands preservative

Table 4.5 Identification of absorption bands in the IR spectra of the initial RBCs suspension

IR absorption spectra of initial erythrocyte suspensions in different ranges (cm^{-1})		Characteristic of bands
Control	Test	
415	415	Vibrational modes of chemical compounds that constitute the preservative for RBCs
435		
	450	
480	480	
510	510	
	520	
525		
580	580	
760–620	780–630	Librational mods of H_2O
1080–1070	1100	Bands of preservative
	1160	
1260		
1290		Fluctuations in the ring structure of the protein and Amide III
	1300	
1400	1400	Bands of preservative
1450		Deformation vibrations of the $\delta(CH_3)$ in the structure of lipids and proteins
	1460	
1550	1550	Valence vibrations of ν(N–H) + ν(C–N) of Amide II
1630	1630	Deformation vibrations of H–O–H
2100		Composite vibration mode of deformation and libration vibrations of H_2O molecules
	2130	
2400		Bands of saline
	2460	
2620		
	2700	
3510–3150	3600–3120	Band of valence vibrations of H–O–H

solution in the IR spectrum in during storage of RBCs appear only on the 28th day (stage V) of observation. Thus, the molecular structure of RBCs is not completely destroyed.

Summary of experiment 1

Analysis of changes occurring in the IR spectra of samples of control and test in the CPD medium was showed that during the first 28 days storage of:

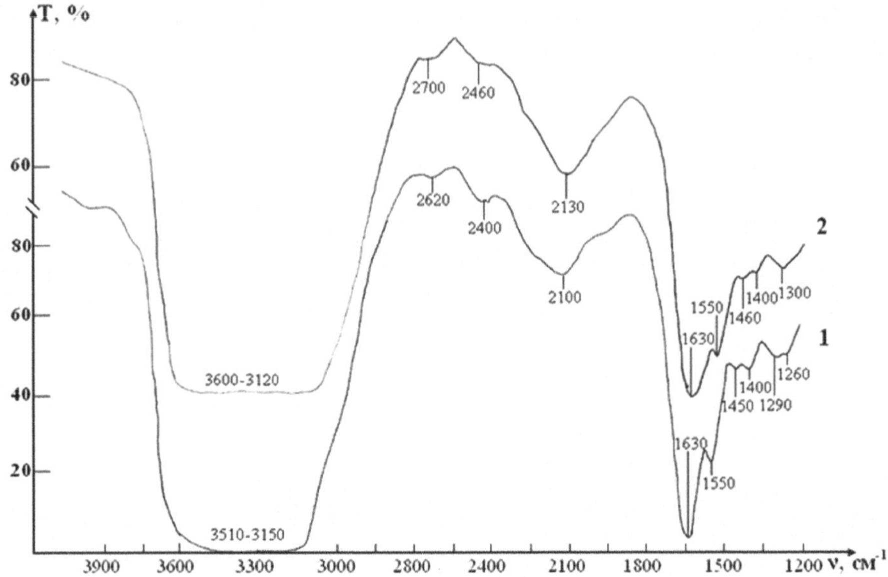

Fig. 4.7 IR absorption spectra of initial erythrocyte suspensions in the range 1200–4000 cm^{-1}: curve 1—control; curve 2—test

1. In the suspension of control of the RBCs, noticeable destructive changes in the molecular structure membrane of erythrocytes at the 14th day (stage III) storage begins. After three weeks, the molecular structure membrane of erythrocytes is completely destroyed.
2. In the sample of test there was a weakening and rupture of molecular bonds only at the 28th day storage of RBCs. Complete destruction of the structure of membrane of erythrocytes occurs at the 35th day (stage VI) storage.

Experiment 2 (preservative CPDA-1)

A sample of control of the RBCs suspension in preservative CPDA-1 with intact saline solution on main stages of the study is presented in the Fig. 4.8.

The IR spectrum of the initial sample contains the same bands of functional groups, which were described above in experiment 1. However, there are some differences in their intensity due to the characteristics of the blood donor from which the sample was obtained. After a three-week storage of RBCs (stage IV) in the spectrum there were no noticeable changes that would indicate disturbance of the molecular structure of the erythrocyte membrane.

After four weeks (stage V), noticeable changes appear in the spectrum (some bands are smoothed, their intensity decreases, small peaks disappear). These changes indicate that the molecular structure of proteins and lipids of erythrocyte membranes begins to break down. After five weeks (stage VI) no significant changes in the spectrum were found. After six weeks of RBCs storage (stage VII) barely noticeable

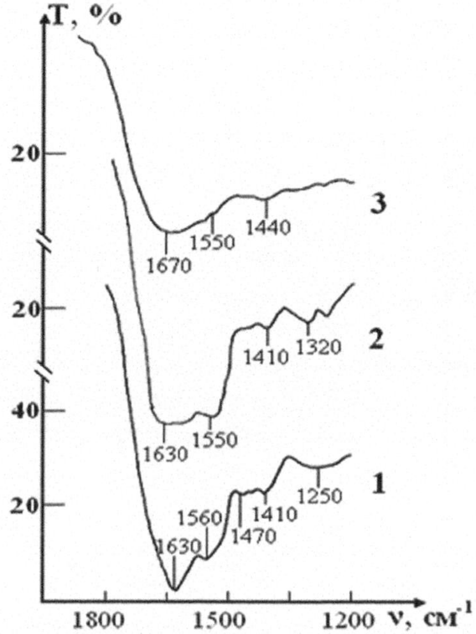

Fig. 4.8 IR absorption spectra of the sample of control consisting of RBCs suspension in native saline solution in CPDA-1 medium on main stages of the study: curve 1—stage I (day 1); curve 2—stage V (day 28); curve 3—stage VII (day 42)

excesses associated with intramolecular oscillations are observed in the spectrum against the background of a wide blurred band in the vibration range O–H (1650–1500 cm^{-1}) H_2O that indicates the complete destruction of the molecular structure of erythrocyte membrane.

A sample of test of the RBCs suspension in preservative CPDA-1 with saline solution that previously was processed by ICNB on main stages of the study is presented in the Fig. 4.9.

The dynamics of changes in the molecular structure of this sample is shown in Fig. 4.9 demonstrates that it remains fully stable for four weeks storage of RBCs. The first signs of a slight decrease in the intensity of the lipid and protein of bands appear after four weeks storage of RBCs. Nevertheless these changes are much less pronounced than in the control sample of this series.

After six weeks, the spectrum changes mainly due to the shift of bands in the high-frequency region and smoothing of small peaks. This indicates a significant weakening of intra-and intermolecular bonds in the structure of RBCs. However, complete destruction of the membrane structure of RBCs is not observed.

After seven weeks of storage, only one weak band of erythrocytes remains in the spectrum, corresponding to the valence vibrations of the most stable bond ν(N–H) + ν(C–N) of Amide II, which previously (in sample of control) experienced a

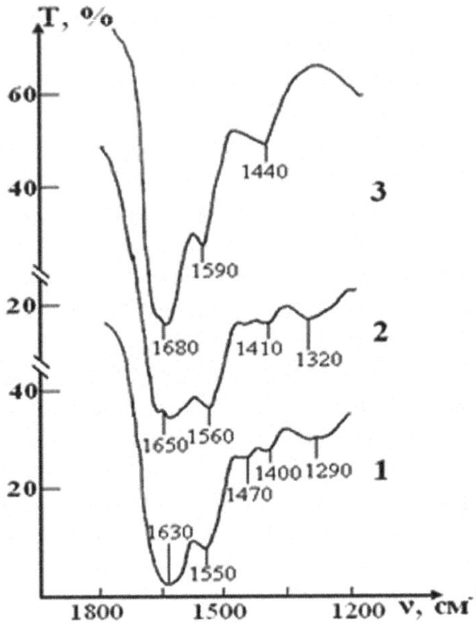

Fig. 4.9 IR absorption spectra of the sample of test consisting of RBCs suspension in preservative CPDA-1 with saline solution that previously was processed by ICNB on main stages of the study: curve 1—stage I (day 1); curve 2—stage V (day 28); curve 3–stage VII (day 42)

significant shift in the high-frequency region due to the weakening of this bond. The bands of deformation vibrations of protein and lipid molecules disappear, i.e. the molecular structure of erythrocytes is disturbed.

Summary of experiment 2

Analysis of changes occurring in the IR spectra of samples of control and test in the CPDA-1 medium was showed that during 49 days storage of:

1. In the suspension of control of RBCs noticeable destructive changes in the molecular structure begins in four weeks, and after six weeks storage the molecular structure of erythrocytes membrane are completely destroyed.
2. In the sample of test, a significant weakening of intra-and intermolecular bonds in the structure of erythrocytes membrane occurs after six weeks. However, the complete destruction of the structure is not observed. After seven weeks storage (stage VIII) of erythrocytes obvious violations of the molecular structure of lipids and proteins that make up the RBCs are visible but some of the strongest compounds still remain.

4.4 Conclusions

The method of infrared spectroscopy made it possible to track the dynamics of changes in all important types of bonds in molecules of erythrocyte membranes at the stages of their storage at positive temperature. The results clearly showed that the presented method of application of nanotechnology significantly increases the storage time of RBCs in different versions of preservatives due to mechanisms to reduce violations of the molecular structure of proteins and lipids in the erythrocyte membranes. In the future, with used nanotechnologies is planned to continue to study the features of metabolic processes of preserved RBCs at storage stages at positive temperature.

Nevertheless, today it is obvious that the presented method of application of nanotechnology is not only safe for use in practice in the Blood Service, Transfusiology and Hematology, but also is the most promising innovation project.

References

1. J.R. Hess, Vox Sang. **91**, 13–19 (2006)
2. C.F. Högman, H.T. Meryman, Transfus. Med. Rev. **13**, 275–296 (1999)
3. J.R. Hess, Transfus Apher Sci. **43**, 51–59 (2010)
4. A.B. Zimrin, J.R. Hess, Vox Sang. **96**, 93–103 (2009)
5. L. Van de Watering, Vox Sang. **100**, 36–45 (2011)
6. D. Wang, J. Sun, S.B. Solomon et al., Transfusion. (2012). https://doi.org/10.1111/j.1537-2995.2011.03466.x
7. S.A. Glynn, Transfusion **50**, 1164–1169 (2010)
8. G.M. D'Amici, S. Rinalducci, L. Zolla, J. Proteome Res. **6**, 3242–3255 (2007)
9. A. Verhoeven, P. Hilarius, D. Dekkers et al., Vox Sang. **91**, 244–251 (2006)
10. R.L. Sparrow, M.F. Veale, G. Healey, K.A. Payne, Transfusion **47**, 966–968 (2007)
11. G.J. Bosman, E. Lasonder, Y.A. Groenen-Döpp et al., J. Proteomics **73**, 396–402 (2010)
12. A.N. Belousov, The use of magnetite nanoparticles in applied medicine. Int. J. Nano Dimens. **2**(1(5)), 25–28. Print Edition ISSN: 2008-8868. Web Edition ISSN: 2228-5059
13. A.N. Belousov, E.Yu. Belousova, The first steps in discovery new mechanisms of cellular regulation in means by nanotechnology preparations, in *X International Conference New Information Technologies in Medicine and Ecology*, Yalta, Gursuf (2002), pp. 420–425
14. A.N. Belousov, E.Yu, Belousova Mechanisms cell regulation by nanotechnology preparations (MCS-B), in *BioNanotech Conference & Expo 2012*, Santa Clura, CA, USA, 18–21 June 2012
15. A.N. Belousov, Effect of Magnet controlled sorbent on parameters of acid-base balance of the blood and the processes of glycolysis in erythrocytes. Pain Anesthesia Intens Care **1**, 263–265 (2000)
16. A.N. Belousov, V.P. Nevzorov Ultrastructure of cells in the kidneys and lungs of rabbits after administration of magnetite, in *International Collection of Scientific Papers IV Scientific and Practical Conference on the Creation and Testing of New Drugs*, Moscow, vol. 4 (1997), pp. 77–87
17. A.N. Belousov, V.P. Nevzorov Ultrastructure of liver cells after administration of magnetite, in *International Collection of Scientific Papers IV Scientific and Practical Conference on the Creation and Testing of New Drugs*, Moscow, vol. 4 (1997), pp. 71–77
18. A.N. Belousov, Spectrum of application magnetite nanopaticles in medicine, in *Nanotech*, vol. 2 (2009), pp. 154–157. ISBN: 978-1-4398-1783-4

19. A.N. Belousov, Effect on hemolysis and transport ATPase activity of erythrocytes by means nanopareticles of magnetit controlled sorbent (MCS-B). Pain Anesthesia Intens. Care **1**, 26–28
20. A.N. Belousov, J. Cell, Mol. Biol. **2**(1), 1–4 (2017)
21. A.N. Belousov, Application magnetite of nanoparticles (ICNB Preparation) as magnetically-resonant contrasting means during visualization of tumours. Clean Technol. Sustain. Ind. Organ. **10**, 379–381. ISBN: 978-1-4822-5819-6
22. A.N. Belousov, Prospect. Med. Biol. **1**, 94–97 (2012)
23. A.N. Belousov, Biomed. Eng. Res. **2**(3), 147–152 (2014)
24. A.N. Belousov, Nanomater. Mol. Nanotechnol. **14–25**, 17–21 (2014)
25. A.N. Belousov, J. Mater. Sci. Eng. A **4**(11), 367–372 (2014)
26. A.N. Belousov, E. Malygon, V.V. Yavorskiy, E. Belousova, New effective method preservation of for red blood cells by means of nanotechnology. Nanomed. Nanotechnol. **3**(3), 000149 (2018). ISSN: 2574-187X. https://doi.org/10.23880/nnoa-16000149
27. A.N. Belousov, E. Malygon, V.V. Yavorskiy, E. Belousova, Int. J. Nanotechnol. Nanomed. **3**(2), 1–8 (2018)
28. A.N. Belousov, E. Malygon, V.V. Yavorskiy, E. Belousova, Int. J. Hematol. Blood Disorders. Symbiosis Group **3**(2), 1–9 (2018)
29. L.A. Kozitsina, N.B. Kupletskaya, *Application of UF, IR and IMR Spectroscopy in Organic Chemistry* (1971), 213p
30. A. Cross, *Introduction to Practical Infrared Spectroscopy* (Publishing House for Foreign Literature, Moscow, 1961), 110p
31. A.A Babushkin, P.A. Bazhulin, F.A. Korolev et.al., *Methods of Spectral Analysis* (Publishing House of the Moscow University, Moscow, 1962), 509p
32. L.A. Gribov, *Introduction to the Theory and Calculation of Vibrational Spectra of Many Molecules* (Ed. LGI, 1965), 134p

Chapter 5
Morphological Changes in Gram-Negative Microorganisms Treated with Silver and Copper Nanoparticles

Ye. Husak, V. Holubnycha, V. Korniienko, P. Myronov, A. Savchenko, A. Yusupova, and V. D. Ivchenko

Abstract Bacterial resistance became the crucial problem in current medicine. Due to this searching for new agents with antibacterial properties is key point. Nanoparticles possess antibacterial as well as many other unique features that can help to solve this problem. The most promising among them is Cu NPs and Ag NPs. The morphological changes in *K. pneumonia* and *E. coli* strains after their treatment with Cu NPs and Ag NPs were examined by scanning electron microscopy (SEM). Silver and cooper nanoparticles cause significant changes in the Gram-negative bacteria structure that are manifested as shortening of the bacteria cells, disruption of the cell wall surface and violation of cell divisions.

5.1 Introduction

Nowadays infections caused by bacteria still stay a main reason of the mortality and morbidity worldwide. *Klebsiella pneumoniae* and *Escherichia coli* are the most common Gram-negative bacteria cause sepsis and hospital acquired infections [1]. Crucial role of antibiotics in fight with infection diseases has been destroyed due to appearance and distribution of antibiotic-resistant bacterial strains. Antibiotic resistance is caused by formation of several protective mechanisms in microorganisms [2].

Nanoparticles have a wide application in medicine due to its unique chemical and physical properties. The main particularities of the nanoparticles action are their low toxicity and few antibacterial mechanisms activation at the same time. The substantial rise of their application as potential antibiotic substitute is indicated previously [2–4].

Ye. Husak · V. Holubnycha (✉) · V. Korniienko · P. Myronov · A. Savchenko · A. Yusupova
Sumy State University, Sumy, Ukraine
e-mail: v.golubnychaya@med.sumdu.edu.ua

V. D. Ivchenko
Sumy National Agrarian University, Sumy, Ukraine

A. D. Pogrebnjak et al. (eds.), *Nanomaterials in Biomedical Application and Biosensors (NAP-2019)*, Springer Proceedings in Physics 244,
https://doi.org/10.1007/978-981-15-3996-1_5

However, the size, structure, shape, surface properties, particle agglomeration and optical characteristics of NPs influence on its antimicrobial properties.

The precise antibacterial mechanism of the metallic NPs action is poorly understood. There are few common hypothesis of NPs antibacterial activity: (1) their influence on bacterial membrane permeability, (2) oxidative stress induction, and (3) activation of intracellular antibacterial effects [3].

The Cu and Ag NPs are a promising alternative to the drugs with antibacterial properties. Moreover, cooper is essential chemical element that stimulates metabolic processes, enzyme activity, and growth of the organism. Some studies reported the mechanism of Cu-NPs bactericidal activity. Karikalan [5] suggested that Cu ions originating from the NPs may interact with phosphorus and sulfur-containing biomolecules such as DNA and protein to distort their structures and thus disrupt intracellular biochemical processes. The mechanisms of Ag NPs antibacterial activity is more often discussed issues rather than antibacterial action of other types of nanoparticles [6]. Two antibacterial mechanisms are widely accepted for it, namely contact killing and ion-mediated killing.

Due to all listed above studying of the antibacterial activity and mechanism of Ag NPs and Cu NPs action against Gram-negative bacteria is a topical issue of the present time.

5.2 Materials and Methods

5.2.1 Materials

Silver nitrate ($AgNO_3$) (impurity > 99.9%, p.a.), polyvinylpyrrolidone (PVP—K25, MW 24,000), ethylene glycol (EG) (impurity > 99.9%), sodium hypophosphite monohydrate ($NaH_2PO_2 \cdot H_2O$), copper sulfate pentahydrate ($CuSO_4 \cdot 5H_2O$), isopropyl alcohol (99% pure, p.a.) were used as raw materials received from Sigma-Aldrich. Distilled water from electric distiller DE 20 was used throughout the experiments. Nutrient broth and nutrient agar were purchased from Hi Media India.

5.2.2 Synthesis of Cu NPs and Ag NPs

Copper and silver nanoparticles were synthesized by the methods described early [9]. After that each type of nanoparticles was treated with low-frequency ultrasound (22 ± 1.65 kHz) for 1 min.

5.2.3 Cu NPs and Ag NPs Characterization

Entire characteristic of the nanoparticles was done with use X-ray diffraction, SEM, transmission electron microscope (TEM), UV-VIS spectroscopy, EDS elemental analysis and total results are reflected in our previous works [7, 8]. Concentration of NPs in the aqua solution was determined by the method of inductively-coupled plasma atomic spectrometry (ICP-AES) using an iCAP 6300 Duo spectrometer (Thermo Scientific Corporation, USA).

5.2.4 Antibacterial Assessment

The mechanism of NPs antimicrobial activity was examined against referents strains of *K. pneumonia* and *E. coli*. The minimum inhibitory concentration (MIC) of NPs obtained in early investigation [7, 8] was measured by tube serial dilution method according to the international recommendations provided by the Clinical and Laboratory Standarts Institute (CLSI). The MIC of the Ag NPs and Cu NPs were 10.0 μg/ml and 0.5 μg/ml respectively. Before investigation of the morphological changes isolates were routinely cultivated overnight in nutrient broth at 37 °C. Then the cultures were diluted with cultivation media to the turbidity equivalent to McFarland 0.5 standard (1.5×10^8 CFU/ml). It was used as an inoculum. Concentrations of NPs equivalent to MIC were added into glass tubes containing 2 ml nutrient broth with microorganisms. The tubes containing growth medium and microorganisms without NPs were used as controls. After incubation at 37 °C for 24 h bacterial suspensions were collected by centrifugation. The bacterial cells were fixed in 2.5% glutaraldehyde with 0.1 M phosphate buffer pH 7.2 twice during 15 min. Then after 3 washes in buffer the samples were dehydrated with series of 40, 50, 60, 70, 80, 85, 90, 95, and 100% ethanol and stored in 100% ethanol. Finally, 2 μl of each sample was placed on 1 mm clean and defatted glass slides. The glass slides were coated with silver and examined under SEM.

The cell morphology of the tested microorganisms was examined by scanning electron microscopy (SEM). Cell dimensions were measured directly from the SEM images to calculate the cell volumes by the following equation:

$$V\left(\mu m^3\right) = \pi W2L/4 + \pi W2R/3,$$

when, W and L mean the width and length, respectively, of the central part of the cylindrical cell, and R was the equatorial radius of the spheroid caps at both ends of the cylinder. Average cellular volumes were calculated by using 30 individual bacteria per population [9].

5.2.5 Statistic

Data were expressed as means ± standard deviation. Student's t-test on unpaired data was used to assess the statistical significance of the difference. Statistical significance was assumed at a confidence level of 95% ($p < 0.05$).

5.3 Results

Scanning electron micrographs of *Klebsiella pneumonia* and *Escherichia coli* are presented on Figs. 5.1 and 5.2 Scanning electron microscopy images *K. pneumoniae* (a) and *E. coli* (b) in control.

As it is shown on Fig. 5.1 the cells in group of control were rod-shaped with smooth and intact cell walls.

At the same time, the morphology of the microorganisms treated with NPs (Fig. 5.2) was changed dramatically and there were a lot of coccobacillus bacteria after treatment with Cu NPs.

As it seen on the picture 2 the cell surface became rough with thickening on the cell ends after cultivation of the microorganisms with Ag NPs. The following results (Table 5.1) provide further evidence of the bacterial cell damage with nonmetals.

As it is reflected in the Tables 5.1 and 5.2 the untreated cells of *K. pneumoniae* and *E. coli* were corresponded normal size of these microorganisms. It was about 1.9994 ± 0.6219 μm and 0.8028 ± 0.1705 μm respectively.

Dramatic shortening of the *K. pneumoniae* cell after Cu NPs and Ag NPs treatment was detected. Length of microorganisms was respectively 0.8028 ± 0.1705 m and 1.2652 ± 0.2665 μm. Moreover, after *K. pneumoniae* incubation with Ag NPs significant decreasing of the of bacteria amount in the slide was found.

Similar changes were revealed after *E. coli* incubation with Cu NPs. The bacteria became shorter than in control group (1.0665 ± 0.3665 μm length). Incubation with Ag NPs did not cause significant influences on size of the cells but surface of

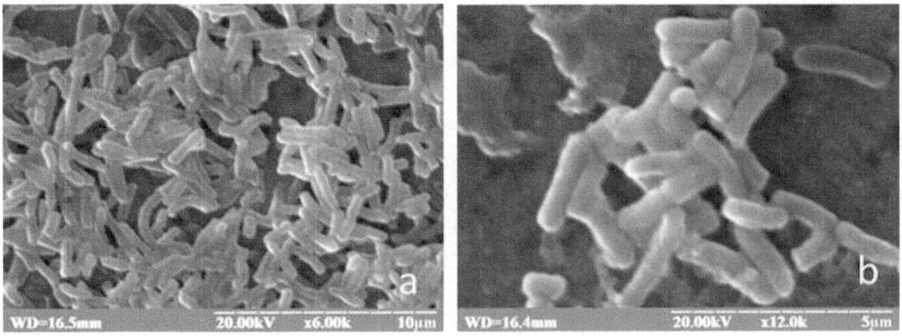

Fig. 5.1 Scanning electron microscopy images *K. pneumoniae* (**a**) and *E. coli* (**b**) in control

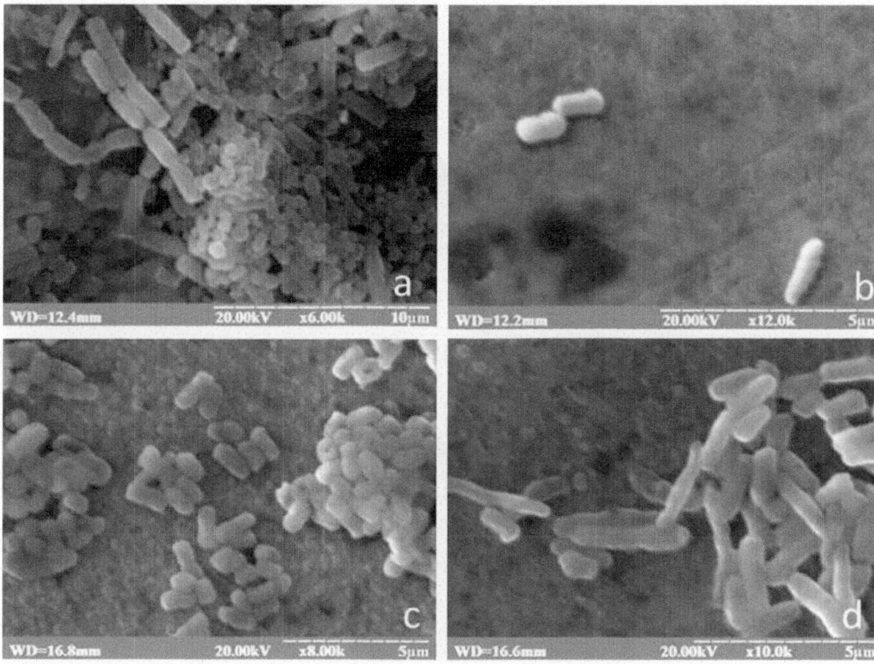

Fig. 5.2 Scanning electron microscopy images of *K. pneumoniae* in the presence of Cu NPs (**a**), Ag NPs (**b**) and *E. coli* in the presence of Cu NPs (**c**), Ag NPs (**d**)

Table 5.1 The mean cell sizes of the K. pneumoniae populations. Mean value ± std. error

	Width (μm)	Length (μm)
Contr	0.4413 ± 0.0571	1.9994 ± 0.6219
NPs Cu	0.3272 ± 0.0939	0.8028 ⊥ 0.1705
NPs Ag	0.5073 ± 0.0737	1.2652 ± 0.2665
	Radius (μm)	Volume (μm^3)
Contr	0.2479 ± 0.0352	0.3561 ± 0.001712
NPs Cu	0.3272 ± 0.0341	0.1041 ± 0.001495
NPs Ag	0.2423 ± 0.0407	0.3208 ± 0.001368

Table 5.2 The mean cell sizes of the *E. coli* populations. Mean value ± std. error

	Width (μm)	Length (μm)
Contr	0.4916 ± 0.0708	1.9350 ± 0.3904
NPs Cu	0.5947 ± 0.1119	1.0665 ± 0.3665
NPs Ag	0.5060 ± 0.0881	1.8103 ± 0.6829
	Radius (μm)	Volume (μm3)
Contr	0.2770 ± 0.0539	0.4371 ± 0.0018
NPs Cu	0.3292 ± 0.0388	0.4179 ± 0.0041
NPs Ag	0.2786 ± 0.0299	0.4385 ± 0.0210

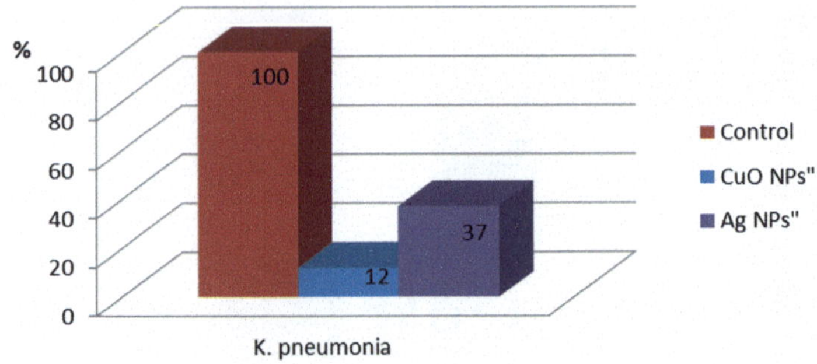

Fig. 5.3 Changes in volume of *K. pneumonia* experimental group compared to control strain, %

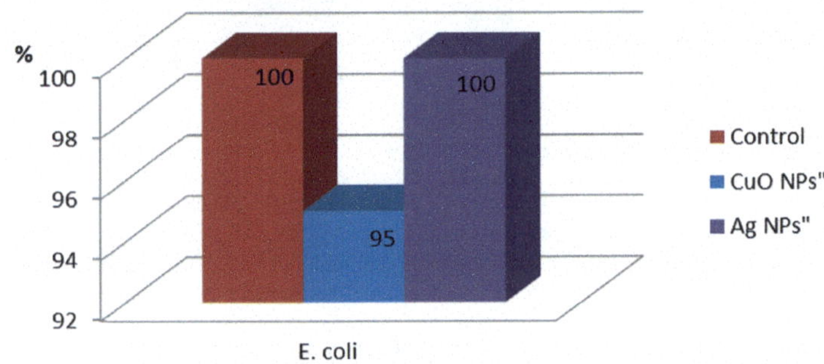

Fig. 5.4 Changes in volume of *E. coli* experimental group compared to control strains, %

the bacteria became more irregular and rough. The cell volumes decreased in both groups with Cu NPs treatment (Figs. 5.3 and 5.4). Moreover, substational drop of the *K. pneumonia* cell volume was revealed.

5.4 Discussion

Nowadays increasing number of nanoparticles using as a new alternative to antibiotics [10]. The numbers of studies on the use of metal nanoparticles with antimicrobial properties rises permanently [11].

The study revealed noticeable changes in the microbial cell morphology after treatment with silver and cooper nanoparticles. These changes included alterations in size and shape. Composition of cell wall determines the cell shape. NPs actively influence on bacteria and damage biological structure (lipids, proteins, nucleic acids) [12]. Copper inhibits and disrupts protein and nucleic acid synthesis [13]. Consequently, the most suitable reason of the morphological changes in *K. pneumoniae* and *E. coli* cell after treatment with NPs is chemical modification of the bacterial cell wall structure.

There was revealed a little number of K. pneumoniae in the slide after their incubation with Ag NPs and the chains formation in the slide with *K. pneumonia* treated with Cu NPs. Obviously, there can be different reasons. On the one hand used concentration of the silver caused whole destruction of the microorganisms and decreased their amount sharply. This assumption corresponds with other scientists data [14, 15] who indicated the violation of the metabolic properties in bacteria cell membranes, affection of the energy processes in the membrane and its barrier function after NPs treatment. On the other hand some authors indicate silver nanoparticles and Cu ions penetrate into the cell and cause further damage by possibly interacting with sulfur- and phosphorus-containing compounds DNA. Due to this we observed the formation of the cell chain.

5.5 Conclusion

Silver and cooper nanoparticles cause significant changes in the Gram-negative bacteria structure that are manifested as shortening of the bacteria cells, disruption of the cell wall surface and violation of cell divisions.

Acknowledgements This research supported by H2020 Marie Skłodowska-Curie Actions Grant NanoSurf 777926 and Ukraine MES Grant "Effectiveness of nanocomposite (chitosan-nanometal) antimicrobial action against multiresistant clinical strains" #0119U003577).

References

1. C. Elliott, A. Justiz-Vaillant, Biol. Biomed. J Spring **4**(2), 72–81 (2018)
2. Michael, A., Naafs, B., Nutri Food Sci. **3**(2), 243–245 (2019)
3. A. Sirelkhatim et al., Nano-Micro Lett. **7**(3), 219–242 (2015)
4. O. Oleshko, V. Deineka, Y. Husak, V. Korniienko, O. Mishchenko, V. Holubnycha et al., Mater. (Basel) **12**(22), 3742 (2019)
5. N. Karikalan, Rasayan J. Chem. **11**(4), 1451–1457 (2018)
6. A. K. Chatterjee, Nanotechnology **25**(13), 135101 (2014)
7. V. Holubnycha et al., in *NAP-2018, IEEE 8th International Conference on —Nanomaterials: Applications & Properties*, 2018, pp. 04NNLS09-1-04NNLS09-4
8. M. Pogorielov, V. Holubnycha, O. Ivashchenko, O. Kalinkevych, B. Peplinska, M. Jarek, V. Korniienko, in *NAP-2017, IEEE 7th International Conference on —Nanomaterials: Applications & Properties*
9. B. Shi, X. Xia, Curr. Microbiol. **46**(2), 120–123 (2003)
10. L. Wang, C. Hu, L. Shao, Int. J. Nanomed. **12**, 227–1249 (2017)
11. Yun'an Qing et al., Int. J. Nanomed. **13**, 3311–3327 (2018)
12. J.R. Morones-Ramirez et al., Nanotechnology **16**(10), 2346–2353 (2005)
13. Q.L. Feng, J. Biomed. Mater. Res. **52**(4), 662–668 (2000)
14. L. Wang, C. Hu, L. Shao, Int. J. Nanomed. **12**, 1227–1249 (2017)
15. Yun'an Qing, L. Cheng, R. Li et al., Int. J. Nanomed. **13**, 3311–3327 (2018)

Chapter 6
Regularities of Obtaining Metal-Filled Polymer Composites

A. N. Kucherenko, V. S. Moravskyi, M. Ya. Kuznetsova, O. N. Grytsenko, A. S. Masyuk, and L. Dulebova

Abstract The example of polypropylene is shown the possibility of using a mechanical activation of a polymeric surface to form a metal coating on it in solutions of chemical metallization. The influence of concentration factors on the process of formation of copper coating on the activated polypropylene surface has been investigated. The obtained results allow to establish optimal conditions and influence the process of polypropylene metallization. Due to change in the concentration of reagents can effectively regulate the amount of reduced copper on the activated polymer surface and the thickness of the formed layer of the metal. Such information is a prerequisite for the creation of highly efficient advanced metal-filled polymer composites.

6.1 Introduction

The need to create new types of polymer composite materials is due to the high demand of modern technology on such materials. Polymer composite materials are becoming more and more practical use in various technological node and components of instrumentation and machine building [1–4]. At present, the production of a wide range of fillers and polymer matrices has already been established, which allows the use of polymer composite materials as both a structural materials [5] and a special purpose materials [6].

The basis of technology for the production of polymer composite materials is almost unlimited possibilities for the regulation of their properties. The use of specific properties of fillers (electrical and thermal conductivity, thermal stability, magnetic properties, etc.) and polymer matrix (high technological and chemical resistance) during the development of technological processes for the production of polymer

A. N. Kucherenko · V. S. Moravskyi (✉) · M. Ya. Kuznetsova · O. N. Grytsenko · A. S. Masyuk
Lviv National Polytechnic University, Lviv, Ukraine
e-mail: vmoravsky@gmail.com

L. Dulebova
Technical University of Kosice, Kosice, Slovak Republic

A. D. Pogrebnjak et al. (eds.), *Nanomaterials in Biomedical Application and Biosensors (NAP-2019)*, Springer Proceedings in Physics 244,
https://doi.org/10.1007/978-981-15-3996-1_6

composite materials allows the creation of materials with unique and predetermined properties.

The use of metal fillers and large tonnage thermoplastics as a polymer matrix is promising when creating polymer composite materials of special purpose. Such combination will allow to obtain materials with the necessary operational, physico-mechanical and physico-chemical properties and low cost.

The presence of metal particles in the polymer matrix contributes to a higher density, better thermal stability of the composite material [7–9] and creates a barrier for high-frequency radiation [10]. Such materials are used as gauges and sensors [11]. Improvement of the performance properties of polymer composites can be achieved by using submicron or nanosized metal particles [12]. However, in this case, the main problem for obtaining high-tech composites is the bad distribution of the metal filler in the polymer matrix. The formation of homogeneous polymer mixtures can be achieved using certain technological techniques: surface modification of the filler [13, 14], metal recovery in the polymer matrix [15], a combination of polymerization and metal recovery processes [16] or use of special mixing equipment [17].

A promising direction in the use of metal-filled polymer composites is their use as highly efficient thermal energy storage system. The problem of accumulation, storage and subsequent use of energy is characteristic of both traditional and non-traditional energy [18]. The introduction of "night tariffs" aims to equalize the daily energy consumption by stimulating the end-user in more intensive energy use at night. One of the most energy-intensive systems, which can be organized using the night tariffs, is the heating system. The introduction of heating systems with energy storage devices will solve a number of problems associated with inconsistencies schedules in production and consumption of energy in both traditional and non-traditional energy.

Polymers are materials with bad thermal conductivity, which imposes certain restrictions on highly efficient heat accumulating systems developed on their basis. Cooling of the surface layers of the polymer in contact with the heat transfer surface affects the significant reduction of the possibility of heat dissipation from the heated inner layers, which reduces the efficiency of the use of the thermal energy storage system. Such disadvantage requires the development of new materials that will enable the rapid and uniform heating of the polymer (charge) and the efficient selection of accumulated heat (discharge) [19, 20]. Thus, there is a need for a developed polymeric heat-accumulating material, which will have high thermal conductivity [21, 22] and high values of the density of accumulated energy (due to the phase transition of crystalline polymers).

In this paper in order to obtain metal-containing composites, the method of introducing a metal filler into a polymer matrix by chemical metallization of the surface of the source polymeric raw material is proposed [23–25]. The formation on the polymeric surface of a metal shell that will collapse during the melting of the polymer will provide a simple introduction and a uniform distribution of metal in the volume of the material [23]. Polypropylene was selected as polymeric base for composite material due to the combination of high technological and operational properties with a low cost, as well as the ability to be processed into products with high-performance methods. In addition, polypropylene is a high-crystal polymer, which will allow it to be used to develop highly efficient heat-storage systems in which the presence of metal particles in the polymer's volume will increase the thermal conductivity. In addition, polypropylene is a high-crystal polymer, which will allow it to be used to develop highly efficient heat-storage systems in which the presence of metal particles in the polymer's volume will increase the thermal conductivity.

6.2 Regularity of Polypropylene Copper Plating

The proposed technology for the production of metallized polymer raw materials involves the use of chemical recovery solutions and the provision of a polymeric surface of catalytic activity. The catalytic activity of the polymer surface was obtained by fixing on it metal particles of the activator-zinc. The chemical recovery solution included copper sulfate, sodium hydroxide, formalin and EDTA-Na_2 [23].

In order to obtain high-tech metal-based composites and to understand the processes occurring during copper recovery, it is necessary to know the basic regularities of forming a metal layer on the activated polymer surface. The kinetics of the formation of a metal layer in solutions of chemical deposition using volumometric method was investigated. The volumometric method for investigating the kinetics of metallization of the activated polymer surface in solutions of chemical precipitation is based on the features of copper recovery in solutions with a complexing agent EDTA-Na_2, whereby one mole of hydrogen whose volume is measured [26] is released per one mol of the recovered copper. The competing reaction of copper recovery, which takes place in the proposed method, occurs without the release of hydrogen and is an exchange reaction with zinc. Thus, the kinetic curves obtained during the study of metallization by volumetric method show only the amount of copper recovered as a result of reaction with formaldehyde.

The most effective factors influencing the rate of copper recovery in the solutions of chemical restoration on the activated polypropylene surface are the concentration of sodium hydroxide, copper sulfate and EDTA-Na_2. Increasing concentrations of sodium hydroxide leads to increase the reaction rate and reduced the induction period and increasing the number of recovered copper (Fig. 6.1).

Reducing the complexing agent concentration effect on the increase of the rate of copper recovery (Fig. 6.2). This is more characteristically for solutions with a EDTA-Na_2 concentration of 47 mmol/l.

This feature can be explained by a certain loss of stability of solutions of chemical deposition, which contributes to the increase of the recovery reaction rate [25].

The study of the effect of changes in the concentration of $CuSO_4$ showed that an increase in the concentration of $CuSO_4$ to 80 mmol/l more significantly affects the rate of recovery of copper and its amount (Fig. 6.3).

Another factor that was investigated to increase the rate of metallization is the effect of a proportional increase in the concentration of all reagents (Fig. 6.4).

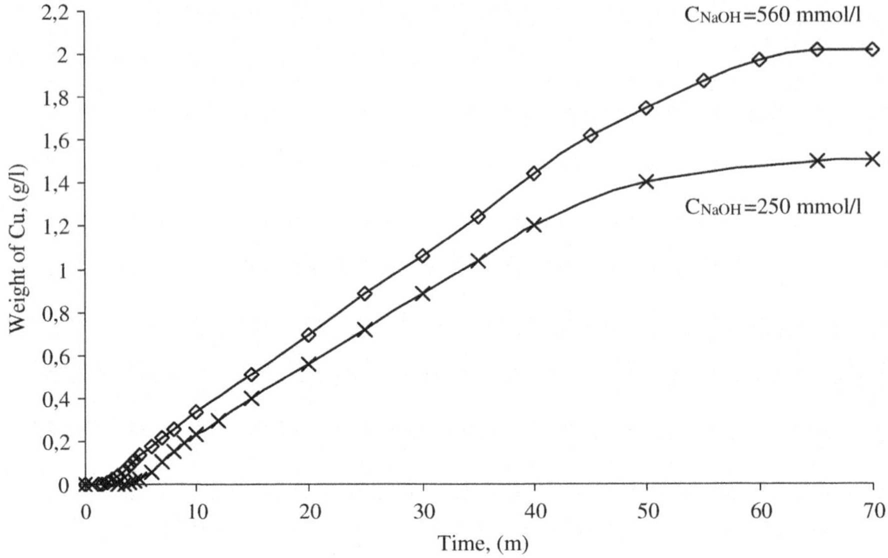

Fig. 6.1 Kinetic curves of the recovery of copper ions on activated polypropylene surface. Concentration (mmol/l): $CuSO_4$—48; EDTA-Na_2—67; Formaldehyde—366

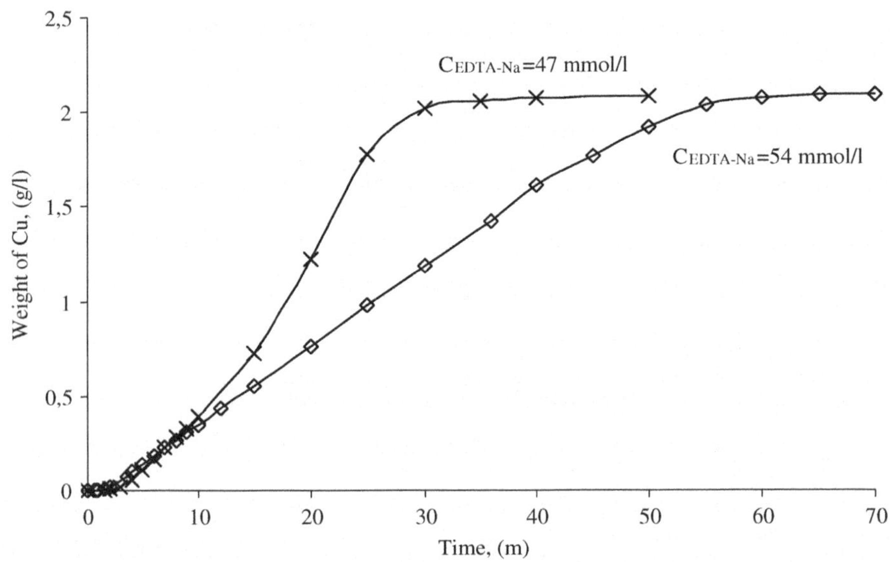

Fig. 6.2 Kinetic curves for the recovery of copper ions on activated polypropylene surface. Concentration (mmol/l): $CuSO_4$—80; NaOH—560; Formaldehyde—366

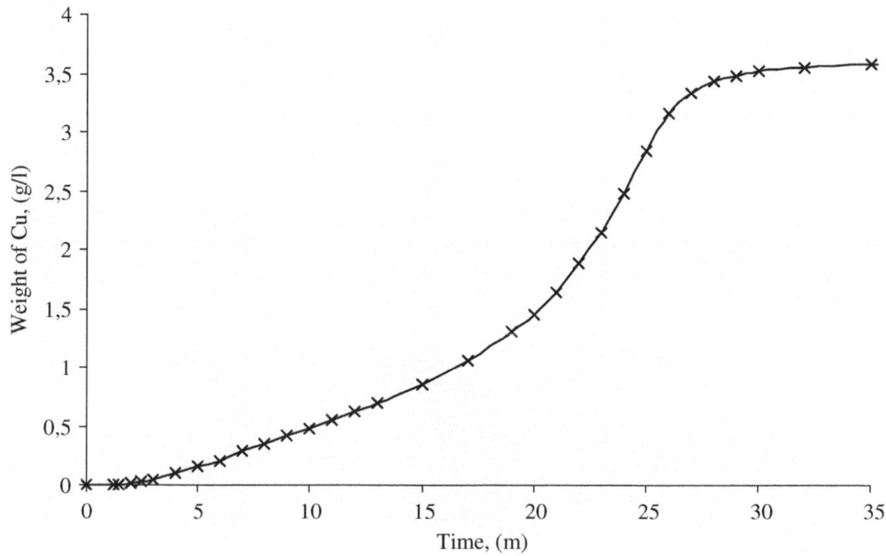

Fig. 6.3 Kinetic curves for the recovery of copper ions on activated polypropylene surface. Concentration (mmol/l): $CuSO_4$—80; EDTA-Na_2—67; NaOH—560; Formaldehyde—366

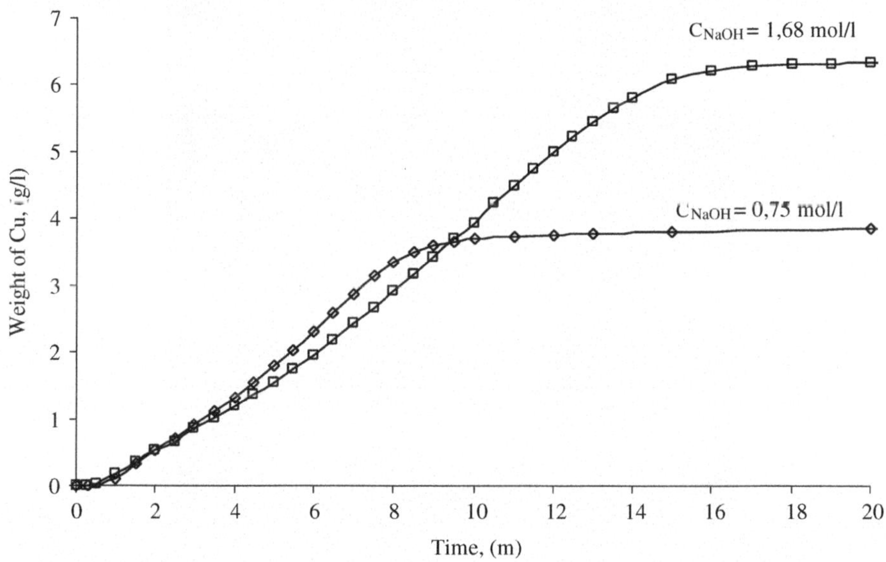

Fig. 6.4 Kinetic curves for the recovery of copper ions on activated polypropylene surface (3-fold increase in the reagents concentration). Concentration (mol/l): $CuSO_4$—0.144; EDTA-Na_2—0.201; Formaldehyde—100.8

One can notice another influence of NaOH concentration on the rate of copper recovery—a certain predominance of the reaction rate at lower concentrations of NaOH.

The use of polypropylene activated by zinc for the production of metallized polymeric raw material showed the high efficiency of this solution. The studies carried out using a scanning electron microscope in contrast mode and the identification of the spectrum of the characteristic X-ray radiation of the metallized surface indicate that the obtained metallized polypropylene is characterized by the formation of copper coating on polymer particles with a high degree of surface filling (Fig. 6.5).

Fig. 6.5 Microphotography of the surface of metallized polypropylene obtained in the contras mode on the average atomic number

6.3 Conclusions

The conducted studies on the influence of concentration factors on the process of metallization of activated polypropylene can highlight the main occurrences in the process of obtaining metallized polypropylene. The change in the concentration of copper sulfate, EDTA-Na_2 and sodium hydroxide can regulate the amount of copper recovered on the activated polypropylene surface and the thickness of the formed layer of the metal. This information will provide the obtaining of polymeric raw materials of the required quality and will allow to control the metal content in the final metal-filled composite. The introduction of metal into a polymer matrix in the form of a metal coating formed on a polymeric surface will provide the production of metal-containing polymer composites, which are characterized by a uniform distribution of metal in the polymer matrix and high technological and operational properties. The production of such metal-filled composites will take place directly during the processing of metallized polymeric raw materials. Properties of the obtained materials can be predicted, as well as change their by content of metal at the stage of metallization of the polymeric surface.

References

1. H. Chen, V.V. Ginzburg, J. Yang, Y. Yang, W. Liu, Y. Huang, L. Du, B. Chen, Prog. Polym. Sci. **59**, 41 (2016)
2. I.K. Bishay, S.L. Abd-El-Messieh, S.H. Mansour, Mater. Design **32**, 62 (2011)
3. Q. Xue, Eur. Polym. J. **40**, 323 (2004)
4. H. Li, J.V. John, S.J. Byeon, M.S. Heo, J.H. Sung, K.-H. Kim, I. Kim, Prog. Polym. Sci. **39**, 1878 (2014)
5. V. Levyts'kyi, Yu. Laruk, L. Bilyi, V. Kochubei, T. Humenets'kyi, Mater. Sci. **51**, 388 (2015)
6. O. Suberlyak, V. Krasinskiy, J. Sikora, A. Krzyzak, Chem. Chem. Tech. **6**(2), 199 (2012)
7. M. Nikzad, S.H. Masood, I. Sbarski, Mater. Des. **32**, 3448 (2011)
8. A.S. Luyt, J.A. Molefi, H. Krump, Polym. Degrad. Stabil. **91**, 1629 (2006)
9. H.J. Park, A. Badakhsh, I.T. Im, M.-S. Kim, C.W. Park, Appl. Therm. Eng. **107**, 907 (2016)
10. H. Pang, L. Xu, D.-X. Yan, Z.-M. Li, Prog. Polym. Sci. **39**, 1908 (2014)
11. O. Grytsenko, E. Spišák, L. Dulebová, V. Moravskii, O. Suberlyak, Mater. Sci. Forum **818**, 97 (2015)
12. T. Tanaka, G.C. Montanari, R. Mulhaupt, IEEE Trans. Dielectr. Electr. Insul. **11**(5), 763 (2004)
13. B. Pukánszky, Eur. Polym. J. **41**(4), 645 (2005)
14. P. Mareri, S. Bastide, N. Binda, A. Crespy, Comp. Sci. Techn. **58**, 747 (1998)
15. O. Suberlyak, O. Grytsenko, V. Kochubei, Chem. Chem. Tech. **9**, 429 (2015)
16. O. Grytsenko, A. Pokhmurska, S. Suberliak, M. Kushnirchuk, M. Panas, V. Moravskyi, R. Kovalchuk, East-Eur. J. Enterp. Technol. **6**, 6 (2018)
17. T. Sakai, Polimery **58**, 847 (2013)
18. S. Prashant, T. Craig, K. Parthiv, Sol. Energy **185**, 494 (2019)
19. L. Cabeza, *Advances in Thermal Energy Storage Systems. Methods and Applications* (Woodhead Publishing, 2015)
20. T. Letche, *Storing Energy* (Elsevier, Oxford, 2016)
21. D. Hansen, R. Tomkiewicz, Polym. Eng. Sci. **15**, 353 (1975)
22. R. Singh, P.K. Sharma, Indian J. Pure Appl. Phy. **49**, 112 (2011)

23. V. Moravskyi, I. Dziaman, S. Suberliak, O. Grytsenko, M. Kuznetsova, in *Abstracts of the 2017 IEEE 7th International Conference Nanomaterials: Applications and Properties (NAP-2017)*, Sumy State University, Odessa, Ukraine, 10–15 Sept 2017
24. V. Moravskyi, I. Dziaman, S. Suberliak, M. Kuznetsova, T. Tsimbalista, L. Dulebova, East-Eur. J. Enterp. Technol. **4**(12), 50 (2017)
25. V. Moravskyi, A. Kucherenko, M. Kuznetsova, I. Dziaman, O. Grytsenko, L. Dulebova, East-Eur. J. Enterp. Technol. **3**(12), 40 (2018)
26. M. Shalkauskas, A. Vashkyalis, *Chemical Metallization of Plastics* (Khimiya, Leningrad, 1985)

Chapter 7
Synthesis, Characterization and Antibacterial Activity of Hydroxyapatite Composite Materials Loaded with ZnO Nanoparticles

A. Yanovska, R. Pshenychnyi, Ye. Husak, V. Korniienko, V. Holubnycha, S. Bolshanina, and T. Dychenko

Abstract Synthesis of hydroxyapatite (HA) composite material loaded with ZnO nanoparticles is proposed. Hydroxyapatite was taken as a basic component due to its biocompatibility with bone tissue and perfect sorption properties. It could be combined with different ions and nanoparticles by simple mixing method due to its high capacity to metal ions retention. The results of studies of the structure, morphology, phase composition and antibacterial activity are presented. The minimum bactericidal concentration against *E. coli* and *S. aureus* was equal 0.034 mg/mL.

7.1 Introduction

Hydroxyapatite (HA) based composite biomaterials are widely used for bone substitution. Acquired infections still stay a crucial complication that leads to failures in open fractures treatment. Incorporating of inorganic metal ions or metal oxide nanoparticles into composite materials could increase their antibacterial properties. Moreover HA has an excellent sorption property so it could be combined with different ions and nanoparticles by simple mixing method due to its high capacity to metal ions retention [1].

There are many different metal and metal oxide nanoparticles that are used for biomedical and biotechnology applications: Fe_2O_3, CuO, TiO_2, MgO, Ag, Cu, ZnO etc. [2]. Zinc oxide nanoparticles are cheap and show antibacterial activity, low toxicity, excellent biocompatibility [2, 3]. Except these, ZnO NPs induce acceleration of wound healing, prevent inflammation and itching [4]. ZnO nanoparticles demonstrate antibacterial activity against most pathogenic and nonpathogenic strains. Wang et al. [5] reported that nanoparticles of ZnO have strong

A. Yanovska (✉) · R. Pshenychnyi · Ye. Husak · V. Korniienko · V. Holubnycha · S. Bolshanina · T. Dychenko
Sumy State University, Sumy, Ukraine
e-mail: yanovskasumdu@gmail.com

A. D. Pogrebnjak et al. (eds.), *Nanomaterials in Biomedical Application and Biosensors (NAP-2019)*, Springer Proceedings in Physics 244,
https://doi.org/10.1007/978-981-15-3996-1_7

antibacterial activity against *E. coli* and the activity increases as the concentration of ZnO nanoparticles increases. Moreover, the incorporation of Zn in an implant material promotes the proliferation and differentiation of osteoblast cells, leading to enhanced osteogenesis [6].

Several studies have reported that the incorporation of zinc into HA is based on the method and condition used for synthesis [1, 7–10]. Synthesis of ZnO nanoparticles could be provided through various methods by controlling synthesis parameters. Among them are sol-gel technique, hydrothermal method, thermal evaporation of ZnO powders, thermal sublimation, polymerized complex method and solution synthesis [11].

In order to decrease the frequency of bacterial complication in patients with trauma we have obtained hydroxyapatite based material with antibacterial properties by adding ZnO nanoparticles and examined its activity against gram-positive and gram-negative bacteria.

7.2 Experiment Details

All reagents were purchased from Sigma-Aldrich and used as received. The chemicals and reagents used were: $Ca(NO_3)_2{\cdot}4H_2O$, $(NH_4)_2HPO_4$, $Zn(CH_3COO)_2{\cdot}2H_2O$, ethylene glycol.

First 23.6 g of $Ca(NO_3)_2{\cdot}4H_2O$ was dissolved in 500 mL of distilled water to obtain 0.1 M solution. 7.92 g $(NH_4)_2HPO_4$ was dissolved in 500 mL of distilled water to obtain 0.06 M solution (Fig. 7.1).

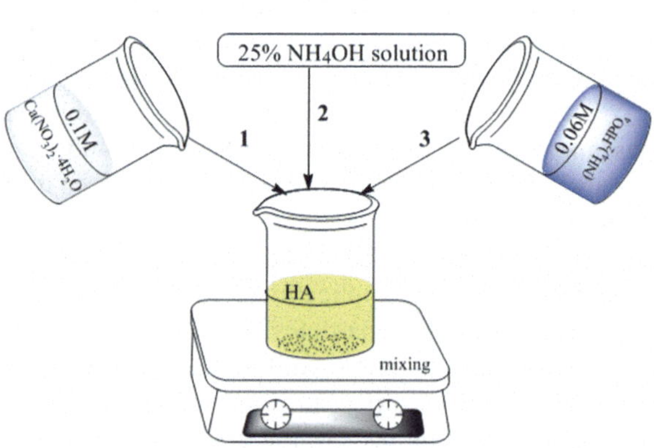

Fig. 7.1 Scheme of hydroxyapatite synthesis: 1—0.1 M $Ca(NO_3)_2{\cdot}4H_2O$ solution was added in a glass, 2—25% ammonia solution was added dropwise (pH = 10), 3—0.06 M $(NH_4)_2HPO_4$ solution was added dropwise to the above mentioned solution

7.2.1 Synthesis of Hydroxyapatite Nanoparticles

$Ca_{10}(PO_4)_6(OH)_2$—stoichiometric HA was synthesized by following scheme:

$$10Ca(NO_3)_2 + 6(NH_4)_2HPO_4 + 8NH_3 \cdot H_2O \rightarrow Ca_{10}(PO_4)_6(OH)_2 + 20NH_4NO_3 + 6H_2O$$

Analytically grade salts $Ca(NO_3)_2$ and $(NH_4)_2HPO_4$ were taken in concentrations 0.1 M and 0.06 M respectively. First 0.1 M $Ca(NO_3)_2$ solution was mixed with ammonia solution, after that 0.06 M $(NH_4)_2HPO_4$ solution was added dropwise to the above mentioned solution (Fig. 7.1).

Solution pH was adjusted to 10 by adding 25% ammonia solution. After aging for two days it was rinsed by distilled water until pH = 7. For obtaining of the composite material it was taken as wet slurry.

7.2.2 ZnO Nanoparticles Synthesis

ZnO nanoparticles were obtained by polyol synthesis. Ethylene glycol was used as a reaction medium. First 2.19 g $Zn(CH_3COO)_2 \cdot 2H_2O$ and 10 mL of ethylene glycol (EG) were placed into 50 mL two-necked bulb with the reflux condenser in the central neck and thermocouple connected with thermo regulator of bulb heater in the right neck.

The obtained mixture was heated in a bulb heater with mixing (3000 rpm) by using Teflon mixer. The mixture was kept 60 min at 160 °C. During synthesis, as a result of decomposition reaction nanostructured ZnO was obtained.

$$Zn(CH_3COO)_2 \cdot 2H_2O \rightarrow ZnO + 2CH_3COOH + H_2O$$

Than the mixture was cooled until the room temperature was reached and the obtained product was separated from organic compound (EG) centrifugation. The remaining of EG was washed by ethanol under intensive mixing with following centrifugation. Washed samples of ZnO was dried 24 h at 60 °C. Schematically this process is shown on the Fig. 7.2.

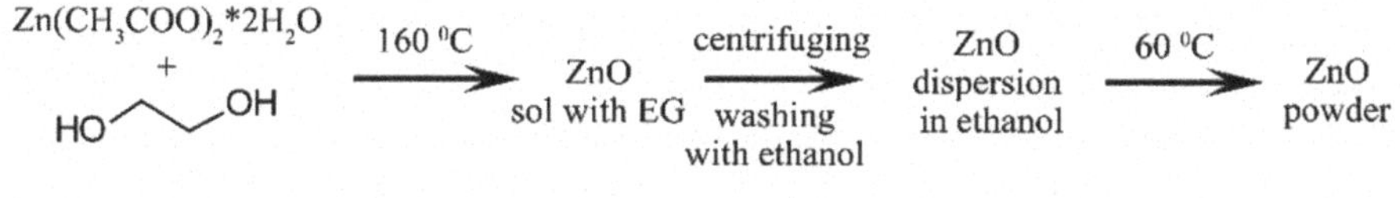

Fig. 7.2 Scheme of ZnO nanoparticles synthesis

HA and ZnO nanoparticles with average size 80 and 10–30 nm respectively were mixed as a wet slurry in relation 5:1 by weight, than obtained composite material was centrifuged at 2500 rpm.

The synthesized HA powders and ZnO nanoparticles phase analysis was done in an X-ray powder diffractometer (XRD) (DRON-3 M, Russia). The diffractometer was fitted with a Ni filter of 0.154 nm and the diffraction patterns were measured using CuKα radiation ($\lambda = 1.5418$ Å) generated at a voltage of 40 kV and current of 40 mA. The samples were scanned in the interval $20° < \theta < 50°$ at a scanning rate 20°/min with a step size of 0.05° in a continuous mode. XRD analysis of the samples was provided with the reference powder diffraction data given by the Joint Committee on Powder Diffraction Standards (JCPDS).

The microstructural study to observe the morphology of the individual nanoparticles and obtained materials were done in Transmission Electron Microscope (TEM-125 K) (SELMI, Ukraine) and Scanning Electron Microscope SEO-SEM Inspect S50-B (FEI Company, Hillsboro, OR, USA). The electrons at 5 kV sources were used to develop relevant information about the samples.

7.2.3 Antibacterial Properties

The antimicrobial activity of HA-ZnO material was examined against Gram-positive (*Staphylococcus aureus*) and Gram-negative (*Escherichia coli*) bacteria. The isolates were cultivated overnight in nutrient broth at 37 °C. The cultures were diluted with nutrient media to the cell concentration equivalent to 10^5 CFU/mL. The antimicrobial activity of HA-ZnO sample was measured by serial dilution method according to the international recommendations provided by the Clinical and Laboratory Standarts Institute (SLCI) with determination minimum bactericidal concentration (MBC). Tubes with the same dilutions of HA in growth medium with microorganisms were used as control. Colony numbers were determined by using streak plate technique. All the measures were triplicate.

7.2.4 Statistic

Statistical analyses were performed using One-way ANOVA with multiple comparisons (GraphPad Prism 8.0 software). Data were expressed as means ± standard deviation. Student's t-test on unpaired data was used to assess the statistical significance of the difference between the results obtained from the tested specimens. Statistical significance was assumed at a confidence level of 95% ($p < 0.05$).

7.3 Results and Discussion

In this work polyol method of synthesis from solutions was applied to obtain nano-sized ZnO particles. HA nanoparticles and ZnO nanoparticles were obtained separately and then mixed with each other as described above in ratio 5:1 respectively. SEM image and XRD diffraction patterns of hydroxyapatite is presented on the Fig. 7.3.

Among synthetic biomaterials there is a strong need for the synthesis of biomaterials in nanoscale range, with a specific morphologies and chemical compositions. The synthesis of nano HA is more desirable than micron-sized one, owing to its great effect on the differentiation of bone cells and mineralization as well as particular affinity towards several adhesive proteins [12].

As it can be observed from the Fig. 7.3a. the average crystal size for HA crystals is nearly 70–80 nm. Formation of HA phase is proved by XRD patterns (Fig. 7.3b).

The pattern reveals the characteristic peaks of HA, confirming the occurrence of no new phase formation. XRD patterns of HA shown in the Fig. 7.3b, shows a broad reflection peak in the range of 31.8–32.50 of 2Θ values, that corresponds to the characteristic peak of apatite phase (according to JCPDS card # 9-0432).

Properties of ZnO nanoparticles are characterized by size, crystallinity, morphology and composition. It is shown [13] that ZnO exhibits significant antibacterial activity when particle size is reduced to nanometer size. Due to the nano-size it can easily interact with bacterial surface or bacterial core and exhibits bactericidal mechanism [13].

TEM microscopy of ZnO nanoparticles is shown on Fig. 7.4. Obtained nanoparticles have spherical shape (Fig. 7.4a) that depends from type of synthesis as reported elsewhere [11].

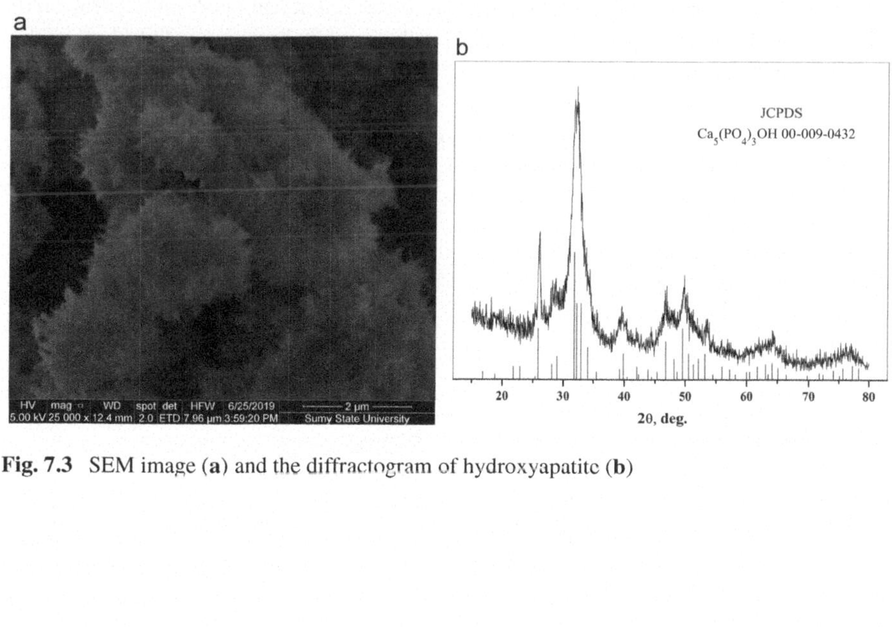

Fig. 7.3 SEM image (**a**) and the diffractogram of hydroxyapatite (**b**)

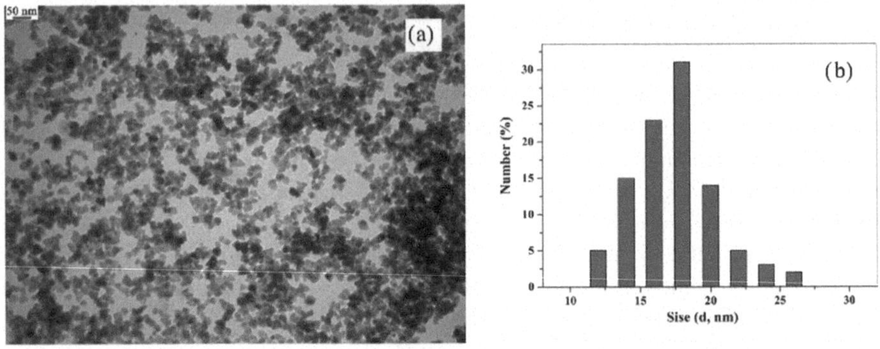

Fig. 7.4 The TEM image of ZnO nanoparticles (**a**) with average size distribution (**b**)

We synthesized nanoparticles with size from 10 to 30 nm with average size 18 nm as well as particles with smaller size provide higher specific surface area and showed higher antibacterial properties as reported [11, 14, 15].

XRD analysis of obtained by polyol method ZnO nanoparticles is given in Fig. 7.5.

The characteristic diffraction peaks for synthesized ZnO nanoparticles (Fig. 7.5a) were observed at position $2\Theta = 32.18°, 34.02°, 36.67°, 47.95°$ corresponds to (100), (002), (101) and (102) crystal planes respectively. It matches with the JCPDS card no. 01-089-1397. Presence of is also proved by electron diffraction (Fig. 7.5b).

The activity of HA-ZnO material against gram-positive and gram-negative bacteria is shown on Fig. 7.6.

It was revealed that pure HA nanoparticles did not demonstrate antibacterial action. However HA-ZnO material possessed antibacterial properties depending on concentration of ZnO NPs in solution against both types of microorganisms. Total growth inhibition of *S. aureus* and *E. coli* was observed at ZnO NPs concentration of 0.034 mg/ml. It can be considered as MBC for these microorganisms. Dilutions of the

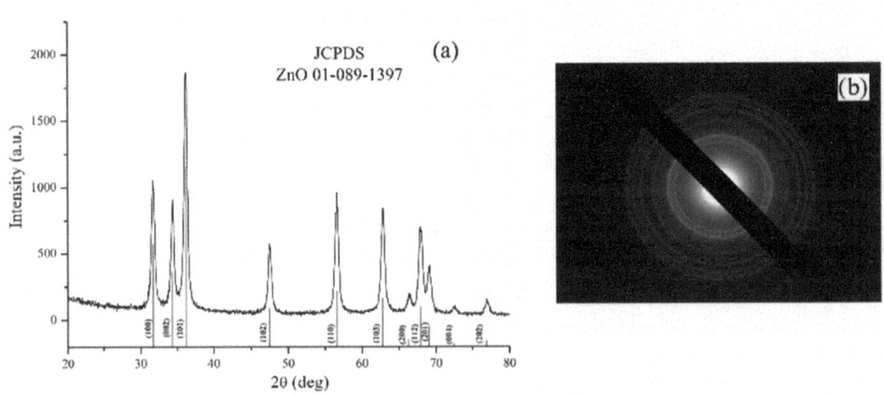

Fig. 7.5 The XRD pattern of ZnO (JCPDS 01-089-1397) (**a**) electron diffraction of ZnO (**b**)

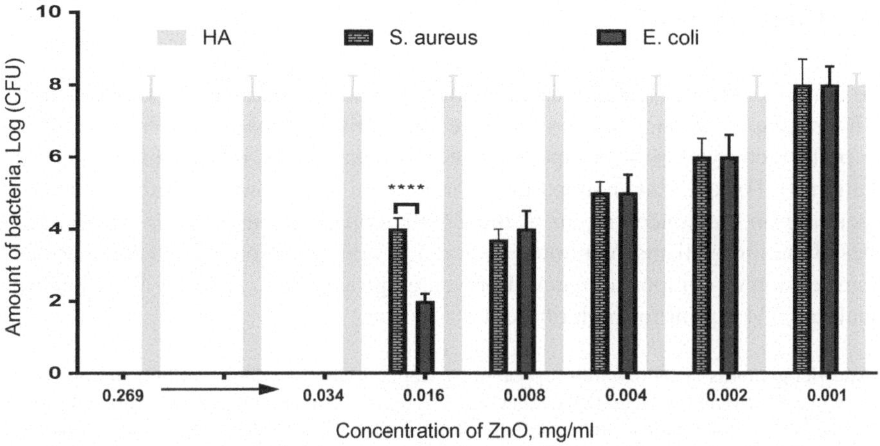

Fig. 7.6 Comparative antibacterial activities of HA-ZnO material against *S. aureus* and *E. coli*

HA-ZnO sample with concentration of ZnO NPs from 0.016 to 0.002 mg/ml caused the inhibition of both microorganisms' growth and differences between amount of microorganisms in equal dilutions HA-ZnO NPs and pure HA were statistically significant ($p < 0.0001$). In addition, ZnO NPs at concentration of 0.016 mg/ml influenced more effectively on gram-negative microorganism than on gram-positive one ($p < 0.0001$).

Currently ZnO NPs application is very broad due to their inexpensiveness, relatively abundance, chemical stability, easy way of preparation, and non toxicity [16–18]. Incorporation of ZnO NPs into different polymers can change their properties as well as give new properties. There are some data about ZnO antimicrobial effects. Yu et al. examined the antibacterial activity of hydroxyapatite whisker/nano zinc oxide biomaterial and revealed the antibacterial action of the composite [19]. However, the effective antimicrobial concentration of ZnO in their work was remarkably higher than our results. Probably such difference can be explained with physic-chemical features of obtained sample.

Possible mechanism of antibacterial activity of ZnO nanoparticles is proposed in the works [11, 12]. High concentrations of nanoparticles produce H_2O_2 and damage the bacterial membrane that causes DNA defragmentation and activates caspases network. Cascade activation of the enzymes plays the main role in the cell death. The increase of ZnO nanoparticles concentrations leads to mitochondrial function disorder as well. Thus, inhibition of bacterial growth is strongly depending on the concentrations of ZnO nanoparticles [5, 11].

7.4 Conclusions

The aim of current research was to synthesize and evaluate antimicrobial activity of hydroxyapatite/nano zinc oxide bone restorative biomaterials. As a result of the performed syntheses, composite material obtained by mixing of as prepared solutions of HA and ZnO nanoparticles (in relation 5:1) shows antibacterial activity depending on the concentration of the ZnO nanoparticles in solution against both types of microorganisms. Minimum bactericidal concentration of ZnO nanoparticles in suspension of composite material prevented growth of *E. coli* and *S. aureus* was equal to 0.034 mg/ml for both of them.

Acknowledgements This research supported by H2020 Marie Skłodowska-Curie Actions Grant NanoSurf 777926 and Ukraine MES Grants "Effectiveness of nanocomposite (chitosan-nanometal) antimicrobial action against multiresistant clinical strains" #0119U003577 and "Biological effectiveness of Plasma Electrolytic Oxidation and sol-gel deposition for development of functional implant surface" #0119U100823).

References

1. C. El Bekkali, H. Bouyarmane, M. El Karbane, S. Masse, A. Saoiabi, T. Coradin, A. Laghzizil, Colloids Surf. A **539**, 364 (2018)
2. V. Gnaneshwar, S.V. Sudakaran, S. Abisegapriyan, J. Sherine, S. Ramakrishna, M.H.Ab. Rahim, M.M. Yusoff, R. Jose, J.R. Venugopal, Mater. Sci. Eng. C **96**, 337 (2019)
3. L. Trandafilović, D. Božanić, S. Dimitrijević-Branković, A. Luyt, V. Djoković, Carbohydr. Polym. **88**(1), 263 (2012)
4. S.A. Aleaghii, E. Fattahy, B. Baei, M. Saghali, H. Bagheri, N. Javid, E.A. Ghaemi, Int. J. Adv. Biotechnol. Res. **7**(3), 1569 (2016)
5. C. Wang, L.L. Liu, A.T. Zhang, P. Xie, J.J. Lu, X.T. Zou, Afr. J. Biotecnol. **11**, 10248 (2012)
6. N. Ohtsu, Y. Kakuchi, T. Ohtsuki, Appl. Surf. Sci. **445**, 596 (2018)
7. X. Xiao, R. Liu, C. Chen, L. Huang, J. Mater. Sci. Mater. Med. **19**(2), 797 (2008)
8. P. Jia, H. Tan, K. Liu, W. Gao, Mater. Lett. **205**, 233 (2017)
9. Z. Beyene, R. Ghosh, Mater. Today Commun. **21**, 100612 (2019)
10. M. Ding, N. Sahebgharani, F. Musharavati, F. Jaber, E. Zalnezhad, G.H. Yoon, Ceram. Int. **44**, 7746 (2018)
11. A. Sirelkhatim, S. Mahmud, A. Seeni, N.H. Mohamad Kaus, L.C. Ann, S.K. Mohd Bakhori, H. Hasan, D. Mohamad, Nano-Micro Lett. **7**(3), 219 (2015)
12. A. Refaat, R.A. Youness, M.A. Taha, M. Ibrahim, J. Mol. Struct. **1147**, 148 (2017)
13. M. Safari Gezaz, S. Mohammadi Aref, M. Khatamian, Mater. Chem. Phys. **226**, 169 (2019)
14. L. Zhang, Y. Jiang, Y. Ding, M. Povey, D. York, J. Nanopart. Res. **9**(3), 479 (2007)
15. O. Yamamoto, Int. J. Inorg. Mater. **3**(7), 643 (2001)
16. A.J. Gimenez, J.M. Yáñez-Limón, J.M. Seminario, J. Phys. Chem. C **115**, 282 (2011)
17. S.K. Gupta, A. Joshi, M. Kaur, J. Chem. Sci. **122**, 57 (2010)
18. J. Yu, W. Zhang, Y. Li, G. Wang, L. Yang, J. Jin, Q. Chen, M. Huang, Biomed. Mater. **10**(1), 015001 (2014)
19. O. Oleshko, V. Deineka, Y. Husak, V. Korniienko, O. Mishchenko, V. Holubnycha et al., Materials (Basel) **12**(22), 3742 (2019)

Chapter 8
Plasma Electrolytic Oxidation of TiZr Alloy in ZnONPs-Contained Solution: Structural and Biological Assessment

O. Oleshko, V. Deineka, Ye. Husak, V. Korniienko, B. Dryhval, J. Dudko, O. Solodovnyk, W. Simka, J. Michalska, O. Mishchenko, K. Grundsteins, and M. Pogorielov

Abstract Titanium (Ti) and its alloys with zirconium (Zr) due of their biological safety, lower elastic modulus and excellent corrosion resistance are the most attractive metallic materials for medical applications. Plasma electrolytic oxidation (PEO) is an environment-friendly process with rapid deposition of anodic oxide coating on the implant surface. PEO coating can incorporate different nanoparticles (NPs) into implant surface such as ZnO that stimulates osteoblast proliferation and mineralization, possesses antibacterial ability. The aim of current research was to evaluate structural and chemical properties of TiZr-alloy with addition ZnO NPs after the PEO in Ca/P solution. The alloy of the Titanium-zirconium system was obtained from Osteoplant R&D (Dębica, Poland). ZnO nanocrystals have been synthesized through the alcoholysis and condensation reaction, known as polyol process. The pretreated samples were subjected to anodization in electrolytic bath, which contained of $Ca(H_2PO_2)_2$ (0.1 Mol/L) and silver nanoparticles (40,000 mg/L). Ready samples were investigated by scanning electron microscope (SEM) which showed surface with oxide layer with large pores in different size. The process of apatite deposition is confirmed the biocompatibility properties of obtained coatings in SBF immersion test. ZnO NPs demonstrated high bacteriostatic activity and absent of cell toxicity. The obtained results showed, that PEO of TiZr alloy in Ca–P based solution with ZnO NPs is a great potential method of surface modification of implants for orthopedics and dentistry.

O. Oleshko (✉) · V. Deineka · Ye. Husak · V. Korniienko · B. Dryhval · J. Dudko · O. Solodovnyk · M. Pogorielov
Sumy State University, Sumy, Ukraine
e-mail: o.oleshko@med.sumdu.edu.ua

O. Mishchenko · M. Pogorielov
Osteoplant Research and Development, 39-200 Dębica, Poland

W. Simka · J. Michalska
Silesian University of Technology, Gliwice, Poland

K. Grundsteins
Institute of Atomic Physics and Spectroscopy, University of Latvia, Riga, Latvia

A. D. Pogrebnjak et al. (eds.), *Nanomaterials in Biomedical Application and Biosensors (NAP-2019)*, Springer Proceedings in Physics 244,
https://doi.org/10.1007/978-981-15-3996-1_8

8.1 Introduction

The one main way in implant design evolution is combined different methods to obtain material with bio-functionality and biocompatibility characteristics. Surface of implant plays the key role of this.

Titanium (Ti) and its alloys is the most attractive metallic materials for medical applications. Its combined such properties as advantages of high strength-to-weight ratio, excellent corrosion resistance and favorable biocompatibility. Especially in orthopedic and dental implants [1–3].

Alloying element such as Zr improves the corrosion resistance by forming stable oxides ZrO_2 [4, 5]. These element was chosen due to their reported biological safety and β-stabilizing effect to obtain a lower elastic modulus close to bone tissue [6].

Nowadays plasma electrolytic oxidation (PEO) is most popular method to enhance surface performance of titanium alloys. PEO is a simple and environment-friendly process with rapid deposition of anodic oxide coating on the implant surface. Moreover, PEO coatings provide the increase surface hardness, roughness, porosity and excellent wear resistance.

During PEO ZnO nanoparticles (ZnONPs) could be incorporated in coating due to promising anabolic effects on bone metabolism by stimulating osteoblast proliferation and mineralization possesses, antibacterial ability [7].

PEO coatings with ZnONPs have thus attracted the interest of the biomedical community. This type of surface treatment could bind the osteogenesis ability and antibacterial effects.

The materials with anti-adhesive polymers or antibacterial agents to inhibit bacterial colonization are demanded by the orthopedic community [6].

The aim of current research was to evaluate structural and chemical properties of TiZr-alloy with addition ZnONPs to achieve both osteogenesis inducing ability and antibacterial effects after the PEO in Ca/P solution.

8.2 Materials and Methods

8.2.1 Materials

The alloy of the Titanium-zirconium system (Zr60Nb21Ti19) that was obtained from Osteoplant R&D (Dębica, Poland) we used in experiment. 6 mm diameter cylindrical samples with a height of 6 mm were prepared for experiment. All chemicals have been purchased from Sigma-Aldrich and used as received: zinc acetate dihydrate ($Zn(CH_3CO_2)_2\ (H_2O)_2$, 99%), poly(N-vinylpyrrolidone) (PVP, Mw = 40,000, K-25), sodium phosphinate monohydrate SPM ($NaH_2{\cdot}PO_2{\cdot}H_2O$, 95%), ethylene glycol EG ($C_2H_6O_2$, 90%), isopropanol (C_3H_8O, 90%).

Human osteoblast (HO) were obtained from collection of Sumy State University.

Staphylococcus aureus B 918, obtained from National collection of microorganism (D. K. Zabolotny Institute of Microbiology and Virology) were used in experiment. All bacteriological media were purchase from HiMedia (India), Alamar blue—from Invitrogen. For the cell culture study all media and reagents were purchased from Gibco®, USA.

8.2.2 Synthesis of ZnO Nanocrystals

ZnO nanocrystals have been synthesized through the alcoholysis and condensation reaction, known as polyol process, of $Zn(CH_3CO_2)_2$ $(H_2O)_2$ (2.2 g) in EG (10 ml) at 160 °C, similar to work. After heating up the solution in a round-bottom flask fitted with a reflux condenser for 4 h under magnetic stirring, the precipitation of ZnO nanocrystals has occurred. The nanocrystals have been recovered by the centrifugation at 4000 rpm for 5 min and washed twice with isopropanol.

8.2.3 Plasma Electrolytic Oxidation

Before starting PEO, the samples were cleaned and degreased. All samples were rinsed with distilled water and ultrasonically cleaned in a deionised water and 2-propanol for 5 min each. The pretreated samples were subjected to anodization in electrolytic bath, which contained of $Ca(H_2PO_2)_2$ (0.1 Mol/L) and ZnONPs (40,000 mg/L). The anodic oxidation was performed under a constant current of 0.1 A cm^{-2} and up to final voltage of either 350 V for 5 min. The anodised specimens were rinsed with distilled water and in a deionised water/2-propanol mixture for 5 min.

The process was performed in a water-cooled electrolysis cell with a titaniummesh cathode and magnetic stirrer. The Zr60Nb21Ti19 alloy served as the anode. A DC power supply (PWR 800H, Kikusui, Japan) was used throughout these treatments.

8.2.4 SEM

To analyzed surface morphology and chemical compound all samples were covered with gold and observed in a scanning electron microscope with EDX (JEOL JSM 5410). Before SEM all samples after PEO and SBF test were covered with gold.

8.2.5 Contact Angle Measurement

Contact angle (CA) measurements experiments were made using a video-based optical contact angle measuring instrument (OCA 15 EC, Data Physics, San Jose, CA city, state, USA) and SCA 20 software for OCA and PCA. The CA data was recorded for ultra-pure water, for at least three parallel samples.

8.2.6 SBF

Bioactivity investigations were performed in a protein-free SBF with an ions concentration simulating that of human blood plasma. The SBF was prepared by dissolving reagent-grade NaCl, $NaHCO_3$, KCl, $K_2PO_4\_3H_2O$, $MgCl_2\_6\ H_2O$, $CaCl_2$, and Na_2SO_4 in distilled water, which was adjusted to a pH of 7.40 using tris (hydroxymethyl) aminomethane and 1 M HCl at 37 °C. This solution has the following ionic composition: 142.0 mmol/L Na^+, 5.0 mmol/L K^+, 1.5 mmol/L Mg^{2+}, 2.5 mmol/L Ca^{2+}, 147.8 mmol/L Cl^-, 4.2 mmol/L HCO^{3-}, 1.0 mmol/L HPO_4^{2-}, 0.5 mmol/L SO_4^{2-}.

After the PEO process, samples were immersed in the SBP for a total of 4 weeks, with daily replacement of the solution. After each 7 days, samples were taken out of solution, then washed with distilled water and dried. The dehydrated samples were covered with gold (The Cressington Sputter Coater 108 auto). And analyzed by observations in a scanning electron microscope (Hitachi S-3400 N).

8.2.7 In Vitro Experiment

8.2.7.1 Bacteria Adhesion Test

S. aureus, strain B 918 was used to describe the adhesive properties of the Zr60Nb21Ti19 disks. Gram positive bacteria grown on nutrient agar at 37 °C for 24 h was suspended in a saline solution (0.9%, w/v NaCl) and re-suspended to a final density of 1×10^8 colony forming units (CFUs)/mL (8 log CFU) in nutrient broth using McFarland standards. The adhesion possibilities was measured by horizontally incubation with 2.0 ml of the bacterial suspension in static conditions in a 24-well plate at 37 °C for 2, 4, 6 and 24 h. Then samples were removed from broth. The washing were conducted tree times with 2.0 ml sterile physiological saline to remove freely-adherent bacteria.

The amount adherent bacteria on surface of TiZr alloy were received after sonication disk in sterile tubes with 1.0 ml of sterile saline solution for 1 min by using an ultrasonic-bath (B3500S-MT, Bransone Ultrasonics Co., Shanghai, China). 10-μl aliquots saline solution from sonicated tubes were inoculation using streak plate technique cultivating onto the solid media for 24 h. Following this, we conducted determining the colony count at each term of incubation. The wells containing disks and tested samples in growth medium without bacterial inoculate were used as a control. All experiments were conducted in triplicate.

8.2.7.2 Cell Culture

Human osteoblast cells were grown in standard conditions: 5% a humidified CO_2, 37 °C air temperature and medium renewal for every 2–3 days in tissue culture flasks. Dulbecco's Modified Eagle Medium/Nutrient Mixture F-12 (DMEM/F-12) with L-glutamine used, containing 100 units/ml penicillin, 100 μg/ml streptomycin, 2.5 μg/ml amphotericin B, 10% Fetal Bovine Serum and 1.0 ng/ml bFGF. After removing medium osteoblasts were seeded on each samples and positive control wells at a cell density of 2×10^4 cells per well.

Alamar blue (Invitrogen) were added in an amount equal to 10% of the volume to each well. The plates were incubated for 4 h at 37 °C, in the dark. The absorbance of medium was measured using a Multiskan FC (Thermo Fisher Scientific) plate reader at wavelengths of 570 nm and 600 nm. The cells were quantified at different time intervals: 1 day, 3 days and 7 days. All experiments were repeated 3 times. The percentage of Alamar blue reduction was performed as Equation according to the manufacturer's protocol.

Cells on the surfaces of TiZr alloy samples at day 7 were fixed with 2.5% glutaric aldehyde and subjected to osmium coating for 1 h, then washed three times in 0.1 M cacodylic buffer. The alloy samples were then dehydrated in increasing ethanol concentrations and covered with gold and observed in the SEM (JEOL JSM 5410).

8.3 Results

8.3.1 SEM, EDX and Contact Angle

After PEO TiZr surface represents by irregular shaped grooves and round pores (Fig. 8.1b). Longitudinal size of grooves sized from 23 ± 4.3 μm to 105 ± 15.7 μm. Pores size ranges from 1.3 ± 0.5 μm to 11.4 ± 3.8 μm. Thus, PEO in complex ZnONPs solution provides formation of mesoporous structures that should be favorable for cell adhesion. Submicron pores can be detected inside the large grooves and completely covers their bottom and lateral walls.

EDX proved formation of oxide layer contained both alloy elements Ti and Zr as well as ions from PEO solution (Fig. 8.1c). Mass concentration of Ca, P and Zn corresponds to 9.92, 17.2 and 24.88 mass%. High concentration of Zn could be toxic for cells that need to clarify in cell culture experiment.

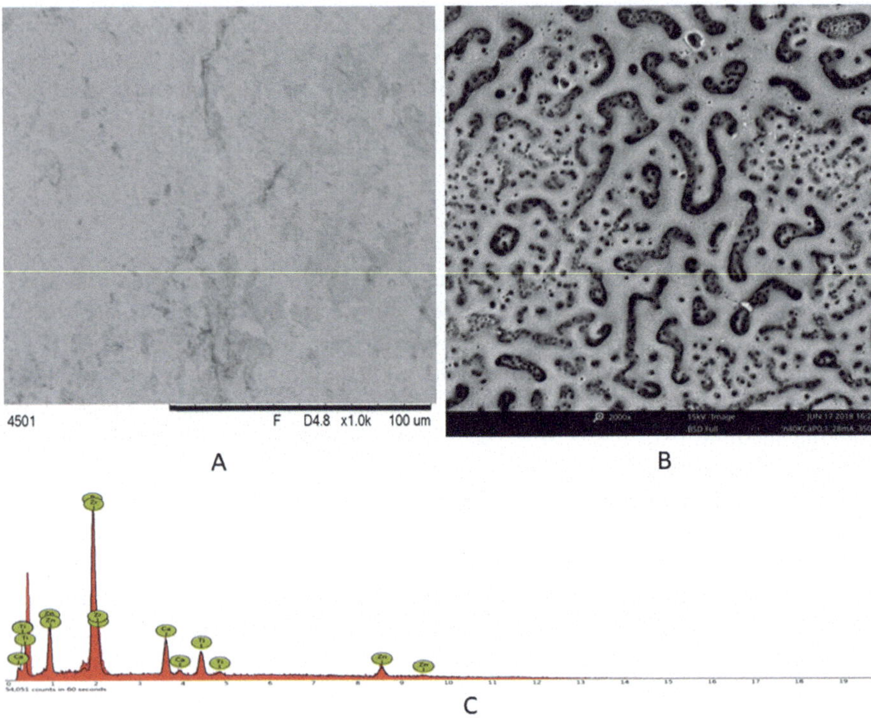

Fig. 8.1 SEM of TiZr samples before (**a**) and after (**b**) PEO in Ca–P–ZnONPs solution with EDX spectra from modified surface (**c**)

Contact angle of TiZr surface after PEO was 58.2 ± 17.6° that significantly low compare the non-treated surface (78.9 ± 6.9°). Hydrophilic surface should provide favorable environment for protein adsorption in first minutes after implantation.

8.3.2 SBF Test

SBF immersion test shown calcium-phosphate deposition from second week of experiment (Fig. 8.2a). In three weeks after immersion whole surface covered by Ca–P substrate (Fig. 8.2b) and we can detect cracks development in four weeks that probably due to Ca–P overload on TiZr surface (Fig. 8.2c). Ca and P adsorption proved high biocompatibility of newly obtained surface. Additionally, high affinity to Ca and P should be advantages after implantation and will facilitate apatite formation on implant surface.

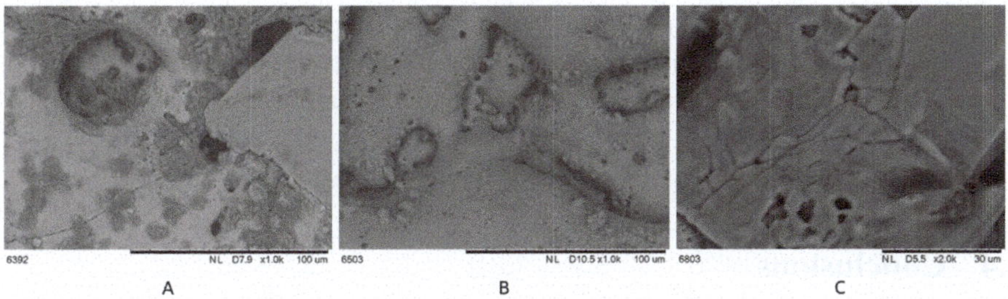

Fig. 8.2 SEM images of TiZr anodized surface after 2 (**a**), 3 (**b**) and 4 (**c**) weeks of SBF immersion

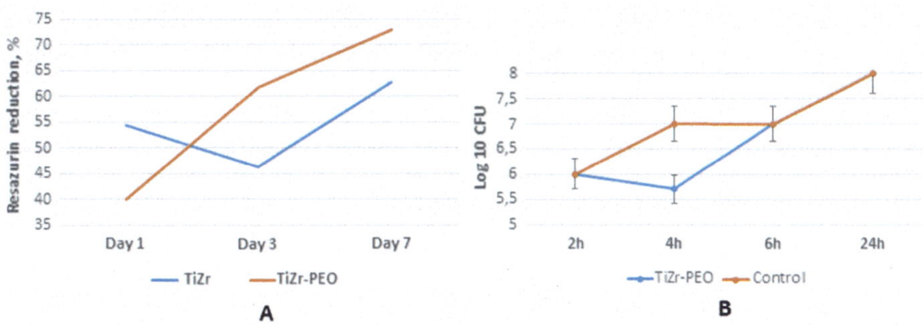

Fig. 8.3 Resazurin reduction during the 7 day of human osteoblast cells cultivation (**a**) and bacteria cell adhesion during the 24 h of co-cultivation (**b**)

8.3.3 Cell Culture

HO cell adhesion on TiZr PEO-treated surface was significant lower ($p \leq 0.01$) compared with pure non-treated TiZr alloy (40.6 ± 7.4% vs. 54.3 ± 3.2%) on day 1 after cell seeding. But next time-point shown progressive cell proliferation on day 3 and 7 compare the slightly cell number decreasing on day 3 (non-treated control). On the last day of culturing osteoblast cell number was significantly higher on PEO-treated surface (Fig. 8.3a). Despite the high Zn concentration after PEO, surface shown no cell toxicity and stimulation of osteoblast proliferation during the 7 day culturing.

8.3.4 Bacteria Adhesion Test

Bacteriological investigation confirmed *S. aureus* adhesion and proliferation during the 24-h experiment (Fig. 8.3b). After 2 hincubation PEO treated surface had the same concentration of bacteria compare the control (polished untreated ZrNb implant). But after 4 h of experiment the microorganism level left stable. While the bacteria concentration of untreated surface shown the significant proliferation during all time point up to 8 Log 10 CFU in 24 h. Present of ZnONPs could negative influence on adhesion process of *S. aureus* or may deter the multiplication bacterial cells on the implant surface.

8.4 Conclusions

Plasma electrolytic oxidation of TiZr alloy in Ca–P based solution with ZnONPs provides formation of mesoporous oxide layer that support osteoblast cell adhesion and proliferation. Additionally, obtained surface decrease bacteria adhesion in first hours after co-cultivation. New surface should be promising for medical implant production.

Acknowledgements This research supported by H2020 Marie Skłodowska-Curie Actions Grant NanoSurf 777926 and Ukraine MES Grant "Biological effectiveness of Plasma Electrolytic Oxidation and sol-gel deposition for development of functional implant surface" (#0119U100823).

References

1. Y. Bai, Y. Deng, Y. Zheng, Y. Li, R. Zhang, Y. Lv, Q. Zhao, S. Wei et al., Mater. Sci. Eng. **59**, 565–576 (2016)
2. P. Kubasiewicz-Ross, M. Dominiak, T. Gedrange, U.U. Botzenhart, Adv. Clin. Exp. Med. **26**(3), 533–537 (2017)
3. I. Gnilitskyi, M. Pogorielov, R. Viter, A.M. Ferraria, A.P. Carapeto, O. Oleshko et al., Nanomed. Nanotechnol. Biol. Med. **21**, 102036 (2019)
4. C. Nune, R.D.K. Misraa, S.J. Li, Y.L. Hao, R. Yang, Dent. Mater. **33**(2), 152–165 (2016)
5. L. White, Y. Koo, S. Neralla, J. Sankar, Y. Yun, Mater. Sci. Eng. B Solid. State. Mater. Adv. Technol. **208**, 39–46 (2016)
6. A. Mohandas, A.G. Krishnan, R. Biswas, D. Menon, M.B. Nair, Mater. Sci. Eng. C Mater. Biol. Appl. **75**, 115–124 (2017)
7. K. Huo, X. Zhang, H. Wang, L. Zhao, X. Liu, P.K. Chu, Biomaterials **34**, 3467–3478 (2013). https://doi.org/10.1016/j.biomaterials.2013.01.071

Chapter 9
Plasma Electrolytic Oxidation of the Titanium-Zirconium Alloy (Zr60Nb21Ti19) for Dental Implant

V. Korniienko, O. Oleshko, Ye. Husak, V. Deineka, V. Holubnycha, O. Mishchenko, W. Simka, and M. Pogorielov

Abstract New metastable beta Ti-based alloys arise as new perspective materials for dental implants due to adjusted chemical compositions that displays a reversible stress-induced martensitic transformation that leads to a superelastic effect (high recoverable strain) and a very significant reduction of the apparent elastic modulus. But it is still require increasing biological response that can achieve by surface modification. Plasma electrolytic oxidation (PEO) can be a perspective method for surface fynctionalisation. ***This research was dedicated*** to application of PEO in $Ca(H_2PO_2)_2$ solution for TIZr alloy surface treatment and investigation of newly formed coating structure and their biological properties. The alloy of the titanium-zirconium system (Zr60Nb21Ti19) was obtained from NanoPrime (Dębica, Poland). The anodic oxidation was performed under a constant current of 0.1 A cm^{-2} and up to final voltage of either 300 (TiZr-300), 400 (TiZr-400) and 450 (TiZr-450) V for 5 min. SEM, EDX and contact angle measurements were used for surface assessment and SBF test, cell culture and bacteriological assay—for biological evaluation. The use of plasma electrolytic oxidation to modify titanium-zirconium dental implants leads to the formation of an oxide layer and a mesoporous structure on their surface. Moreover, the peculiarities of morphology directly depend on electric voltage disturbance. Cultivation of cells on the surface of the samples showed a positive effect of adhesion of osteoblasts, compared with the control alloys, and confirmed that PEO at 450 V gives the best conditions for cell reproduction. Bacteriological examination showed that the experimental surface is less favorable for bacterial contamination compared to the control, but not significantly different from the voltage of the electric current. Therefore, this method of modification of titanium-zirconium alloys is promising for further research and opens new opportunities in medical practice.

V. Korniienko · O. Oleshko (✉) · Ye. Husak · V. Deineka · V. Holubnycha · M. Pogorielov
Sumy State University, Sumy, Ukraine
e-mail: o.oleshko@med.sumdu.edu.ua

O. Mishchenko · M. Pogorielov
Osteoplant Research and Development, 39-200 Dębica, Poland

W. Simka
Silesian University of Technology, Gliwice, Poland

A. D. Pogrebnjak et al. (eds.), *Nanomaterials in Biomedical Application and Biosensors (NAP-2019)*, Springer Proceedings in Physics 244,
https://doi.org/10.1007/978-981-15-3996-1_9

9.1 Introduction

The most of the implants used in orthopedics and dentistry are made from Ti–6Al–4V alloy [1] with higher elastic stiffness and a relatively high elastic modulus (110 GPa) that in 4–6 time higher the cortical bone (30 GPa). It is well known that decreasing of elastic modulus induces a better stress transfer along the bone–implant interface and could limit the crestal bone loss. Moduli mismatch could lead to excessive micromotion between implants and bone, which prohibits bone formation and contrarily facilitates fibrous tissue ingrowth [2]. Additionally, Ti–6Al–4V alloy can release Al and V ions during corrosion (after long term implantation) that should make an adverse tissue reaction.

New metastable beta Ti-based alloys (contained nontoxic elements such as Nb, Ta, Mo, and Zr) can partially solve this problem due to adjusted chemical compositions that displays a reversible stress-induced martensitic transformation that leads to a superelastic effect (high recoverable strain) and a very significant reduction of the apparent elastic modulus [3]. Ti–Ni alloy was considered as alternative material over last two decades due to super-elasticity and high damping but possible release and accumulation of Ni ions limit their wide application [4]. Based on the d-electron alloy design method, many new Ti alloys with low elastic modulus have been developed, such as the Ti–Nb–Zr series with the lowest modulus of 38.8 GPa [5], the Ti–35Nb–4Sn–6Mo–9Zr alloy with the lowest modulus of 65 GPa [6], and the Ti–Nb–Zr–Ta–Si series alloys with the compression elastic modulus of 37–48 GPa [7]. Additionally, alloying titanium with more thermodynamically stable elements such as Zr and Nb makes the implant more corrosion resistant, biocompatible, and promotes cell adhesion [8]. Two-component Ti–Zr alloy was first established in 1995 when Kobayashi et al. investigated this new alloy for its suitability as a biomedical implant [9]. Since that time a lot of in vitro, animal and clinical trials shown excellent biocompatibility and clinical effectiveness of Ti–Zr alloy for dental and orthopedics implants [10, 11].

Additionally to stress-shielding syndrome, bone-implant interaction still play a major role in postoperative complications due to pure bone development around implant and bacterial contamination. Incomplete osteointegration represents a major contribution toward aseptic loosening and still plays main role in other postoperative complications [12]. The development of functional surface is clearly important to ensure an optimal, long-term functionality of medical implants. Several inexpensive methods, such as grit blasting or acid etching used within many years with clinical success but number of complication still not decrease a lot [13]. Chemical coating using biomimetic calcium phosphate/hydroxyapatite has represented a main focus of research for more than 30 years now [14] and originally achieved by plasma spray and electrodeposition methods. Poor mechanical stability of thick calcium phosphate coatings were later reported [15] and new method for formation of biomimetic coating still required.

Plasma electrolytic oxidation (PEO) is a method to ceramized metallic alloys and in contrast to conventional methods, the new layer is not separate from the substrate, but rather an inert part of the ceramized alloy [16, 17]. There are a lot of evidence of improving of corrosion and wear resistance [18] as well as biocompatibility and bioactivity [19, 20] of standard Ti alloy after PEO treatment in Ca and P solutions. There are some preliminary studies using PEO for Nb and Zr based alloys. Pereira B.L. and co-authors suggested that PEO with electrolytes containing phosphorous and/or calcium on Nb alloy has positive characteristics for osseointegration processes [21]. Jung O. et al. made 15–20 μm thick coating on TiZr alloy that was biocompatible and resistant to corrosion [22]. There are some research for PEO application in silicate-based electrolytes for Zr and Zr alloy treatment but they are not concentrated in Ca–P solutions [23, 24]. In our previous research we proved formation of porous structures on pure zirconium with satisfied corrosion resistance after PEO in Ca–P solution [25] but it is required additional experiment for TiZr system treatment.

This research was dedicated to application of PEO in $Ca(H_2PO_2)_2$ solution for TIZr alloy surface treatment and investigation of newly formed coating structure and their biological properties.

9.2 Materials and Methods

9.2.1 Materials

The alloy of the titanium-zirconium system (Zr60Nb21Ti19) was obtained from NanoPrime (Dębica, Poland) we used in experiment as a 6 mm diameter cylindrical samples with a height of 6 mm.

All chemicals for PEO have been purchased from Sigma Aldrich and used as received. Gram-positive bacterium (S. aureus, strain B 918) obtained from National collection of microorganism (D. K. Zabolotny Institute of Microbiology and Virology, Ukraine) was used for evaluation the adhesive surface properties of the samples. Required bacterial culture media were procured from HiMedia (India), Alamar blue—from Invitrogen (USA). The media and reagents for the cell culture study were purchased from Gibco®, USA. Human osteoblast (HO) were obtained from collection of Sumy State University.

9.2.2 Plasma Electrolytic Oxidation

All samples were rinsed with distilled water and ultrasonically cleaned in a deionised water and 2-propanol for 5 min. The pretreated samples were subjected to anodization in electrolytic bath, which contained of $Ca(H_2PO_2)_2$ (0.5 Mol/L). The anodic oxidation was performed under a constant current of 0.1 A cm^{-2} and up to final voltage of either 300 (TiZr-300), 400 (TiZr-400) and 450 (TiZr-450) V for 5 min. The surface area of the face that was modified was equal to 1.89 cm^{-2} (diameter = 6 mm).

The anodized specimens were rinsed with distilled and than—in a deionised water for 5 min. A DC power supply (PWR 800H, Kikusui, Japan) was used throughout these treatments. The process was performed in a water-cooled electrolysis cell with a titanium mesh cathode and magnetic stirrer. The ZrNb alloy served as the anode.

9.2.3 SEM, EDX and Contact Angle Measurements

SEM investigation is used to assess sample morphology after PEO process and SBF biocompatibility tests. Before SME all samples after PEO and SBF test were covered with gold (The Cressington Sputter Coater 108 auto). Surface analysis was carried out with a 5,000 times magnified image in Tescan (Brno–Kohoutovice, Czech Republic) and Phenom Pro X (Eindhoven, The Netherlands) scanning electron microscopes with an EDX system. Contact angle (CA) measurements experiments were made using a video-based optical contact angle measuring instrument (OCA 15 EC, Data Physics, Filderstadt, Germany). The CA data was recorded for ultra-pure water, for at least three parallel samples.

9.2.4 SBF

Bioactivity investigations were performed in a protein-free SBF with an ions concentration simulating that of human blood plasma. This solution has the following ionic composition: 142.0 mmol/L Na^+, 5.0 mmol/L K^+, 1.5 mmol/L Mg^{2+}, 2.5 mmol/L Ca^{2+}, 147.8 mmol/L Cl^-, 4.2 mmol/L HCO^{3-}, 1.0 mmol/L HPO_4^{2-}, 0.5 mmol/L SO_4^{2-}.

The SBF was prepared by dissolving reagent-grade NaCl, $NaHCO_3$, KCl, K_2PO_4-$3H_2O$, $MgCl_2$-6 H_2O, $CaCl_2$, and Na_2SO_4 in distilled water, which was adjusted to a pH of 7.40 using tris (hydroxymethyl) aminomethane and 1 M HCl at 37 °C. After the PEO process, the specimens were immersed in the SBF for a total of 4 weeks. SBF solution was refreshed every 24 h during of experiments. At the end of each week, samples were taken out of solution, then washed with distilled water and dried at room temperature during 24 h and proceed for SEM.

9.2.5 Bacteria Adhesion Test

The bacterial strain (S. aureus) was cultivated on nutrient agar at 37 °C for 24 h. The disks were incubated horizontally in a 24-well sterile plastic plate with 2.0 ml of the bacterial suspension with a final density of 1×10^5 colony forming units (CFUs)/mL (5 log CFU) in static conditions at 37 °C. After 2, 4, 6 and 24 h the specimens were washed three times with 2.0 ml sterile physiological saline to purge out freely-adherent bacteria. Then, disks were sonicated for 1 min in sterile tubes with 1.0 ml of sterile saline solution using an ultrasonic-bath (B3500S-MT, Bransone Ultrasonics Co., Shanghai, China) to extract adherent bacteria from the surfaces of the specimens. 10-μl aliquots saline solution from sonicated tubes were grown on the solid media for 24 h using streak plate technique to determine the colony count at each term of incubation. As a control we used disks and tested samples in growth medium without bacterial inoculate. All experiments were repeated in triplicate.

9.2.6 Cell Culture

Primary human osteoblast cell culture were cultured in Dulbecco's Modified Eagle Medium/Nutrient Mixture F-12 (DMEM/F-12) with L-glutamine used, containing 100 units/ml penicillin, 100 μg/ml streptomycin, 2.5 μg/ml amphotericin B, 10% Fetal Bovine Serum and 1.0 ng/ml bFGF. Cells were maintained at 37 °C in a humidified incubator with 5% CO_2 for 24 h, until a monolayer, with greater than 80% confluence, was obtained. Osteoblasts were seeded on each samples and positive control wells at a cell density of (2×10^4 cells per well). Cell viability was analyzed by the colorimetric Alamar blue assay in which resazurin is a blue dye reduced to the pink colored resorufin only by viable mitochondria. Alamar blue were added in an amount equal to 10% of the volume to each well. As negative control Alamar Blue solution was added to the medium without cells. As positive control Alamar blue solution was added to the medium from wells contains only cells without samples. The plates were incubated for 4 h at 37 °C in the dark. The medium was transferred to another 96-well plate and absorbance was measured using a Multiskan FC (Thermo Fisher Scientific) plate reader at wavelengths of 570 and 600 nm. The cells were quantified at different time intervals: 1, 3 and 7 days. All experiments were repeated 3 times. The calculation of the percentage of Alamar blue reduction was performed as Equation according to the manufacturer's protocol.

9.2.7 Statistics

Data was expressed as means ± standard deviation. Student's t-test on unpaired data was used to assess the statistical significance of the difference. Statistical significance was assumed at a confidence level of 95% ($p < 0.05$).

9.3 Results

9.3.1 SEM with EDX

The use of plasma electrolytic oxidation (PEO) to modify the surface of TiZr samples leads to the formation of a solid oxide layer with structure depending on applied voltage. TiZr-300 regimen provide formation of round, randomly distributed pores with 1–5 μm in diameter. Numerous cracks start from big pores and distribute to small ones, that may be as a factor of low oxide layer resistance to the physiological solutions. Increasing in voltage leads to formation of mesoporous coatings with big flat valley (30–70 μm in diameter) and porous interconnected hills between them. Pores size varies from 5 to 30 μm that is more optimal for cell adhesion. Flat valleys have less pores but Ca–P crystals covers most of them. Interconnected cracks identified on ceramic surface after TiZr-400 regimen and their number and size significantly decreased with voltage rise up to 450 V.

The SEM was equipped with an integrated energy-dispersive X-ray spectrometer (EDX), which was utilized to obtain the elemental composition of the films. The table of element distribution (Fig. 9.1) demonstrates the relative changes in the concentration of the surface elements in the oxide layers. Coatings were composed of oxygen, titanium, and zirconium with small amounts of Cl. We did not include Oxygen in calculation because of too small atomic weight of this element to quantify it by EDX.

According to data, calcium was incorporated into coatings much easier than phosphorus. Difference of incorporation amount of the phosphorus and calcium was not

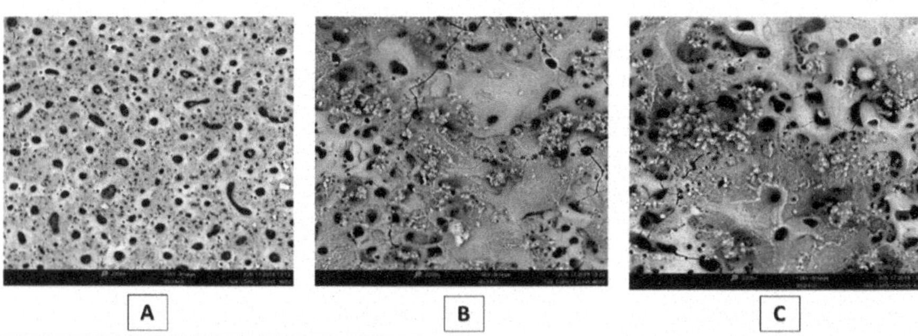

	Atomic percentage		
	300V	400V	450V
Ca	47,96±5.2	49,29±2.1	47,93±12.5
P	40,63±9.9	43,66±8.8	42,03±5.9
Ti	5,13±1.4	3,84±1.3	4,02±1.1
Zr	4,62±2.1	3,21±0.6	3,84±1.5
Cl	0,90±0.3		2,18±0.6
I	0,75±0.4		

Ca/P atomic ratios of PEO coating			
	300V	400V	450V
Ca/P	1,18	1,13	1,14

Fig. 9.1 SEM images and EDX of TiZr samples after PEO in 0.5 M $Ca(H_2PO_2)_2$ solution with applied voltage 300 V (**a**), 400 V (**b**) and 450 V (**c**) V. Tables of EDX atomic percentage of ceramic layer and Ca/P rations

significant within increasing voltage. In addition less zirconium and titanium were detected as result of oxide coating formation. The elemental composition of the coatings was closely bound up with that of the solutions regardless voltage level. The composition of the electrolytes was elected to achieve the Ca/P atomic ratios of 1.67 as atomic ratio of stoichiometric hydroxyapatite, Nb is not detected with EDX.

9.3.2 SBF

SBF experiment is used as biomimetic mineralization test for mineralization surfaces. SBFs comprised ions at similar concentrations in blood plasma. This coating technique is performed under biological conditions in terms of temperature, pressure, and pH. Carbonated apatite is formed on a substrate with resembling chemical composition and material properties (crystallinity and dissolution rate) to bone mineral.

PEO coatings can release ions such as calcium. These released ions contribute to nucleation and subsequent surface mineralization, thus forming bone-like mineral apatite. Mineral formation is governed by the surface characteristics of the materials and the immersion parameters, such as the composition of the SBF, ionic strength, pH, temperature, and immersion time.

The CaP layers formed on the all sample surface. The deposition layer had similar view to all samples. It may depend to similar elemental composition of the coatings. The square of apatite depended on the immersion time. The volume of this coating had more value after immersion three weeks in SBF solution compares the same samples immersion two weeks (Fig. 9.2).

9.3.3 Cell Culture

All obtained surfaces shown biocompatibility and ability for osteoblast adhesion and proliferation. There are no significant differences in cell attachment in day 1 between TiZr-300 and TiZr-400 samples, but osteoblast adhere much better for TiZr-450 surface ($p \leq 0.05$). Between day 1 and day 3 we can see significant cell proliferation up to 60% on TiZr-300 samples and 78%—on TiZr-450 ones. Cell proliferation slows down from day 3 and 7 on TiZr-300 and TiZr-400 samples that can indicate disability of long time osteoblast proliferation. Other way, on TiZr-450 surface osteoblast actively proliferates up to 99% ($p \leq 0.01$) of resazurin redaction. It should be noted, that osteoblast adhesion to TiZr-400 and TiZr-450 samples significantly higher compare the TCP control both in day 3 and 7 (Fig. 9.3).

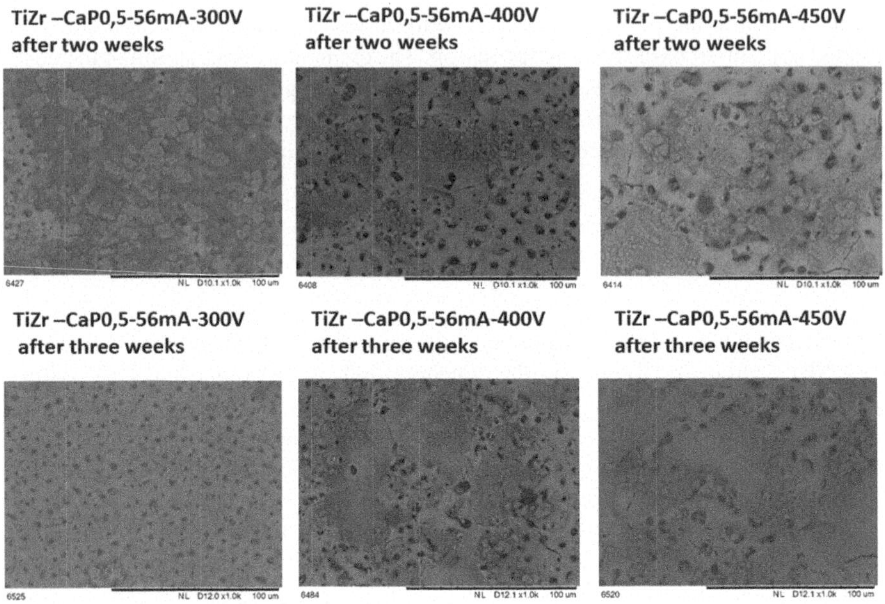

Fig. 9.2 SEM images of TiZr samples after 2 and 3 weeks of SBF test

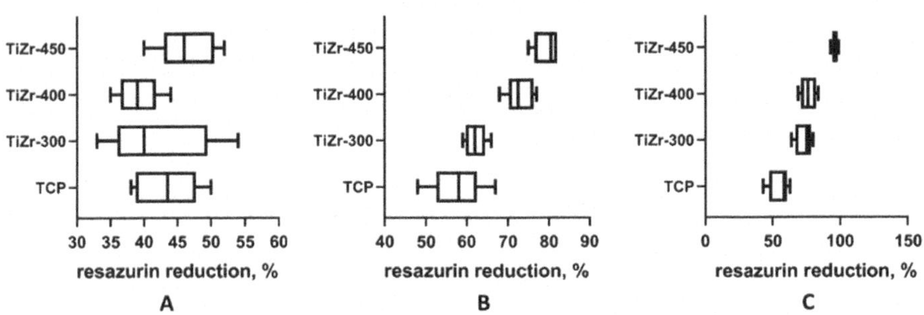

Fig. 9.3 Osteoblast adhesion properties of the experimental samples and TCP control in day 1 (**a**), 3 (**b**) and 7 (**c**)

9.3.4 Bacteria Adhesion Test

Bacteriological study after 2 h incubation TiZr-450 implant shown less adhesive properties compared to TiZr-300 and TiZr-400 ones (Fig. 9.4). Thus, amount of adherent bacteria on the TiZr-450 surface was to 2 time less after 2 h and 30% less after 4 h in comparison with TiZr-400 disk respectively. But after 6 and 24 h incubation implant TiZr-450 displayed loss of antiadhesive properties and amount of bacteria on its surface exceeded 7 Log CFU in 24 h.

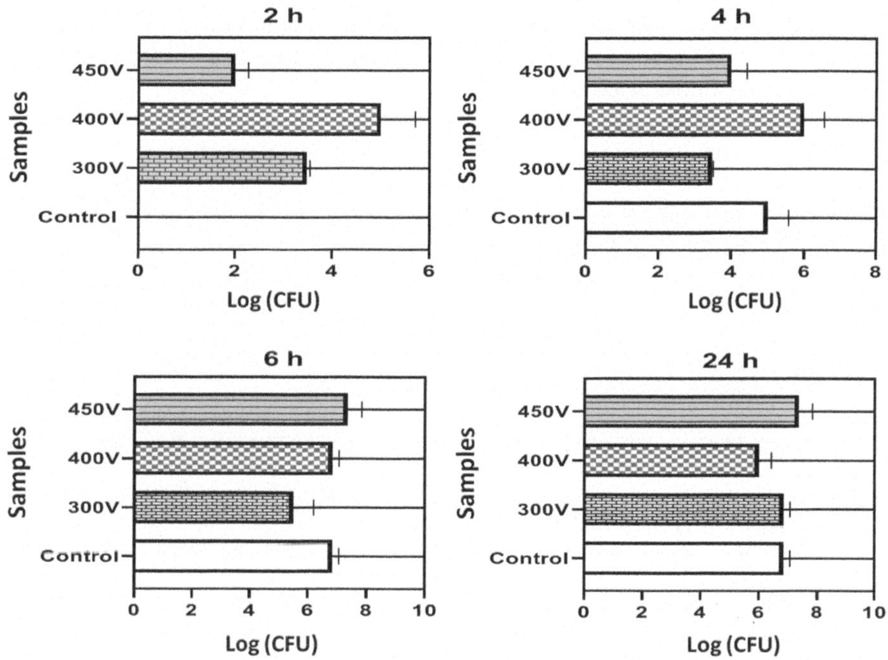

Fig. 9.4 The amount of attached bacteria on the surfaces of TiZr samples during 24 h incubation

Sample TiZr-300 remained stable rate of adhesiveness after 2 and 4 h and shown the least amount of bacteria (5 log10 CFU) compare rest samples after 6 h but long-time incubation led to intensive bacteria attachment up to 7.6 log10 CFU after 24 h. Adherent properties of TiZr-400 sample were almost the same during all time of experiment (from 5 log10 CFU to 6.6 log10 CFU).

Long-time co-cultivation leads to loss of antiadhesive properties and all implants demonstrated intensive bacteria growth on surface up to 6–7, 6 log10 CFU in 24 h.

9.4 Conclusion

The use of plasma electrolytic oxidation to modify titanium-zirconium dental implants leads to the formation of an oxide layer and a mesoporous structure on their surface. Moreover, the peculiarities of morphology directly depend on electric voltage disturbance. Cultivation of cells on the surface of the samples showed a positive effect of adhesion of osteoblasts, compared with the control alloys, and confirmed that PEO at 450 V gives the best conditions for cell reproduction.

Bacteriological examination showed that the experimental surface is less favorable for bacterial contamination compared to the control, but not significantly different from the voltage of the electric current.

Therefore, this method of modification of titanium-zirconium alloys is promising for further research and opens new opportunities in medical practice.

Acknowledgements This research was supported by H2020 MSCA, grant NanoSurf 777926 and Ukraine MES grant #0119U100770.

References

1. M. Bönisch, M. Calin, J. Van Humbeeck, W. Skrotzki, J. Eckert, Mater. Sci. Eng. C **48**, 511–520 (2015)
2. J. Duyck, K. Vandamme, L. Geris, H. Van Oosterwyck, M. De Cooman, J. Vandersloten, R. Puers, I. Naert, Arch. Oral Biol. **51**(1), 1–9 (2006)
3. D.M. Gordin, R. Ion, C. Vasilescu, S.I. Drob, A. Cimpean, T. Gloriant, Mater. Sci. Eng. C **44**, 362–370 (2014)
4. D.N.A. Shri, K. Tsuchiya, A. Yamamoto, Mater. Sci. Eng. C Mater. Biol. Appl. **43**, 411–417 (2014)
5. L. You, X.P. Song, Scr. Mater. **67**(1), 57–60 (2012)
6. S.J. Dai, Y. Wang, F. Chen, X.Q. Yu, Y.F. Zhang, Trans. Nonferr. Metals Soc. China **23**, 3027–3032 (2013)
7. Y.H. Li, C. Yang, F. Wang, H.D. Zhao, S.G. Qu, X.Q. Li, W.W. Zhang, Y.Y. Li, Mater. Des. **85**, 7–13 (2015)
8. R. Bhola, S.M. Bhola, B. Mishra, D.L. Olson, Trends Biomater. Artif. Organs **252011**, 34–46 (2011)
9. E. Kobayashi, S. Matsumoto, H. Doi, T. Yoneyama, H. Hamanaka, J. Biomed. Mater. Res. **29**(8), 943–950 (1995)
10. B. Wen, F. Zhu, Z. Li, P. Zhang, X. Lin, M. Dard Clin. Oral Impl. Res. **25**(7), 1–7 (2013)
11. I. Gnilitskyi, M. Pogorielov, R. Viter, A.M. Ferraria, A.P. Carapeto, O. Oleshko et al., Nanomedicine: nanotechnology. Biol. Med. **21**, 102036 (2019)
12. C. Mas-Moruno, M. Espanol, E. B. Montufar, G. Mestres, C. Aparicio, F. J. Gil et al., Biomater. Surf. Sci. 337–374 (2013)
13. D.D. Deligianni, N. Katsala, S. Ladas, D. Sotiropoulou, J. Amedee, Y.F. Missirlis, Biomaterials **22**, 1241 (2001)
14. F. Barrere, C.M. van der Valk, G. Meijer, R.A. Dalmeijer, K. de Groot, P. Layrolle, J. Biomed. Mater. Res. Part B **67B**, 655 (2003)
15. J. Lee, L. Rouhfar, O. Beirne, J. Oral Maxillofac. Surg. **58**, 1372 (2000)
16. A. Kopp, T. Derra, M. Müther, L. Jauer, J.H. Schleifenbaum, M. Voshage, O. Jung, R. Smeets, N. Kröger, Acta Biomater. **15**(98), 23–35 (2019)
17. O. Jung, R. Smeets, A. Kopp, D. Porchetta, P. Hiester, M. Heiland et al., Vivo **30**(1), 27–33 (2016)
18. L. White, Y. Koo, S. Neralla, J. Sankar, Y. Yun, Mater. Sci. Eng. B Solid. State. Mater. Adv. Technol. **208**, 39–46 (2016)
19. S. Durdu, M. Usta, A.S. Berkem, Surf. Coat. Technol. **301**, 85–93 (2016)
20. O. Oleshko, V. Deineka, Y. Husak, V. Korniienko, O. Mishchenko, V. Holubnycha et al., Mater. (Basel) **12**(22), 3742 (2019)
21. B.L. Pereira et al., J. Mech. Behav. Biomed. Mater. **77**, 347–352 (2018)
22. Ole Jung et al., Mater. (Basel) **12**(9), 1403 (2019)

23. R.G. Farrakhov, V.R. Mukaeva, A.R. Fatkullin, M.V. Gorbatkov, P.V. Tarasov, D.M. Lazarev et al., IOP Conf. Ser. Mater. Sci. Eng. **292**, 12006 (2018)
24. M. Sandhyarani, N. Rameshbabu, K. Venkateswarlu, D. Sreekanth, C. Subrahmanyam, J. Alloys Compd. **553**, 324–332 (2013)
25. M. Sowa, W. Simka, Mater. (Basel) **11**(5), 723 (2018)

Chapter 10
Nanostructured Hemostatic Sponges Made from Chitosan: Structural and Biological Evaluation

J. Radwan-Pragłowska, V. Korniienko, Ye. Husak, V. Deineka, Ł. Janus, D. Matysek, V. Holubnycha, O. Oleshko, M. Piątkowski, and M. Pogorielov

Abstract Local hemostatic agents are a necessary element of surgical work that reduces blood loss and mortality. Despite the variety of materials, there is no perfect styptic. The most promising hemostatic agents are based on biomaterials therefore they can be left in the wound. The aim of the following research was to obtain a new type of biocompatible chitosan-based agents with high hemostatic activity. The biomaterials were obtained under microwave radiation using L-aspartic acid, L-glutamic acid as crosslinking agents. Chitosan sponges showed high porosity and antioxidant activity, and also high antibacterial activity against both *S. aureus* and to *E. coli*. This material was absolutely non-toxic on cell culture and showed good hemostatic properties after interaction with blood in vitro.

10.1 Introduction

Nowadays hemorrhage remains the key reason of combat death and it is in the place second among the reasons leading to death after traumatic brain injury in the civilian sector [1]. Hemostasis is an essential complexion of surgical practice as can lead to serious clinical problem and consequences [2].

Hemostatic agents and dressings for managing hemorrhage have evolved since ancient times. Despite the main approach to hemostasis has not changed noticeably in modern medicine but renewed and more effective methods have been introduced. Hemostatic dressings can be categorized according to their mechanism of action [3] into 3 groups: (1) factor concentrators (fast absorption of the water content of blood

J. Radwan-Pragłowska · Ł. Janus · M. Piątkowski
Faculty of Chemical Engineering and Technology, Cracow University of Technology, Kraków, Poland

D. Matysek
Faculty of Mining and Geology, Technical University of Ostrava, Ostrava, Czechia

V. Korniienko (✉) · Ye. Husak · V. Deineka · V. Holubnycha · O. Oleshko · M. Pogorielov
Sumy State University, Medical Institute, Sumy, Ukraine
e-mail: vicorn77g@gmail.com

A. D. Pogrebnjak et al. (eds.), *Nanomaterials in Biomedical Application and Biosensors (NAP-2019)*, Springer Proceedings in Physics 244,
https://doi.org/10.1007/978-981-15-3996-1_10

lead to concentration of its cellular and protein components results in clot formation); (2) mucoadhesive agents (strong adherence to the tissues cause physically block bleeding from wounds); (3) procoagulant supplementors (delivering procoagulant factors to the hemorrhagic wound). Nevertheless here are several requirements to hemostatic agent including be able to stop large-vessel arterial and venous bleeding within 2 min of application, and to be delivered through a pool of blood when applied; be ready to use with no pre-application preparation; be simple to use with minimal training; have lightweight and durable properties; have minimum 2-year shelf-life in extreme environmental conditions; be safe to use with no risk of further injury to tissues or transmission of infection; and be inexpensive [4]. Additionally absence of toxic effects, lack of anaphylactic potential, and ease of removal at the time of surgery [5] are vital features of hemostatic agent. In case of non-sterile environmental conditions when traumatic wounds exposed to dirt and pathogens, ideal hemostatic agent should have antimicrobial properties [6].

Various materials have been studied for control of certain kind of bleeding. According to their forms these materials may be classified as hemostatic dressings [7], solid particles, powders, sponges [8] and fibers [9], hydrogels [10], liquid tissue sealants [11], and dispersions [12]. Proteins (fibrinogen, thrombin, collagen, gelatin, album), polysaccharides (chitosan, chitin) and cellulose, polyester, inorganic minerals, ceramics and their combinations are typical types of biomaterials [13].

Chitosan is obtained from chitin via deacetylation by enzymatic or alkaline treatment. Chitin and chitosan have a lot of useful biological properties such as biodegradality, bio- and cytocompatibility, antioxidant and anti-inflamatory activity, analgestic and haemostatic action and mucoadhesive properties [14–21].

The hemostatic mechanism of chitosan is based on vasoconstriction and the rapid mobilization of red blood cells, clotting factors, and platelet to the site of the damage [22]. Chitosan is able to promote platelet adhesion and aggregation, leading to the interaction of platelets with damaged tissues and promoting wound healing effect of chitosan [23]. Finally, chitosan contributes blood clot formation due to the interactions with negatively charged thrombocytes and erythrocytes [24]. Futhermore, chitosan possesses antibacterial properties against both Gram-negative and Gram-positive bacteria strains due to free amino groups on its surface [25, 26].

Biocompatibility, biodegradability, properly mechanical strength, porous hydrophilic structure of hydrogels [27], water retention ability due to their hydrophilic structure allow to apply hydrogels as hemostasis bandages [28]. There are two main types of hydrogels depending on their preparation method. They may Hydrogels are obtained by the method of physical or chemical crosslinking. Physically cross-linked hydrogels are less durable and possess reversible nature after pH or temperature changes. While chemically cross-linked hydrogels are more resistant to ambiance and maintain their mechanical properties for a long time. Unfortunately, these hydrogels can be toxic because of the some crosslinking agents using such as glutaraldehyde or photoinitiators [29]. Nevertheless, implementability hydrogels is still limited because of their low mechanical strength and fragile nature.

Thus, the aim of the current research was to create novel chitosan hydrogels with stronger and more stable properties including antibacterial which would be flexible and conformable enough to produce effective hemostasis wounds that have irregular depth and geometry and in junctional regions as well.

10.2 Materials and Methods

For the hemostatic agents obtainment chitosan with 85% and 95% deacetylation degree was used prepared from shellfish. L-aspartic acid (L-Asp), L-glutamic acid (L-Glu), 1,2-propanodiol, isopropanol, methanol, 2,2-diphenyl-1-picrylhydrazyl, ethanol 95%, were purchased from Sigma Aldrich, Poland. NaOH, NaCl, $NaHCO_3$, KCl, KH_2PO_4•$3H_2O$, $MgCl_2$•$6H_2O$, $CaCl_2$, Na_2SO_4 were also purchased from POCH, Poland.

10.2.1 Hemostatic Agents Synthesis

For the obtainment of the Sample 1 0.5 of chitosan with 95% deacylation degree was dissolved in the aquatic solution of L-aspartic (0.15) and L-glutamic acid (0.75). For the preparation of Sample 2 0.5 g of chitosan with 95% deacetylation degree was dissolved in the same amounts of aminoacids. Next, 10 ml of propylene glycol were added, and reacting mixture was placed in the domestic microwave oven. The crosslinking process was carried out for 3 min total. After that, the ready product was washed out from acid residues and freeze-dried.

10.2.2 FT-IR Analysis

All FT-IR/ATR analyses were performed using IR Thermo Nicolet Nexus X 470 spectrometer (diamond crystal ATR), USA. The range was between 400 and 4000 cm $^{-1}$ with 32 scans and 4 cm $^{-1}$ resolution.

10.2.3 Porosity and Density

The density and porosity of the obtained chitosan materials were determined by isopropanol displacement because it does not wet the sample. Investigated biomaterials were placed into the previously measured volume of isopropanol. After fixed time (5 min) the change in volume of the alcohol-impregnated aerogel was measured. Then studied chitosan scaffold was removed from the isopropanol. In the last step the difference in isopropanol volume was measured. Basing on the obtained data density (Eq. 10.1) and porosity (Eq. 10.2) was calculated using following equations:

$$d = \frac{W}{(V_2 - V_3)} \tag{10.1}$$

$$p = \frac{(V_1 - V_3)}{(V_2 - V_3)} \cdot 100\% \tag{10.2}$$

where

d density, g/cm^3
p porosity, %
W weight of the investigated sample
V_1 initial volume of isopropanol, cm^3
V_2 volume of isopropanol with immersed sample, cm^3
V_3 volume of isopropanol after sample removal, $cm.^3$

All experiments were repeated 3 times.

10.2.4 Swelling Properties

Swelling properties were determined in distilled water, phosphate buffer solution (PBS) and simulated body fluid (SBF). To determine this ability, each dried and weighed sample was placed in the swelling medium for 24 h. The swelling degree was calculated using the following equation:

$$SD = \frac{(W_t - W_0)}{W_0} \tag{10.3}$$

where

SD swelling degree
W_0 initial weight
W_t the weight after time $t = 24$ h.

All experiments were repeated 3 times.

10.2.5 Water Vapor Transmission Rate

To determine water vapor transmission rate (WVTR) samples were fixed using polymeric glue to polystyrene well onto its opening (area 1 cm^2). Each plate contained 5 ml of distilled water. The studies were carried out for 24 h at 37 °C. The WVTR was measured basing on the amount of water loss. Water vapor transmission rate was calculated using following equation:

$$\mathrm{WVTR} = \frac{(W_t - W_O)}{(tA)\left(g \cdot m^{-2} \cdot d^{-1}\right)} \quad (4)$$

where

W_O the initial weight,
W_t the weight after time t,
t the measuring time
A the area of the opening of polystyrene well.

All experiments were repeated 3 times.

10.2.6 Antioxidant Activity

Antioxidant properties of the prepared chitosan scaffolds were investigated by a standard DPPH method. For this purpose, a solution of DPPH in methanol was prepared so that the solution absorbance was 1.0 at 517 nm using Aligent 8453 spectrophotometer. To determine the ability of free radicals scavenging, 0.10 g of each sample was placed in 5 ml of DPPH solution and left in darkness for 1 h with constant shaking. Then, the absorbance of each solution was measured at 517 nm. The percentage of the free radicals removed was calculated using following Eq. 10.5:

$$\%S = \frac{(A_s - A_c)}{A_c} \quad (5)$$

where

%S the % of the free radicals which were neutralized
A_c the absorbance of the DPPH solution without the sample
A_s the absorbance of the DPPH solution containing sample

All experiments were repeated 3 times.

10.2.7 SEM Analysis

SEM microphotographs and X-ray fluorescence analysis were performed using Scanning Electron Microscope SEM FEI QUANTA 650 FEG.

10.2.8 Time-Depending Antimicrobial Assay

E. coli and *S. aureus* were used as model of Gram-negative and Gram-positive bacteria for this study. Sterilized under the UV light 4 mg samples of the sponges were utilized. The bacterial cultures were incubated in nutrient broth overnight. Samples wereincubated for 2, 4, 6, 8, 10, 24 h at 37 °C in 2 ml of microbial inoculums with microorganism concentration equivalent to 10^4 CFU/ml (4 log10 CFU). The tubes containing tested samples in 2 ml of growth medium without bacterial inoculums were used as a control. 100-μl aliquots from each tube were applied onto Petri dish with solid media using streak plate technique and incubated at 37 °C for 24 h. Determining the colony count were carried out counting viable organisms in log10 (Fig. 10.1).

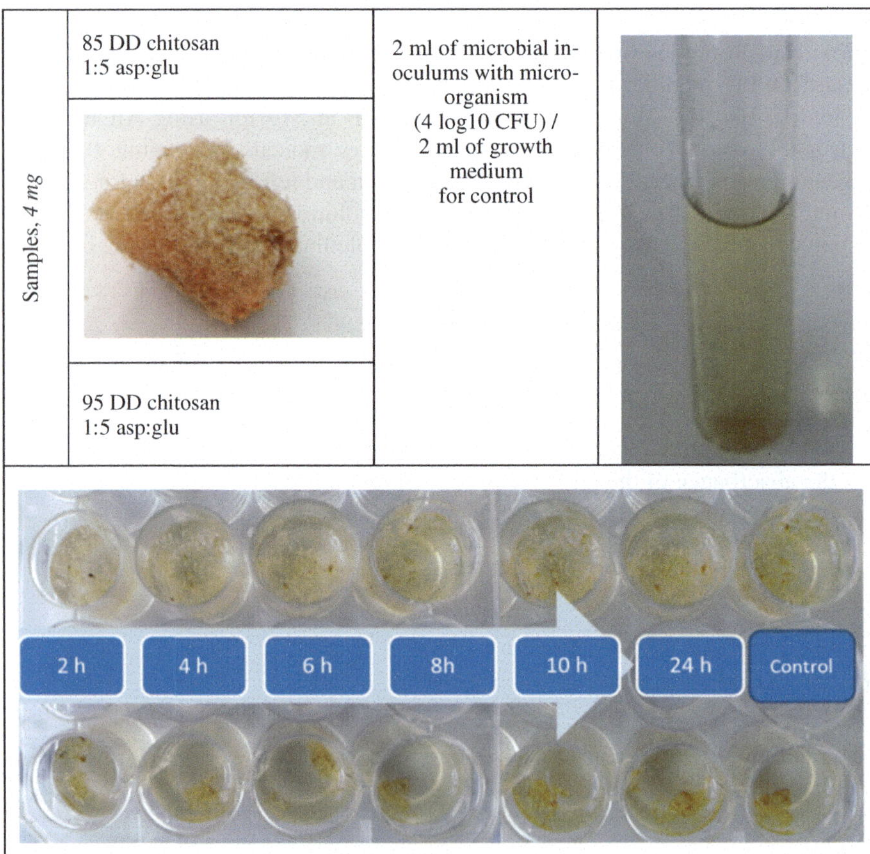

Fig. 10.1 Scheme of time-depending antimicrobial assay

10.2.9 Blood Clotting Tests

50.0 mL of blood obtained from two human subjects volunteered was drawn by a registered nurse. The study was previously approved by the Ethic Committee on Medical Research of Medical Institute Sumy State University. An appropriate informed consent was received from all volunteers. The samples of chitosan sponge weighted 50 mg were placed in individual Becton Dickinson Vacutainer for each filled with 2 mL of human blood. To ensure good interaction between material and blood vacutainers were shaken constantly during the 10 min. All samples were removed, weighted and blood sorption (BS) rate was calculated as follows:

$$BS = W_2 - W_1$$

where

W_1 initial weight (50 mg)
W_2 weight after the blood clotting test, mg.

The remaining blood was used for a completed blood count (CBC) test that was performed in Medical Centre "Floris". CBC was performed using by hematology analyzer CELL-DYN 3700 (ABBOTT, Irving, TX, USA) with reagents DIAGON (Budapest, Hungary). The analysis carried out the following parameters: platelet count (PLT, × 109/L), platelet distribution width (PDW, %) and mean platelet volume (MPV, fL).

10.2.10 Scanning Electron Microscopy (SEM)

After extraction from blood the samples were immediately placed in 2% glutaraldehyde for 2 h, then dehydrated samples were placed in ethanol and dried. To avoid surface charge accumulation in the electron-probe, samples were covered with a thin (30–50 nm) layer of silver in the vacuum set-up VUP-5 M (SELMI, Sumy, Ukraine). The SEM images of sponges were observed by FEI Inspect S50B (FEI, Brno, Czech Republic) with the Everhart–Thornley secondary electron detector.

10.2.11 Cell Culture

For the exposure experiments, cell culture medium was prepared using Dulbecco's Modified Eagle Medium/Nutrient Mixture F-12 (DMEM/F-12) with L-glutamine contained 100 units/mL penicillin, 100 μg/mL streptomycin, 2.5 μg/mL amphotericin B, 10% fetal bovine serum and 1.0 ng/mL bFGF (basic Fibroblast Growth Factor). All reagents were obtained from Gibco™. Primary human fibroblasts were grown in 75 cm^2 tissue culture flasks under standard culture conditions with 5% humidified CO_2 in the air at 37 °C with medium renewal every 2–3 days.

The cells were seeded in 24-well plates at a cell density of 5×10^4 cells per well and volume of medium 1.0 mL/well and pre-incubated overnight. Thereafter in each well the chitosan strip weighed 50 mg was added. Traditional cell culture on plate (TCP) with no chitosan sponge was used as a positive control. Cell growth and proliferation were assessed using EVOS XL Core cell imaging system (Thermo Fisher Scientific, Waltham, MA, USA). Chitosan sponge can change the optical density after interaction with medium accordingly resazurin or tetrazolium assays were unable to be carried out.

10.2.12 Statistics

A One-way ANOVA (GraphPad Prism 8.0 software) wasconducted to compare the antibacterial effect of chitosan sponges on different bacterial strains. All experiments were performed independently in triplicatefor each time point. The data were expressed as means ± standard deviation. Statistical significance was considered at $p < 0.05$.

10.3 Results and Discussion

10.3.1 FT-IR Analysis

Figure 10.2 shows FT-IR spectra of the pure chitosan Sample 1 and Sample 2. Spectrum of the raw chitosan shows some typical bands coming from free hydroxyl (3359 cm^{-1}) and amino groups (1591 cm^{-1}, 1150 cm^{-1}), aliphatic groups (2923 cm^{-1}, 28711 cm^{-1}) amide bonds of quite low intensity (1650 cm^{-1}), glycosidic bonds (1065 cm^{-1}) and finally glucopyranose ring (891 cm^{-1}). The spectrum of Sample 2 shows changes which occured due to the microwave radiation such as formation of free carboxyl groups due to the slight surface degradation (sample 1: 3268 cm^{-1}; sample 2: 3215 cm^{-1}). Moreover, one may noticed that the new amide bonds were formed in each sample which resulted due to the reaction between free carboxyl groups coming from L-Asp and L-Glu and amino groups of chitosan (sample 1 1651 cm^{-1}; sample 2 1654 cm^{-1}). At the same time, free amino groups bands (sample 1 1581 cm^{-1}; 1150 cm^{-1}; sample 2 1574 cm^{-1}, 1150 cm^{-1}) are still visible in each spectrum due to the fact, that aminoacids which were incorporated into the polymeric chain introduced new NH_2 into the structure. Finally, it can be noticed that no chain degradation appeared since bands coming from glycoside bonds between mers (sample 1 and 2 1066 cm^{-1}) as well as pyranose rings are still present (sample 1 and 2 894 cm^{-1}).

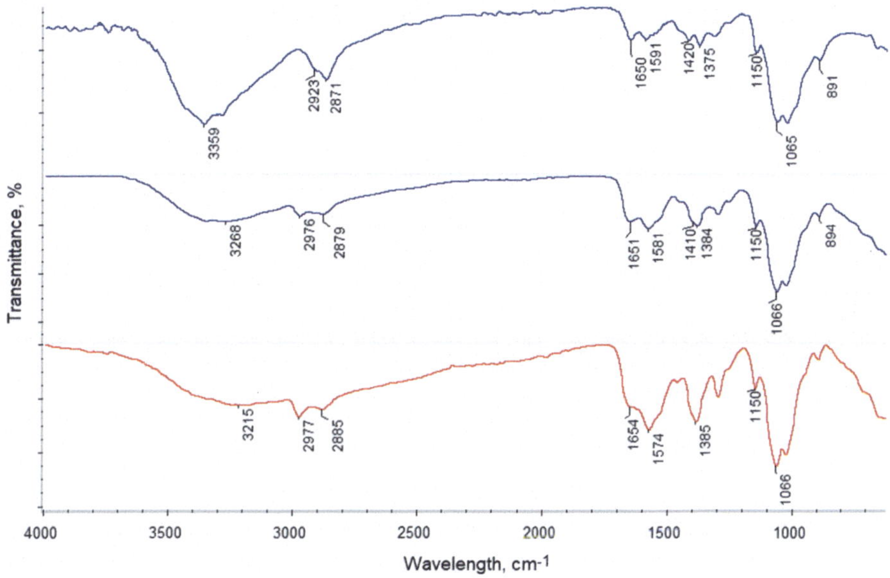

Fig. 10.2 FT-IR of the prepared samples (blue line—chitosan, dark blue sample 1, red—sample 2)

10.3.2 Morphology and Elemental Composition Study

Figure 10.3 presents microphotographs of the prepared samples. It can be noticed that both samples are highly porous. The biomaterials have open cellular structure. The pores size is around 300–500 μm. In the structure of both samples interconnecting

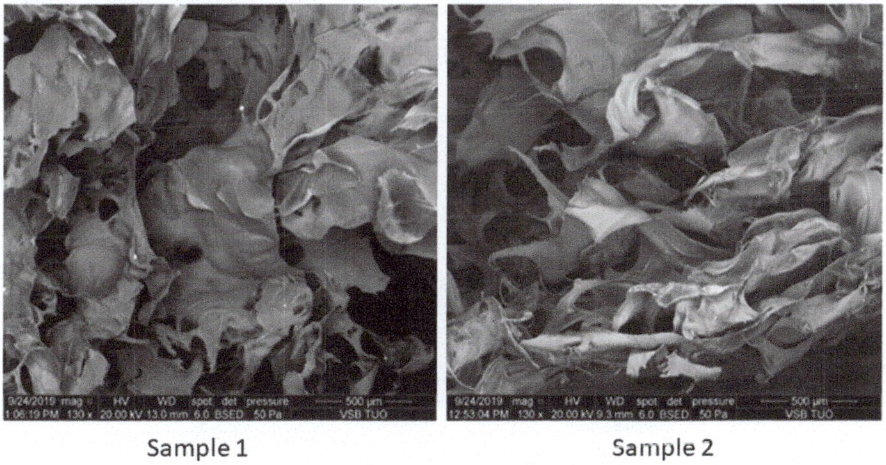

Fig. 10.3 SEM microphotographs of the prepared samples

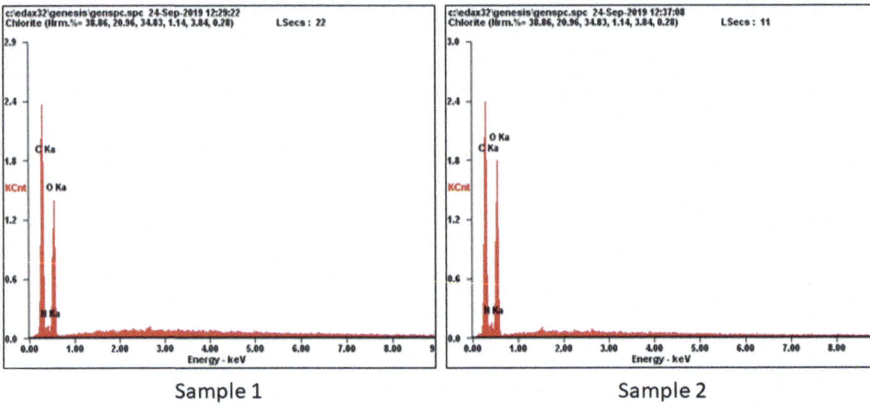

Fig. 10.4 The elemental composition of the prepared samples

channels are visible. The pores' edges are curved. Such morphology is excellent for the sorption of aquatic solutions and body fluids to the possibility of biomolecules migration inside the three-dimensional structure.

Figure 10.4 presents the elemental analysis of the samples. It may be noticed that only three elements are present that is carbon, oxygen and nitrogen which confirms the preliminary biosafety of the samples due to the lack of any toxic metals or residues. Such results confirm successful purification of the ready products.

10.3.3 Swelling Properties

Figure 10.5 presents results of swelling properties study which are crucial for the biomaterials application as hemostatic agents. One may observe that both samples

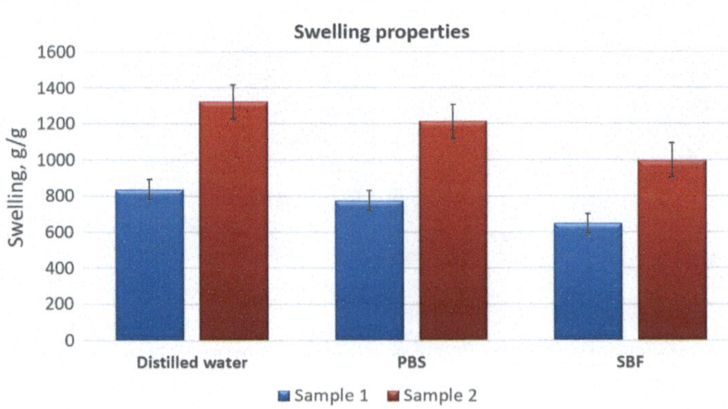

Fig. 10.5 The swelling abilities the prepared samples

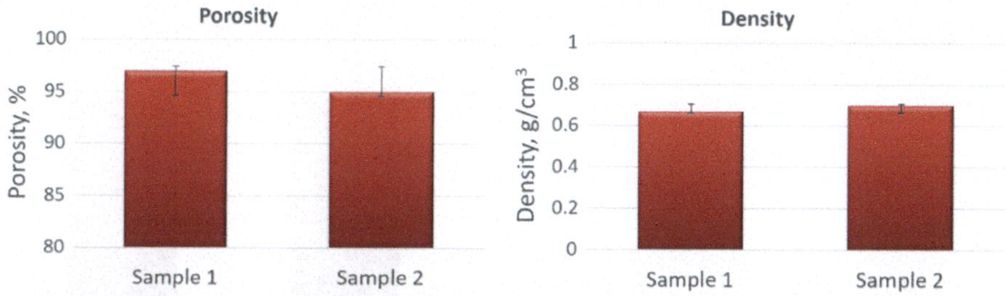

Fig. 10.6 The porosity and density of the prepared samples

have excellent sorption abilities. This could be assigned to the highly porous morphology shown in Fig. 10.2 and the presence of hydrophilic functional groups such as hydroxyl, amino and carboxylic. It may be observed that swelling ability is slightly better in the case of sample 2 which may be assigned to the higher porosity. Moreover, without a doubt the sorption ability depends on the swelling medium composition, the more ions present in the solution, the lower medium intake.

10.3.4 Porosity and Density Study

Figure 10.6 presents the results of porosity and density study. It can be noticed that the data presented in the charts corresponds to the SEM microphotographs and swelling ability evaluation since both samples are characterized by very high porosity (above 90%) and very low density typical for aerogels. It can be noticed that the higher porosity provides better sorption abilities.

10.3.5 Antioxidant Properties Study

Figure 10.7 shows the results of antioxidant properties study. Raw chitosan is known of its ability of free radicals removal. However, only when it is dissolved in the acidic

Fig. 10.7 The antioxidant activity of the prepared samples

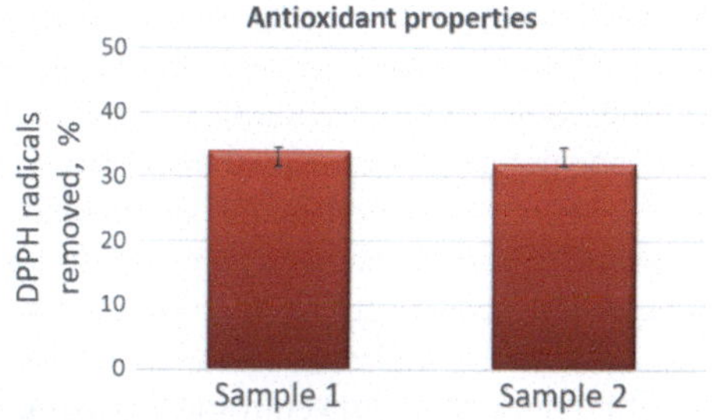

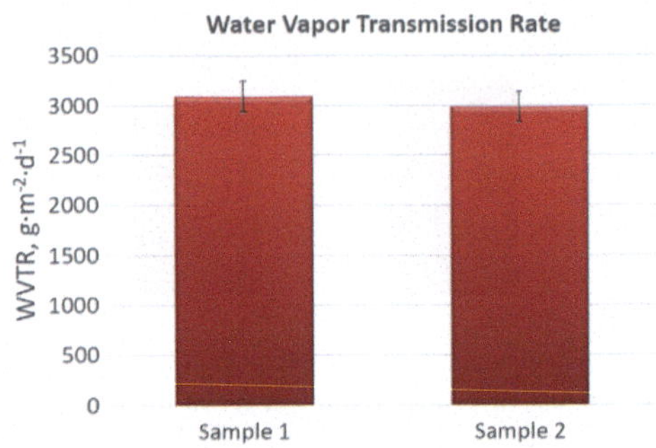

Fig. 10.8 Water vapor permeability of the prepared samples

medium. The obtained samples exhibit very good antioxidant activity against DPPH radicals even though they are crosslinked. The ability of free radicals scavenging can be explained by the presence of high number of free amino groups as well as pyranose rings. The data suggests that the biomaterials will be able to neutralize Reactive oxygen species (ROS) which are very often generated when wounds are created.

10.3.6 Water Vapor Transmission Rate (WVTR)

Figure 10.8 presents the results of water vapor permeability rate. The hemostatic dressings which are used in the case of superficial wounds must let moisture pass through so to enable skin regeneration. It can be observed that both samples have a very high WVTR value which can be assigned to the high porosity and hydrophilic nature of the biomaterials. Again, WVTR depends on the number of pores which leads to the conclusion that the hemostatic agents physical properties are determined mostly on their morphology.

10.3.7 Time-Depending Antimicrobial Assay

It is clearly evident from the assays that both tested sponges showed more prevalent antibacterial activity against *S. aureus* compared to *E. coli*. Thus, the results indicated significant difference between the amount of bacterial stains throughout all time points of cultivation ($p < 0.0001$).It is interesting to note that sponge85 DD chitosan sponge demonstrated better bactericidal activity due to microbial inoculum of *S. aureus* was killed in 2 h of incubation (Fig. 10.9). Whereas sample 95 DD chitosan 1:5 asp:glu showed lesser inhibitory actions against Gram-positive bacteria and caused total killing of microorganism in 4, 6 and 8 h. The amount of *S. aureus* was significantly higher 2, 10 and 24 h of incubation with sponge 95 DD chitosan 1:5 asp:glu than with 85 DD chitosan 1:5 asp:glu ($p < 0.0001$). There were no significant

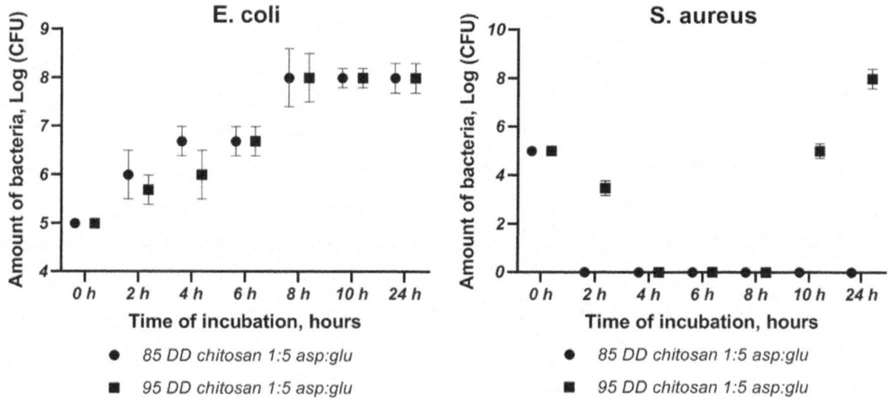

Fig. 10.9 Time-depending antibacterial effect of chitosan sponges against *E. coli* and *S. aureus*

differences in the CFU of *E. coli* between the samples during experiment in spite of revealed time dependence of antimicrobial properties of both types of sponges.

Chitosan has been reported to have bacteriostatic and bactericidal activities on both gram-negative and gram-positive bacteria [30]. However, antibacterial action of chitosan as antimicrobial agents is dependent on the origin of microorganism. According to [31] the outer membrane of gram-negative bacteria consists basically of lipopolysaccharides containing phosphate and pyrophosphate groups that explain higher surface density of negative charges higher to that observed for gram-positive ones that facilitates electrostatic interaction between positively charged chitosan molecules and negatively charged microbial cell membranes [32]. Moreover, hydrophilicity in gram-negative bacteria is significantly higher than in gram-positive bacteria, making them more sensitive to chitosan [33]. On the other hand, the degree of acetylation (DA) of chitosan is inversely proportional to its antimicrobial properties and haemostatic properties as well [34]. Besides the ionic surface interaction, the inhibition of the mRNA and protein synthesis via the penetration of chitosan into the nuclei of the microorganisms can be involved to justify the mechanism of chitosan antimicrobial action [35]. Current research pointed out that proposed biomaterials are more effective against *S. aureus* and formulation with lower DD of chitosan (85 DD chitosan 1:5 asp:glu) showed the best antibacterial properties. It should be admitted that 95 DD chitosan sponge showed antibacterial properties against gram-positive bacteria depending on time point of incubation that can be elucated by time depending peculiarities of sponges biodegradation. This result can be interpreted considering the method of chitosan development and such peculiarities as difference in crosslinking time while the composition was the same.

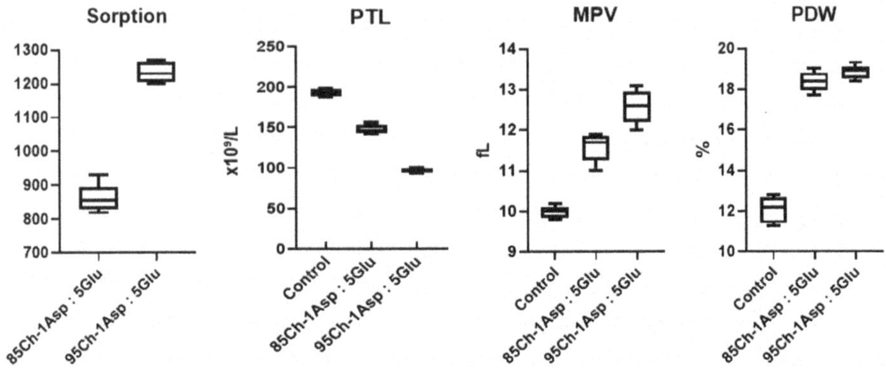

Fig. 10.10 Blood sorption by chitosan sponges (Sorption), namber of platelet (PLT), mean platelet volume (MPV), platlet distribution width (PDW) after Blood cloing test

10.3.8 Blood-Clotting Experiment and Scanning Electron Microscopy (SEM) Analysis

95Ch-1Asp:5Glu sponge showed high sorption ability comparatively 85Ch-1Asp:5Glu ($p \leq 0.05$) which reached to 1300% from the initial weight. And number of platelet for both samples significantly lower ($p \leq 0.005$) in comparison with control and 95Ch-1Asp:1Glu was significantly lower 85Ch-1Asp:5Glu ($p \leq 0.005$). But ordinary one-way ANOVA test did not shows significant differences between samples for MPV and do not significantly differ from the control ($p \geq 0.68$). Interaction of chitosan sponges with blood leads to significant increases PDW parameters (ANOVA, $p \leq 0.05$) but chitosa grups did not differ from the control. The results indirectly indicates that proposed chitosan biomaterials have possible hemostatic properties (Fig. 10.10).

SEM image shown erhytrocyte and plateletes adgesuon over the sponge surface after blood contact. We can detect amorphous substance aver cells and between erhytrocytes and sponge. Based on previous research we can coclude that is probably blood proteins, including thrombin that is the main component of blood clot (Fig. 10.11).

10.3.9 Cell Toxicity Experiment

Both chitosan sponge samples showed no toxic effect. Cell morphology and proliferation rate did not differ between groups. On day 7 of incubation, the confluence of cells in the wells with sponges and TCP has reached 70–80%. During the experiment in the medium, the chitosan sponge began to fragment but there was no indication of its resorption fibroblasts (Fig. 10.12).

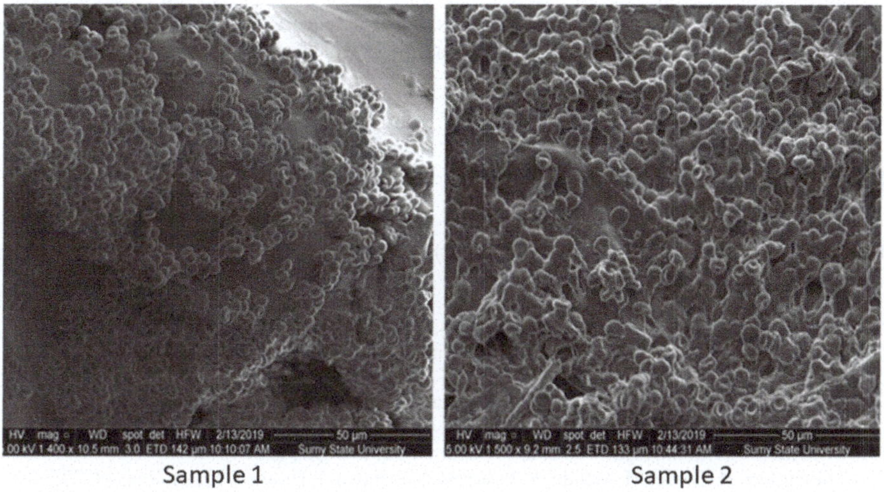

Fig. 10.11 Scanning electron microscopy of chitosan sponge after blood clotting test with adhered blood cells

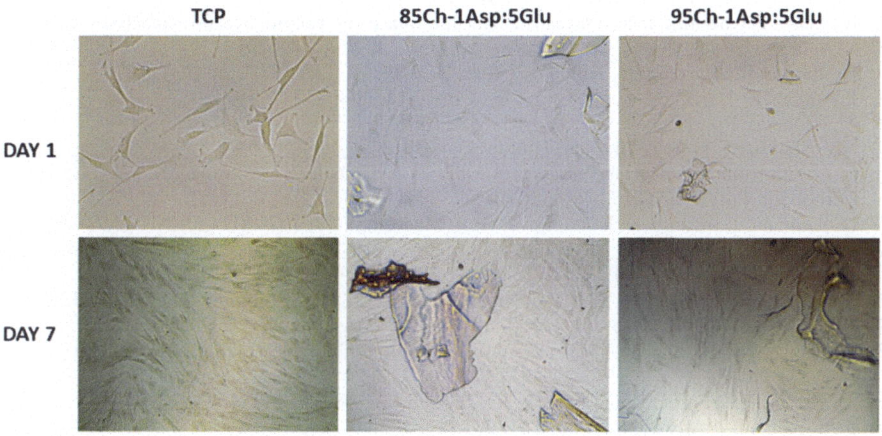

Fig. 10.12 Optical image of human cell primary fibroblasts at one and seven days after cultivation with chitosan sponges. (TCP—tissue culture plastic). Magnification ×200

10.4 Conclusions

Sponges made from chitosan in L-aspartic and L-glutamic acids without acetic acid shown excellent hemostatic properties which are confirmed by blood sorption and affect platelet aggregation. These biomaterials had high porosity and very good antioxidant properties. Additional antibacterial properties and non-toxic nature make them promising to clinical application.

Acknowledgements This research support by National Science Centre, Poland, grant number 2016/23/N/ST8/01273 and H2020-MSCA-RISE grant 777926 NanoSurf.

References

1. B.L. Bennett, Wilderness Environ. Med. **28**(2), 39–49 (2017)
2. O.E. Ogle, J. Swantek, A. Kamoh, Dent. Clin. North Am. **55**(3), 433–439 (2011)
3. B.L. Bennett, L. Littlejohn, Mil. Med. **179**(5), 497–514 (2014)
4. H. Khoshmohabat, S. Paydar, H.M. Kazemi, B. Dalfardi, O Trauma Mon. **21**(1), e26023 (2016)
5. J. Granville-Chapman, N. Jacobs, M.J. Midwinter, Injury **42**(5), 447–459 (2011)
6. P. Nakielski, F. Pierini, Acta Biomater. **84**, 63–76 (2019)
7. S. Pendharkar, J. Guo, Biodegradable hemostatic wound dressings, European Patent Application No. EP1481694 A1 (2004)
8. H.B. Alam, Z. Chen, A. Jaskille et al., J. Trauma **56**, 974–983 (2004)
9. T. Sugamori, H. Iwase, M. Maeda, Y. Inoue, H. Kurosawa, J. Biomed. Mater. Res. **49**, 225–232 (2000)
10. D. Jensen, G. Machicado, K. Hirabayashi, J. Trauma **57**, S33–S37 (2004)
11. D. Albala, Surg. **11**, S5–S11 (2003)
12. E. Hick, A. Morey, R. Harris, M. Morris, J. Urol. **173**, 1801–1804 (2005)
13. Tao Peng, Artif. Organs **24**(1), 27–68 (2010)
14. M. Dash, F. Chiellini, R.M. Ottenbrite, E. Chiellini, Prog. Polym. Sci. **36**, 981–1014 (2011)
15. J. Shaji, V. Jain, S. Lodha, Int. J. Pharm. Appl. Sci. **1**, 11–28 (2010)
16. F. Shahidi, R. Abuzaytoun, Adv. Food Nutr. Res. **49**, 93–135 (2005)
17. I. Aranaz, R. Harris, A. Heras, Curr. Org. Chem. **14**, 308–330 (2010)
18. M. Rinaudo, Prog. Polym. Sci. **31**, 603–632 (2006)
19. S.V. Madihally, H.W.T. Matthew, Biomaterials **20**, 1133–1142 (1999)
20. R. Logith Kumar, A. Keshav Narayan, S. Dhivya, A. Chawla, S. Saravanan, N. Selvamurugan, Carbohydr. Polym. **151**, 172–188 (2016)
21. D. Archana, J. Dutta, P.K. Dutta, Int. J. Biol. Macromol. **57**, 193–203 (2013)
22. Y. Okamoto, R. Yano, K. Miyatake, I. Tomohiro, Y. Shigemasa, S. Minami, Carbohydrate Polym. **53**, 337–342 (2003)
23. T.C. Chou, E. Fu, C.J. Wu, J.H. Yeh, Biochem. Biophys. Res. Commun. **302**, 480–483 (2003)
24. J.M.D. Lima, R.R. Sarmento, J.R.D. Souza, et al. BioMed Res. Int. **6** (2015)
25. Y. Ma, L. Xin, H. Tan, M. Fan, J. Li, Y. Ji, Z. Ling, Y. Chen, X. Huc, Mater. Sci. Eng., C **81**, 522–531 (2017)
26. M. Kong, X. Chen, K. Xing, H. Park, Int. J. Food Microbiol. **144**, 51–63 (2010)
27. E.M. Ahmed, J. Adv. Res. **6**, 105–121 (2015)
28. A.S. Hoffman, Adv. Drug Deliv. Rev. **64**, 18–23 (2012)
29. M. Piatkowski, Ł. Janus, J. Radwan-Pragłowska, D. Bogdał, D. Matysek, B Biointerfaces **164**, 324–331 (2018)
30. M. Kong, X.G. Chen, K. Xing, H.J. Park, Int. J. Food Microbiol. **144**(1), 51–63 (2010)
31. P. Sahariah, V.S. Gaware, R. Lieder, S. Jónsdóttir, M.Á. Hjálmarsdóttir, O.E. Sigurjonsson, M. Másson. Mar Drugs. **12**(8), 4635–4658 (2014)
32. F. Devlieghere, A. Vermeulen, J. Debevere, Food Microbiol. **21**(6), 703–714 (2004)
33. Y.C. Chung, Y.P. Su, C.C. Chen, G. Jia, H.L. Wang, J.C. Wu, J.G. Lin, Acta Pharmacol. Sin. **25**(7), 932–936 (2004)
34. C. Chatelet, O. Damour, A. Domard, Biomaterials **22**(3), 261–268 (2001)
35. F.S. Kittur, A.B. Vishu Kumar, R.N. Tharanathan, Carbohydrate Res **338**(12), 1283–1290 (2003)

Chapter 11
Composite Ultrafiltration Membrane Incorporated with Dispersed Oxide Nanoparticles

L. M. Rozhdestvenska, O. I. V'yunov, L. N. Ponomarova, A. V. Bilduykevich, T. V. Plisko, Y. G. Zmievskii and V. D. Ivchenko

Abstract Organic-inorganic membranes containing the nanoparticles of hydrated zirconium dioxide and $BaFe_{12}O_{19}$ magnetic nanoparticles were obtained. The nanoparticles were inserted into polymer matrices, they form aggregates, a size of which is up to 20 nm (active layer) and up to 2 μm (macroporous fibrous support). Larger aggregates are formed in absence of the magnetic constituent (up to 5 μm). The membranes were tested for filtration of sugar beet juice. Due to smaller particle size, the membrane containing also $BaFe_{12}O_{19}$ shows the liquid flux of 4.3×10^{-7} to 5.7×10^{-7} m^3 m^{-2} s^{-1} at 2 bar and rejection towards vegetable protein of 55–87%. Regarding the membranes including no magnetic nanoparticles, these values are 3.8×10^{-7} to 5.5×10^{-7} m^3 m^{-2} s^{-1} and 38–77%.

11.1 Introduction

Ultrafiltration technology is widely used for removal of colloidal particles from ground and brackish water, wastewater, sea water [1]. This stage of water treatment is before reverse osmosis to prevent membrane fouling. Ultrafiltration is also applied

L. M. Rozhdestvenska · O. I. V'yunov
Vernadsky Institute of General and Inorganic Chemistry, Ukrainian National Academy of Sciences, Kiev, Ukraine

L. N. Ponomarova (✉)
Sumy State University, Sumy, Ukraine
e-mail: l.ponomarova@chem.sumdu.edu.ua

A. V. Bilduykevich · T. V. Plisko
Institute of Physical Organic Chemistry, National Academy of Science of Belarus, Minsk, Belarus

Y. G. Zmievskii
Ministry of Education and Science of Ukraine, National University of Food Technologies, Kiev, Ukraine

V. D. Ivchenko
Sumy National Agrarian University, Sumy, Ukraine

A. D. Pogrebnjak et al. (eds.), *Nanomaterials in Biomedical Application and Biosensors (NAP-2019)*, Springer Proceedings in Physics 244,
https://doi.org/10.1007/978-981-15-3996-1_11

to beverage industry, for instance, for milk skimming and effluents treatment etc. The main problem of filtration is a decrease of membrane permeability due to fouling with organics. This is especially important, when liquids of biological origin are processed.

In general, species of organic substances, microorganisms, iron oxide and silicon dioxide significantly decrease the time of filtration. The membranes need chemical regeneration that involves aggressive reagents. Frequent regeneration reduces a lifetime of the membranes. One of the ways to overcome these disadvantages is to enhance hydrophilicity of polymer membranes. As a rule, nanoparticles of inorganic ion-exchangers are used for modification of polymers. A number of inorganic compounds are applied to modifying [20]: zirconium hydrophosphate [15] (the attempt to use these materials as a filler for electromembrane processes is known [4], moreover, they are used for modifying of ion exchange resins [6]), silica [18], hydrated zirconium [15, 16] or iron oxide [17]. This approach allows one to enhance liquid permeability and anti-fouling ability without sufficient changes of membrane structure.

Magnetic particles are another type of modifier that improves functional properties of polymer membranes. The membrane containing magnetic Fe_3O_4 nanoparticles and graphene oxide particles shows high flow of pure water and high degree of rejection (up to 83.0%) [11]. Membranes containing iron nanoparticles can be used to remove copper and lead ions from wastewater [9, 11]. Adsorption capacity increases due to improved hydrophilicity on the one hand and nucleophilic functional groups on the surface of nanoparticles on the other hand. Nanocomposite membrane exhibits minimal interaction with whey protein due to its higher hydrophilicity, which leads to a polar-non-polar interaction between membrane surface and protein. This depresses membrane fouling [19].

The membranes modified with magnetic nanoparticles shows an increase in water flow due to changes in the average pore radius, porosity and hydrophilicity of the membranes. The membrane surface roughness and hydrophilicity are considered to be main factors, which minimize membrane fouling.

The aim of the work was to obtain organic-inorganic membranes containing inorganic modifier, particularly magnetic one, and to establish the effect of the filler on separation ability of the composite membranes and their stability against fouling.

11.2 Experiment Details

11.2.1 Membrane Modifying

Ultrafiltration membranes (produced by the Institute of Physico-Organic Chemistry of the National Academy of Science of the Republic of Belarus) were used for investigations as a polymer substrate. These materials consist of macroporous substrate (non-woven polyester) and ultrathin active layer (polysulfone (PS) or polyacrylonitrile (PAN)). Further the membranes were marked according to the polymer forming the active layer. PS and PAN rejects globular proteins, molecular mass of which is 100 and 50 kDa, respectively.

Magnetic nanoparticles (MNP) $BaFe_{12}O_{19}$ were synthesized according to [2]. In order to provide their fixation in membrane pores, hydrated zirconium dioxide (HZD) was used. First of all, sol of insoluble zirconium hydroxocomplexes was obtained from a 0.25 M $ZrOCl_2$ solution similarly to [8]. MNP were dispersed in zirconium sol and treated with ultrasound at 30 kHz. The membranes were degassed in deionized water under vacuum conditions at 343 K, and impregnated with suspension of MNP in sol. Then HZD and MNP were coprecipitated directly in the polymer with a 0.1 M NH_4OH solution. The membrane was dried at 50 °C and cleaned with ultrasound to remove the precipitate from its outer surface. This approach, which involves impregnation of a membrane with the suspension of insoluble compounds followed by precipitation, was applied earlier to modification of ceramics [7]. For comparison, the membranes containing only HZD were obtained. In this case, the polymer matrix was impregnated with zirconium-containing sol.

Morphology of the membranes was investigated using scanning electron microscopy (SEM). Fractal dimension of aggregates in macroporous support was determined with methods of cube counting, triangulation, and power spectrum analysis similarly to [3].

Before the application of transmission electron microscopy (TEM), the active layer was separated from the macroporous substrate, and milled in the medium of liquid nitrogen.

11.2.2 Membrane Testing

The experimental set-up for filtration consisted of typical elements for baromembrane separation (magnetic pump, manometer, rotameter). A divided two-compartment flow-type cell was used. An effective area of the membrane was 2.82×10^{-3} m^2. Before the measurements, the membrane was pressed by means of pumping deionization water at 4 bar. The effluent volume was measured after predetermined time. Filtration was stopped, when the constant flow rate through the membrane was achieved.

Tap water containing 1 and 0.2 mol dm^{-3} Ca^{2+} and Mg^{2+} respectively was used for testing. The content of ions in permeate was determined by means of atomic absorption technique. Filtration was carried out at 2 bar. Sugar beet juice (PC "Salyvonkivskyy sugar factory") was also applied to investigations. Before testing, juice was diluted in 10 times. The content of vegetable proteins was determined in permeate using such dye as Coomassie brilliant blue G-250 [12]. Selectivity of membranes (φ), i.e. rejection of species was estimated via [1]:

$$\varphi = \left(1 - \frac{C_p}{C_f}\right) \times 100\%. \tag{1}$$

where C_p and C_f are the concentration of feeding solution and permeate, respectively.

11.3 Results and Discussion

11.3.1 Morphology of Membranes

As an example, typical SEM images of the pristine PAN membrane are given in Fig. 11.1a, b. It is seen that the microporous support consists of sprung fibers, a size of which is 10–20 μm. Active layer is attached to the support forming the membrane that is able to reject colloidal particles. During HZD precipitation, the particles, a size of which is up to 5 μm, are formed in the support, when MNP are absent (Fig. 11.1c).

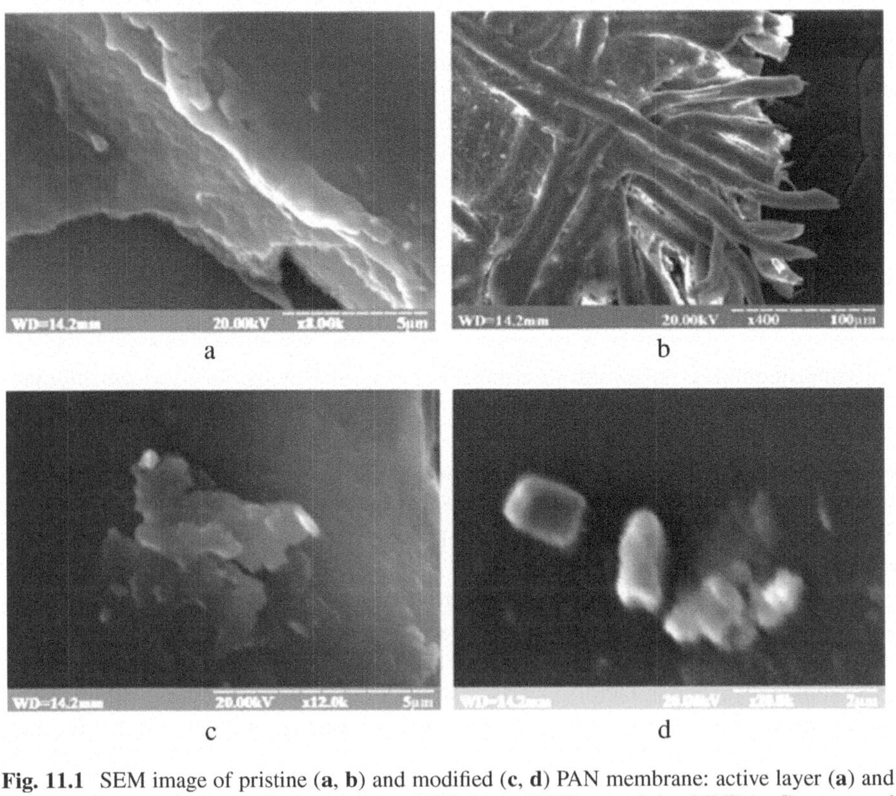

Fig. 11.1 SEM image of pristine (**a**, **b**) and modified (**c**, **d**) PAN membrane: active layer (**a**) and macroporous support (**b**–**d**). One-component HZD (**c**) and HZD containing MNP (**c**, **d**) were used as a modifier

In the case of MNP in sol, the size of aggregate is up to 2 μm (Fig. 11.1d). Analysis of fractal dimension gives 2.4–2.7 indicating diffusion as a limiting stage during aggregate formation. The mechanism involves sticking of particles to a small cluster (DLA model) [14]. In our case, MNP particles are evidently additional precipitation centers.

Indeed, the flux of particles (J) during precipitation is determined by Fick's law:

$$J = DgradC, \tag{2}$$

where D and C are the diffusion coefficient and the concentration of particles being formed. Formation of smaller particles causes increase of their concentration gradient, which moves from the outer sides of a membrane to its middle together with a precipitator. When deposition occurs, higher concentration gradient is realized for smaller particles. On the other hand, magnetic nanoparticles provide local magnetic fields inside membrane. As found for solutions of NaCl, KCl, $CaCl_2$ and Na_3PO_4, their conductivity increases under the influence of magnetic field [10]. The reason is suggested to be structuring water in hydrate shells of ions, this results in increase of their diffusion coefficient. It is possible to assume that bonded water around the particles is also structured promoting faster diffusion. Enhancement of particle movement affected by magnetic field leads to formation of smaller aggregates comparing with the case of MNP absence.

TEM image of the active layer (Fig. 11.2a) shows very small aggregates of nanoparticles (up to 20 nm). Dark contrast spots evidently correspond to MNP, grey traces are related to HZD.

For comparison, the image for MNP is also given. The shape of nanoparticles is seen to be close to globular. A size of the primary particles is about 10 nm. The size of aggregates embedded to the active layer corresponds to pore size of the polymer according to its rejection ability towards proteins, molecular mass of which is 50 kDa.

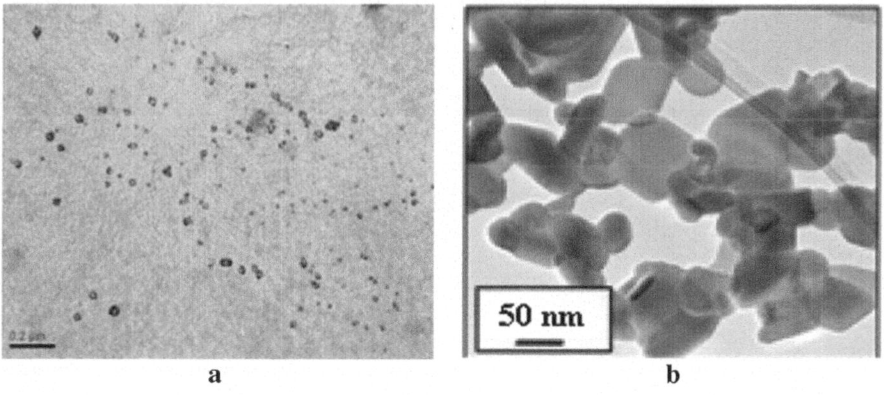

a b

Fig. 11.2 TEM image of active layer of PAN membrane containing HZD and MNP (**a**). The image of MNP that are outside the membrane is also given (**b**)

11.3.2 Water Filtration. Secondary Active Layer

Figure 11.3 illustrates a volume of permeate (V) as a function of time of water filtration (τ).

As seen, the dependencies are linear. This allows us to estimate water flux (J) as:

$$J = \frac{dV}{d\tau}\frac{1}{A}, \quad (3)$$

where A is the membrane area. The calculation were made from the slopes of the lines to the abscissa axis, the results are given in Table 11.1.

As seen, the PAN membrane containing HZD and MNP shows lower permeate flux than the pristine membrane. due to filling of the polymer pores. At the same time, rejection of hardness ions becomes higher due to decrease of pore size in the active layer and charge effect. The φ values are similar for the membranes containing MNP and free from them. At the same time, the PS membranes show higher values

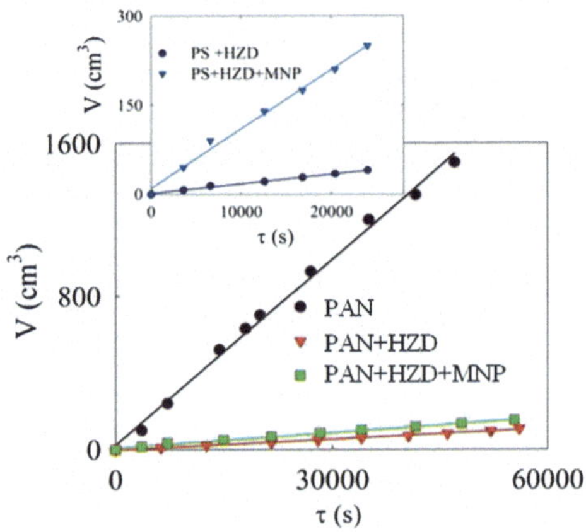

Fig. 11.3 Permeate volume as a function of time of water filtration through the pristine PAN membrane (PAN), PAN membrane containing HZD (PAN + HZD), PAN membrane containing HZD and MNP (PAN + HZD + MNP)

Table 11.1 Filtration of liquids at 2 bar

Membrane	Water		Sugar beet juice	
	J (m^3 m^{-2} s^{-1})	φ (%), Ca^{2+}, Mg^{2+}	J (m^3 m^{-2} s^{-1})	φ (%), VP
PAN	1.1×10^{-5}	6–7	1.1×10^{-6}	16–26
PAN + HZD	5.9×10^{-7}	8–19	3.8×10^{-7}	58–77
PAN + HZD + MNP	9.5×10^{-7}	7–20	4.3×10^{-7}	78–87
PS	2.3×10^{-5}	2–3	2.4×10^{-6}	6–13
PS + HZD	5.7×10^{-7}	5–7	5.5×10^{-7}	38–49
PS + HZD + MNP	3.6×10^{-6}	5–6	5.7×10^{-7}	55–60

of fluxes and lower rejection of Ca^{2+} and Mg^{2+}. It means that the PAN polymer membrane, which is characterized by smaller holes in active layer, is more attractive for modifying.

The inorganic particles form "secondary active layer" inside the polymer pores: this layer determines water flux and rejection ability of the membrane. Its thickness (l) was calculated from Kozeny–Carman equation [13]:

$$\frac{\Delta P}{l} = \frac{180\mu(1-\varepsilon^2)}{\Phi^2 d^2 \varepsilon^3} J. \tag{4}$$

where ΔP is the pressure drop, ε is the porosity (0.33 for compact bed of globules), Φ is the particle sphericity (it is assumed that $\Phi = 1$), d is the particle diameter ($\approx$10 nm), μ is the dynamic viscosity (9×10^{-3} Pa s at 298 K). The calculations give $l \approx 0.58$ µm both for the PS and PAN membranes containing HZD. This value is comparable with a thickness of active layer of the membrane. Regarding the membranes containing also MNP, $l = 95$ nm (PAN) and 0.87 µm (PS). Thus, thinner "secondary active layer" is formed in the polymer matrix containing smaller pores. In the case of PS, the nanosized inorganic particles are dispersed through the polymer active layer.

11.3.3 Filtration of Sugar Beet Juice

In all cases, filtration of sugar beet juice caused its clarification. Vegetable proteins partially penetrated into the permeate (see Table 11.1). Modifying was shown to improve ability of the membranes to reject this valuable component. Higher φ values were found for the membranes containing MNT. As seen from Fig. 11.4, the V-τ dependences can be fitted with linear functions (organic-inorganic membranes) or tend to plateau formation (pristine membrane).

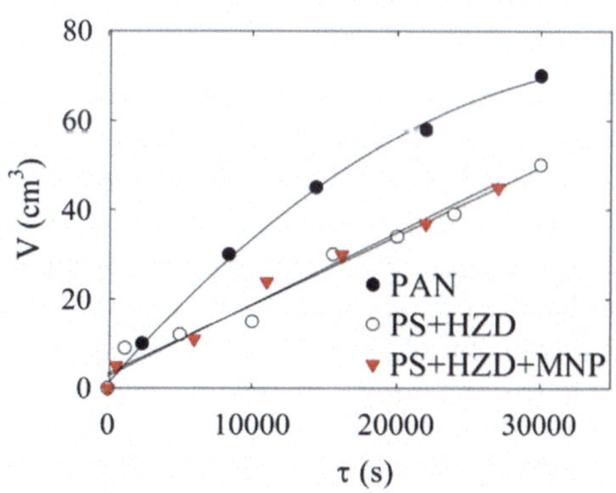

Fig. 11.4 Permeate volume *vs* time of filtration of sugar beet juice

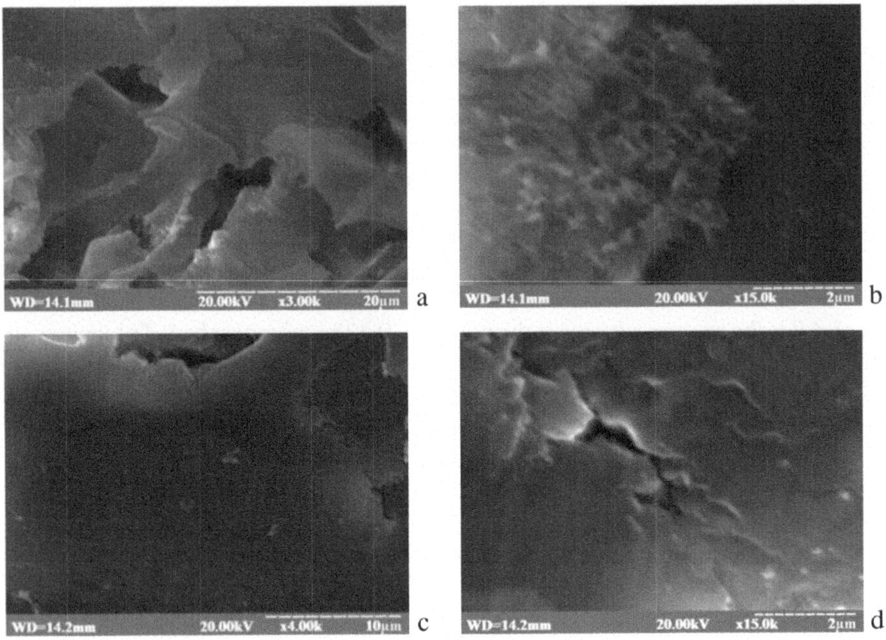

Fig. 11.5 SEM images of the samples after filtration of sugar beet juice: pristine PAN membrane (**a**, **b**), membrane modified with HZD (**c**), HZD and MNT (**d**)

The depression of filtration is due to fouling with organics: the precipitate is seen as a web-like patina (Fig. 11.5). At the same time, the outer surface of the organic-inorganic membrane remains clean.

11.4 Conclusions

When hydrated zirconium dioxide is deposited in ultrafiltration polymer membranes, aggregates of the nanoparticles are formed both in macroporous support and active layer. MNP provides formation of smaller HZD particles. Fractal analysis shows the DLA model of particle formation. The function of MNP is assumed to accelerate diffusion of HZD nanoparticles being precipitated. This depresses enlargement of the aggregates, As a result of modifying, the composite membrane shows slight improvement of rejection of hardness ions and much higher rejection of vegetable proteins comparing with pristine membranes. The modifying effect is most expressed for the PAN polymer membrane, which is characterized by smaller pores through its active layer comparing with the PS membrane. The composites also demonstrate stability against fouling with organics due to additional hydrophilization of polymer support. The membranes can be recommended for water treatment and processing of feedstock and wastes of food industry. The composites also demonstrate stability against fouling with organics due to additional hydrophilization of the polymer support similarly to [5, 15].

Acknowledgements This work was performed within the framework of the project entitled "Baro- and electromembrane processes in technologies for purification of liquid media produced by food technologies" (grant number 0117U001247 supported by the Ministry of Education and Science of Ukraine). The work was also supported by the joint Ukrainian-Belorussian project that is called "Development of composite ultra- and nanofiltration membranes with predetermined functional properties for complex processing of wastes of food industry" (supported by the National Academy of Science of Ukraine and by the National Academy of Science of Belarus). The authors thank Dr. S. Scherbakov (M.G. Kholodnii Institute of Botany of the NAS of Ukraine) for his support of investigation using electron microscopes.

References

1. S. Adham, *Development of a Microfiltration and Ultrafiltration Knowledge Base* (AWWA Research Foundation and American Water Works Association, Denver, CO, 2005)
2. A.G. Belous, O.I. V'yunov, E.V. Pashkova, V.P. Ivanitskii, O.N. Gavrilenko, J. Phys. Chem. B **110** (2006)
3. C. Douketis, Z. Wang, T. Haslett, M. Moskovits, Phys. Rev. B **51** (1995)
4. Y. Dzyaz'ko, L. Rozhdestvenskaya, A. Pal'chik, Russ. J. Appl. Chem. **78** (2005)
5. Y. Dzyazko, L. Rozhdestvenskaya, Y. Zmievskii, A. Vilenskii, V. Myronchuk, L. Kornienko, S. Vasilyuk, N. Tsyba, Nanoscale Res. Lett. **10** (2015)
6. Y. Dzyazko, L. Ponomaryova, Y. Volfkovich, V. Trachevskii, A. Palchik, Microporous Mesoporous Mater. **198** (2014)
7. Y. Dzyazko, A. Rudenko, Y. Yukhin, A. Palchik, V. Belyakov, Desalination **342** (2014)
8. Y. Dzyazko, Y. Volfkovich, V. Sosenkin, N. Nikolskaya, Y. Gomza, Nanoscale Res. Lett. **9** (2014)
9. N. Ghaemi, S. Madaeni, P. Daraei, H. Rajabi, S. Zinadini, A. Alizadeh, R. Heydari, M. Beygzadeh, S. Ghouzivand, Chem. Eng. J. **263** (2015)
10. L. Holysz, A. Szczes, E. Chibowski, J. Colloid Inter. Sci. **316** (2007)
11. Y. Huang, C. Xiao, Q. Huang, H. Liu, J. Hao, L. Song, J. Membrane Sci. **548** (2018)
12. H. Mæhre, L. Dalheim, G. Edvinsen, E. Elvevoll, I.-J. Jensen, Foods **7** (2018)
13. W. MacCabe, P. Harriott, J. Smith, *Unit Operations Of Chemical Engineering* (McGraw-Hill Higher Education, Boston, MA, 2005)
14. P. Meakin, Physica D: Nonlinear Phenomena **86** (1995)
15. V. Myronchuk, Y. Dzyazko, Y. Zmievskii, A. Ukrainets, A. Bildukevich, L. Kornienko, L. Rozhdestvenskaya, A. Palchik, Acta Periodica Technologica (2016)
16. R. Pang, X. Li, J. Li, Z. Lu, X. Sun, L. Wang, Desalination **332** (2014)
17. Z. Rahimi, A.A. Zinatizadeh, S. Zinadini, J. Appl. Res. Water Wastewater **1** (2014)
18. J. Shen, H. Ruan, L. Wu, C. Gao, Chem. Eng. J. **168** (2011)
19. L. Upadhyaya, M. Semsarilar, A. Deratani, D. Quemener, J. Membrane Sci. Res. **3** (2017)
20. M. Zahid, A. Rashid, S. Akram, Z. Rehan, W. Razzaq, J. Membrane Sci. Technol. **8** (2018)

Chapter 12
The Laser-Induced Coagulation Method of Biological Tissues

I. M. Lukavenko, V. V. Andryushchenko, and O. V. Yazykov

Abstract The results of the study of the laser-induced radiation method effectiveness for the treatment of different sizes cysts are presented. It is shown that under the influence of laser radiation, the effects are observed: anti-inflammatory action, accelerating the regeneration of damaged tissues and improving the blood supply to organs. It is confirmed that during the short pulse of heating the tissue is faster than during continuous radiation, which eliminates the possibility of thermal damage to surrounding tissues, contributes to a significant reduction in the possibility of complications and shortening treatment time. The effectiveness of scleroobliteration of breast cysts using the laser-induced coagulation method was 98.4%, with the use of chemical sclerosis—82.7%. The incidence of cyst recurrence was found to be dependent on the cavity volume and decreased with the use of laser-induced sclerotherapy.

12.1 Introduction

The integration of radiological techniques into practical mammology has led to the emergence of new surgical techniques that are performed using ultrasound, magnetic resonance imaging and radiography, and allow for the removal or destruction of nodal lesions with minimal tissue injury. For this purpose, sclerotherapy, vacuum biopsy, focused ultrasound, high-intensity laser radiation are also used [1–3]. The use of fiber lasers is extremely promising in medicine because they meet the key requirements: use ease, reliability, a wide choice of parameters and the possibility of their exact control, the absence of special requirements for the premises, mobility and minimal operating costs.

The main characteristic that determines the result of laser thermal exposure is the biological tissue temperature and the character of its spatial distribution. When laser

I. M. Lukavenko (✉) · V. V. Andryushchenko · O. V. Yazykov
Department of Surgery, Traumatology, Orthopedics and Phthisiatry, Sumy State University, Sumy, Ukraine
e-mail: doctorlv@ukr.net

A. D. Pogrebnjak et al. (eds.), *Nanomaterials in Biomedical Application and Biosensors (NAP-2019)*, Springer Proceedings in Physics 244,
https://doi.org/10.1007/978-981-15-3996-1_12

radiation acts on a biological tissue, it is absorbed in the tissue volume. Depending on the radiation power, reflection and absorption coefficient, a certain power is absorbed in a unit volume of the biological tissue. The absorption coefficient depends on the tissue type and the laser radiation wavelength. The spatial distribution of temperature is determined by the following factors: the amount of absorbed energy decreases with depth (according to a exponential law), therefore, the temperature in the tissue interior is less; scattering of radiation leads to the fact that a significant part of the radiation is absorbed away from the original direction of propagation of the beam; at the same time, heat is removed due to heat conduction and blood flow; the temperature distribution depends on the geometry of the laser beam. Thus, under the action of laser radiation on biological tissue, a temperature gradient arises, both in the depth of the tissue (in the direction of the action of the radiation beam) and in the directions perpendicular to it.

The purpose was to investigated the physicochemical bases of the laser-induced coagulation method from the point of view of its application in mammology and comparative analysis of the effectiveness of laser-induced coagulation and chemical sclerotherapy as methods of breast cyst obliteration.

12.2 Research Methodology

The technique of using high-intensity laser radiation is based on photometric and ablation effects, which are irreversible (coagulation and denaturation) of tissue changes, which leads to the formation of necrosis zones, with its replacement by connective tissue [4]. In laser surgery ablation is the elimination of a living tissue section under the laser radiation photons action. In this case, we are talking about the effect manifested in the course of irradiation, when the irradiated tissue section remains in place after the cessation of laser exposure, and its gradual elimination occurs as a result of a local biological reactions series in the irradiation zone. The energy characteristics and ablation performance are determined by the properties of the irradiated object, the radiation characteristics, and the laser beam parameters: the reflection, absorption and scattering coefficients of this type of radiation in a given tissue type. Figure 12.1 shows the physical processes that occur in the biological tissue when interacting with laser radiation.

When the temperature of the biotissue increases, various processes occur in it: enzyme activation, membrane change (37–45 °C); protein denaturation, onset of coagulation and necrosis (60 °C); collagen denaturation, membrane defects (80 °C); dehydration (100–150 °C); charring (above 150–300 °C); ablation and gas formation (over 300 °C). The dynamics of tissue temperature change under the influence of laser radiation is shown in Fig. 12.2.

The result of exposure to radiation on the biotissue also depends on the time during which the tissue is heated. This time is determined by the time of exposure to radiation and by the processes of redistribution of heat during cooling upon exposure cessation.

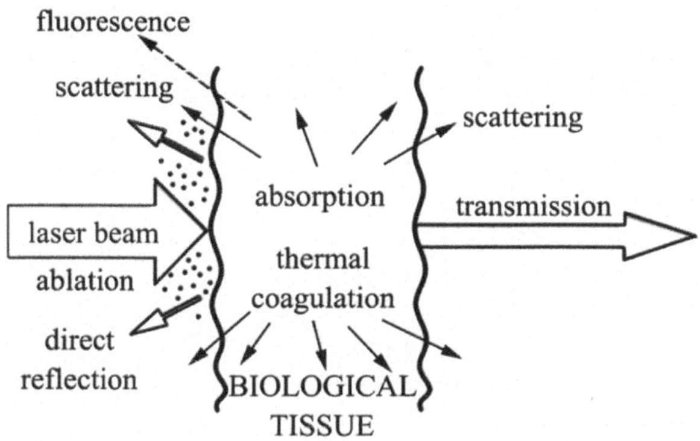

Fig. 12.1 Physical and chemical processes in biological tissue under the action of laser radiation

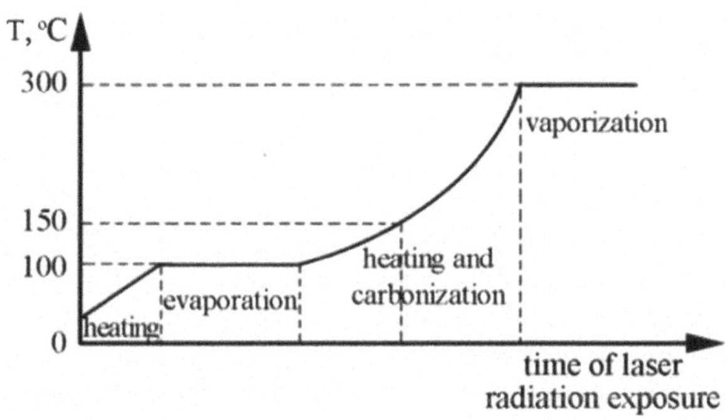

Fig. 12.2 The dynamics of temperature changes in the of biological tissue under the action of laser radiation

The properties of the irradiated biological object include: the ratio of the liquid and the dense components, their chemical and physical properties, the nature of the intra- and intermolecular bonds, the thermal sensitivity of cells and macromolecules, the blood supply to the tissue, etc. Under the influence of laser radiation, there are effects: anti-inflammatory action, accelerating the regeneration of damaged tissues and improving the blood supply to organs.

In the treatment of fibrous-cystic mastopathy we used the laser-induced coagulation method (device—laser coagulator “Lika-surgeon”, firm “Photonics-plus”, Cherkasy, Ukraine) for surgical treatment of nodular entities breast or residual cystic cavities. The operation principle of the device is based on the laser radiation generation by semiconductor lasers of the laser-optical module and output through the optical output connector (SMA-905) to connect it to the optical fiber and deliver laser radiation to the site of the operation. The set of operating modes and parameters of laser radiation is accomplished by the keypad block buttons, which are displayed on the indicator of the display unit and are fed to the control unit that provides control of

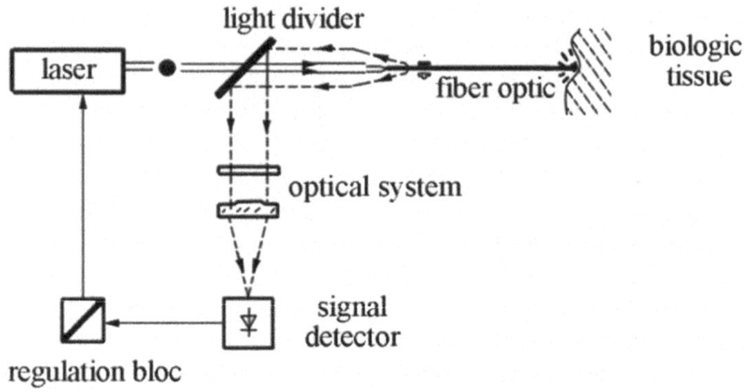

Fig. 12.3 Scheme of laser-induced coagulation

the laser radiation parameters. The control unit consists of driver boards, the number of which depends on the output power of the laser radiation.

The method of intra-tissue laser coagulation is as follows. The laser beam is delivered to the damaged tissue by fiber through a subcutaneous needle with simultaneous visual inspection of the exposure site. Under the ultrasound control, the cyst is injected into the cavity under a puncture needle, then a flexible fiber light guide is placed on the lumen of the puncture needle. At least 50% of the contents of the cyst are evacuated through a puncture needle before laser exposure. Exposure is carried out by continuous radiation with a guidance beam wavelength of $\lambda = 650$ nm, a power $P = 5.0$–10.0 W, irradiation time $\tau = 0.1$–10 s. The laser-induced coagulation scheme is shown in Fig. 12.3.

The method allows to reduce the trauma due to indirect thermal effects on the cyst walls, to reduce the time of surgery, to ensure the prevention of recurrence. Invasive interventions on the breast were performed under real-time ultrasound control using 7.5–12 MHz linear sensor. For anesthesia, local infiltration anesthesia was performed with retromammary blockade. A quartz electrode with a diameter of 600 μm was inserted into the cavity of the brush through a 14 G plastic catheter. The apparatus radiation is transmitted to the surgical manipulation place by means of flexible fiber-optic fibers of small diameter. The length of the flexible section is $D = 2$ m, the optical fiber diameter is $d = 400$–440 μm. For comparative analysis, studies were also performed by chemical sclerotherapy using 96% ethanol.

12.3 Results and Discussion

In the practical activity of the mammal surgeon, there are often situations where major pathological changes in the breast are caused by cystic transformation of its tissue. There is no single algorithm for the patients treatment of regarding single and multiple breast cysts [4–6]. Treatment tactics are based on the differential diagnosis results

Table 12.1 Distribution of cysts depending on their volume

Cysts volume (sm^3)	Number of surveys			
	Laser-induced coagulation method		Chemical obliteration method	
	Kists (pieces)	%	Kists (pieces)	%
1.5–3.0	62	51.6	42	56.0
3.1–6.0	37	30.9	24	32.0
6.1–12.0	12	10.0	5	6.7
12.1–16.5	9	7.5	4	5.3
Total	120		75	

of cancer and benign breast dysplasia. The most informative method of breast cyst detection is the ultrasonic research which sensitivity makes 46–59%, specificity—98%. The combination of ultrasound imaging and methods of physical and chemical destruction of biological tissues allows the removal of breast cysts with minimal traumatization of surrounding tissues.

The study involved patients from 30 to 58 age years, except for breast cysts with a diameter of less than 1.5 cm, atypical breast cysts, cysts with inflammation signs, polycystosis of breast. After aspiration biopsy under the ultrasound control cytological examination of punctate was performed. Patients over 40 age years underwent mammograms according to the standard method.

The study involved several steps. The first stage is the selection of patients who have been diagnosed with breast cysts for laser-induced or chemical scleroobliteration. The criteria for inclusion in the study were: the presence of primary and (or) recurrent breast cysts with a diameter of more than 1.5 cm. After the patients selection with cystic disease, breast cysts were divided according to the cavity volume (Table 12.1).

The second stage is scleroobliteration. The total number of cysts obliterated by laser was 120, and chemically obliterated cysts—75. When using the laser-induced obliteration method, the evacuation of the cavity contents was performed before the introduction of the fiber. Due to the laser radiation effects, the cyst is emptied before its walls fall and the laser seam is formed, which ensures complete comparison of the cyst sockets (Fig. 12.4).

This, in turn, helps to prevent relapses and improve the treatment effectiveness. During the short pulse of heating the tissue is faster than in continuous radiation, with the spread of heat to be minimal, which eliminates the possibility of thermal damage to the surrounding tissues, contributes to a significant reduction in the possibility of complications and, consequently, shortening the healing time.

Make a comparative analysis. Consider first the chemical obliteration method. In the presence of breast cysts with volume 12.1–16.5 cm^3 relapses occurred in 7 (9.3%) patients; 6.1–12.0 cm^3 in 4 (5.3%), 3.1–6.0 cm^3 in 2 (2.7%). During chemical cysts obliteration of 1.5–3.0 cm^3 recurrence was not. Thus, the efficacy of breast cysts scleroobliteration using the chemical method was 82.7%.

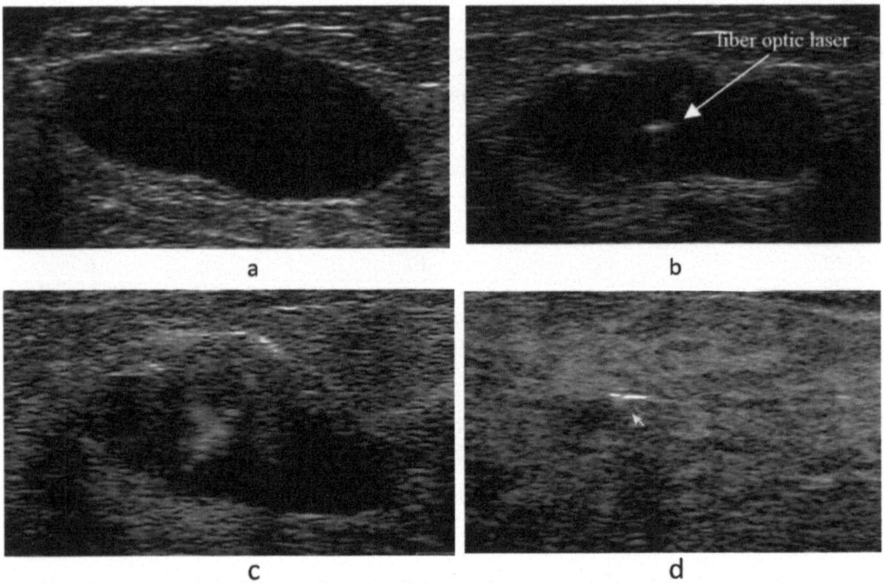

Fig. 12.4 Stages of laser-induced scleroobliteration of the breast cyst (BC) . Sonograms: **a** is a typical cyst in the initial state; **b** the optic laser injection; **c** the laser irradiation effect (gas microbubbles at the fiber end); **d** after the laser beam (the marker shows the location of the laser coagulation cyst, the inner wall—in the form of a hyperechogenic strip)

Laser-induced treatment of breast cysts showed the following results. The cysts recurrence of occurred only in 2 (1.6%) patients with cysts of 6.2–16.5 cm^3. Complications of laser coagulation were manifested in the form of inflammatory infiltrate on day 2 after the procedure, which disappeared after 4–5 days after antibacterial therapy, use of anti-inflammatory agents and topical treatment. At laser irradiation, breastcysts of 1.5–6.0 cm^3 did not relapse. Thus, the efficacy of scleroobliteration of breast cysts using the laser-induced coagulation method was 98.4%. This indicating greater efficiency of the laser method for the treatment of large breast cysts.

12.4 Conclusions

It is shown that the use of laser tissue destruction modern methods is a safe alternative to surgical treatment and is extremely promising in medicine, because they meet the key requirements: use ease, reliability, a wide choice of parameters and the ability to accurately control them, the lack of special requirements for the room, mobility and minimal operational costs.

The use of the laser-induced coagulation method is found to be effective for the treatment of large-sized breast cysts—from 6.1 to 16.5 cm^3. It was determined that the scleroobliteration effectiveness of breast cysts using methods of laser-induced coagulation and chemical obliteration, which were 98.4 and 82.7% respectively.

References

1. K. Shokrollahia, E. Raymondb, M.S.C. Murisona, J. Surg. **2**(1), 28 (2004)
2. B.S. Santosh, H.K. Aiyappan, P. Choudhary, S. Varsha, H. Harikeerthy, M. Islam, D. Daniel, J. Adv. Med. Dental Sci. Res. **6**(9), 46 (2018)
3. V. Gafiychuk, I. Lubashevsky, B. Datsko, Phys. Rev. E **72**, 051920-1 (2005)
4. I.M. Lukavenko, V.V. Andryushchenko, A.V. Yazykov, J. Nano-Electron. Phys. **10**(6), 06031-1 (2018)
5. I.M. Lukavenko, V.V. Andryushchenko, V.Iu. Garbuzova, A.V. Yazykov, Georg. Med. News **1** (238), 12 (2015)
6. A.V. Yazikov, V.V. Leonov, V.V. Andryushchenko, I.M. Lukavenko, Georg. Med. News **2**(239), 11 (2015)

Chapter 13
Fullerene C_{60}-Containing Hydroxyapatite/Polymer Polyelectrolyte Composite for Dental Applications

L. B. Sukhodub, M. A. Kumeda, L. F. Sukhodub, and Yu. I. Prylutskyy

Abstract The novel bioactive Fullerene C_{60}-containing apatite/polymer polyelectrolyte composite for dental applications based on hydroxyapatite (HA), sodium alginate (Alg) and chitosan (CS) in form of thing membrane was synthesized by the "wet chemistry" method and investigated by various experimental techniques. XRD data indicate the formation under influence of MW irradiation of only one phase of HA (JCPDS 9-432) with average crystallites size of 20 nm and small carbonate content. The including of fullerene to the samples leads to a decrease in the liquid absorption degree for of about 60% and liquid retention ability for of about 50%. The fullerene-containing samples have higher shape stability after oscillations for 10 days in the shaker (rpm 80) at 37 °C. HPLC studies have shown the ability to use the material as a system for prolonged drug release. The addition of C_{60} accelerates and equals the kinetics of chlorhexidine (CHX) release during 48 h. The components ratio in the composite affects the thickness and mechanical properties of the film, its solubility and release of the antibacterial agent. The biodegradable and biocompatible with native surroundings material can be used in surgical and conservative dentistry for the treatment of periodontal diseases. Material can stimulate the bone tissue repair as an additional source of calcium and phosphorus ions, do not cause side effects, have antimicrobial properties and act as a system for prolonged drug release.

L. B. Sukhodub (✉) · M. A. Kumeda · L. F. Sukhodub
Sumy State University, Sumy, Ukraine
e-mail: l.sukhodub@gmail.com

Yu. I. Prylutskyy
Taras Shevchenko National University of Kyiv, Kiev, Ukraine

A. D. Pogrebnjak et al. (eds.), *Nanomaterials in Biomedical Application and Biosensors (NAP-2019)*, Springer Proceedings in Physics 244,
https://doi.org/10.1007/978-981-15-3996-1_13

13.1 Introduction

Periodontitis is an infection that includes the inflammatory process and an immune response. It destroys the alveolar bone, cement, periodontium and gum due to the accumulation of germs [1]. The periodontal restoration process involves osteogenesis and the formation of functionally oriented periodontal fibers both in newly formed cement and in alveolar bone [2]. This process is problematic in the absence of ancillary supportive medicine.

Medicinal treatment of periodontitis, at present, is one of the most commonly used methods. However, the effect of this approach is reduced by the fact that the therapeutic concentration of antibiotics is difficult to maintain for a long period in problem areas. Therefore, it is promising to create and research materials for local application containing antibiotics or growth factors and capable to prolong their release up to one month in order to improve the regeneration process [3]. A wide range of new drug delivery systems such as fibers, strips, inserts, implants, films, gels, plates, sprays, micro- and nanoparticles were developed for local treatment and prevention of various diseases in the oral cavity.

Hydroxyapatite, having the chemical composition $Ca_{10}(PO_4)_6(OH)_2$, is the most frequently used biomaterial for bone tissue regeneration. HA is the main mineral constituent of natural hard tissue and its excellent bioactivity and osteoinductivity are well-known [4]. In order to improve the weak HA mechanical properties, its structure can be modified by carbon nanoparticles, which are one of the hardest materials in the world. The most widespread among them is fullerene C_{60}. Aqueous suspensions of fullerenes exhibit relatively strong antibacterial activity [5], and over the last decade they have been studied as delivery and release systems for drugs [6] because of their ability to cross cell membranes [7].

Alginate is a natural anionic linear polysaccharide extracted from brown algae, such as kelp. Alg has several unique properties that allow it to be used as a matrix for the capture and delivery of various drugs [8].

Chitosan is a linear polysaccharide composed of (1→4)-2-acetamino-2-deoxy-β-D-glucan units—a hydrophilic biopolymer produced by alkaline deacetylation of chitin. CS is a major component of shells of crustaceans, and has beneficial properties such as non-toxicity, biocompatibility, bioadhesivity, biodegradation, excellent mucoadhesive and penetrating influence on biological surfaces [9].

CS has antimicrobial activity. It forms a complexes with negatively charged functionalities of the carboxymethylcellulose, citrates, pectin, agar, sodium caprylate, stearic acid, glutaraldehyde, sodium tripolyphosphate, lactic and alginic acids [10]. In the medical applications, chitosan is often crosslinked with glutaraldehyde, but glutaraldehyde is very harmful even at low concentrations. In this work sodium tripolyphosphate is used as a crosslinking agent instead of glutaraldehyde.

Chlorhexidine is a chemical antiseptic and widely used as antibacterial agent for dental applications. It is a positively charged hydrophobic and lipophilic molecule, that interacts with phospholipids and lipopolysaccharides on a cell membrane of bacteria [11].

In this study, Fullerene C60-containing apatite/polymeric polyelectrolyte composites in form of thing membrane, loaded with CHX were investigated as potential system for prolonged drug release. The dependence of the term of release on the content of fullerene is analyzed.

13.2 Experiment Details

13.2.1 Materials

The next materials and chemicals were used: calcium nitrate tetrahydrate $Ca(NO_3)_2{\cdot}4H_2O$, diammonium dihydrophosphate $(NH_4)_2HPO_4$, ammonium hydroxide NH_4OH, sodium alginate (low viscosity, E407, China), chitosan (m.m. 300 kDa Acros organics, USA), pharmaceutical 0.05% chlorhexidine. All reagents were analytically grade and used as received.

13.2.2 Composite Material Obtaining

HA hydrogel HA hydrogel was obtained according to the "wet chemistry" method. Briefly, for HA synthesis 50 ml of $Ca(NO_3)_2{\cdot}4H_2O$ (0.167 M) and 50 ml of $(NH_4)_2HPO_4$ (0.1 M) were used. Ammonium hydrophosphate was added drop wise to the calcium nitrate tetrahydrate. The pH of about 10.5 was obtained by the addition of ammonia solution. 100 ml of the obtained suspension was transferred to the consumer microwave oven Samsung M1712NR for its microwave irradiation (MW) for 3 min. Then, the product was cooled at room temperature in the closed, but not sealed volume. After being washed, the solid fraction of the sample was separated by centrifugation. The moisture content of the resulting HA hydrogel was about 90% by weight.

HA–Alg scaffold formation The hydrogel (p.a) was dispersed in a 0.05% aqueous sodium hydroxide solution under the influence of ultrasound, followed by the addition of sodium alginate powder in such a way that the ratio of CaP:Alg equaled 1:1, based on dry matter. Propylene glycol (10% by volume) and the TWEEN-80 emulsifier were added to the mixture. The resulting mixture was poured into a 5 mm layer with subsequent freezing and lyophilic drying for 6 h using a laboratory device ALPHA 1-2 LDplus. As a result, a high-porous intermediate HA–Alg scaffold was formed.

Fullerene C60 loaded polyelectrolyte composite formation 30 ml of 1% chitosan in ascorbic acid was prepared and divided into 3 parts by 10 ml: fullerene—free chitosan, chitosan containing 100 mg/l of fullerene and chitosan containing 200 mg/l of fullerene. After adding the required amount of colloidal fullerene to the chitosan solution, the homogeneous suspension was obtained using ultrasound. Then experimental HA-Alg scaffolds, obtained according to sect. HA-Chitosan scaffold formation, were immersed into prepared chitosan solutions for 12 h, followed by rinsing in distilled water and drying at 37 °C. The obtained samples for further research were named: HA-Pol, HA-Pol-100Full, HA-Pol-200Full. The chlorhexidine introduced into HA-Pol, HA-Pol-100Full and HA-Pol-200Full by the method of impregnation, followed by drying.

13.2.3 Analytical Methods

SEM study The morphology investigations carried out using a scanning electron microscope (FEI Inspect S 50). Microphotographs of the samples surface were made in the secondary electron mode with HV = 10.00 kV.

XRD analysis The crystallinity and structure of all precipitates were examined using X-ray diffractometer DRON 3 (RPE "Burevestnik") connected to a computer-aided system for the experiment control and data processing.

The Ni-filtered CuKα radiation (wavelength 0.154 nm) with a conventional Bragg-Brentano θ-2θ geometry was used. The data were collected over the 2θ range 5.0°–60.0° with step of 0.02° and counting time of 2 s. Identification of the crystal phases was performed using the JCPDS (Joint Committee on Powder Diffraction Standards) card catalog. The average crystallite sizes (L; nm) along [002] directions were estimated from the corresponding peak broadening by the Scherrer equation:

$$L = k\lambda/\beta_m \cos\theta \qquad (1)$$

where k is the constant, dependent on the crystallite shape ($k = 1$ is taken), λ is the wavelength of X-ray radiation [in nm], β_m is an integral width of (002) peak, in which physical broadening occurs only due to the small sizes of the CSR (coherent scattering region), θ is the diffraction angle. Calculation of the hexagonal lattice parameters a and c was held by the corresponding formulas, using the Miller indexes [12].

Water uptake and retention abilities The swelling behavior was quantified by measuring the changes in sample weight as a function of sample immersion time in simulated body fluid (SBF) at pH = 7.33 prepared according to the protocol described by Kokubo et al. [13] The osmolarity and ion concentrations of the solutions match those of the human body fluid.

Dried composites with mass of 0.1 g (W_o) were immersed in SBF for 1 h. The samples were then carefully removed from the solution and placed on a filter paper for 5 min, followed by weighing (W_t) to determine the absorption of water (swelling) [14]

$$S_w = (W_t - W_o)/W_o * 100\% \qquad (2)$$

where W_o is the initial sample weight, and W_t is the final weight of the swollen sample.

To measure the water retention ability, the wet samples were transferred to centrifuge tubes with filter paper at the bottom, centrifuged at 600 rpm for 5 m and weighed immediately (W_t). The percentages of water retention (E_r) of the beads at equilibrium were calculated using following equation [15]:

$$E_r = \left(W_j' - W_t\right)/W_t * 100\% \quad (3)$$

Determination of drug release kinetics The material composition and fullerene content influence on the drug release was investigated in our experiment. The pharmaceutical 0.05% Chlorhexidine solution used as a model drug. The experimental samples were saturated with Chlorhexidine solution for 1 h followed by drying at 37 °C. For the drug release test, the 0.15 g of each sample, namely HA-Pol, HA-Pol-100Full, HA-Pol-200Full, saturated with Chlorhexidine, was placed into 6.0 ml of PBS (pH = 7.4) and incubated at 37 °C with continues shaking of 80 rpm. The rate of drug release from composites was determined by taking of 600 μl aliquots of PBS from each experimental tube daily for 120 h. An equal volume of the fresh medium was added back to maintain a constant initial volume in the tubes. The drug release from experimental samples was studied using high-performance liquid chromatography (HPLC; Agilent Technologies 1200, detector with UV-Vis Abs, detection at λ = 280 nm, column C18 (Zorbax SB-C18 4.6 × 150 mm, 5 μm)). The data acquisition and processing made with software Empower 2. The mobile phase was: 0.05 M potassium hydrogen phosphate buffer with 0.2% triethylamine (pH = 3.1 at 210 °C)—68%; acetonitrile—32%. Isostatic treatment was applied at a rate of elution eluent 2 ml/min and a temperature of 40 °C analytical column.

13.3 Results and Discussion

The ability of sodium alginate to gel formation, as well as antibacterial properties, biocompatibility, biodegradation of chitosan has been used to create the composite material. Present in the CS polymer chain the amino groups are protonated in acidic media at pH <6.5 to form NH_3^+ groups. In this case, CS macromolecules acquire a positive charge, which facilitates their binding to the anionic functionality of the sodium alginate to form a polyelectrolyte complex. Being localized near the negatively charged surfaces of microbial cells, chitosan has an antimicrobial effect by binding to the phospholipids of the membrane shell, resulting in a violation of metabolic processes and cell death [16].

The structure of the obtained apatite/polymer hydrogel includes nanoparticles of up to 100 nm with a characteristic reactivity for such particles. The nanoparticles are immobilized in an alginate matrix, the limiting pore size of which allows them to remain a nanosized in the suspension for a long time before and after sublimation.

Fig. 13.1 Morphology of the HA-Pol-100Full sample at different magnification

This fact significantly differs them from polymer-free powders, where calcium phosphate particles are agglomerated.

The dispersion of calcium phosphate nanoparticles, made during synthesis, is an important process in obtaining a homogeneous HA polymer complex. Usually, nanoparticles have a large surface area and large free surface energy, so they are easily agglomerated to reduce surface energy. In this work, Alginate acts as dispersant for HA nanoparticles. Being an anionic polysaccharide, Alg forms a polyelectrolyte complex with CS polycation.

SEM study Figure 13.1 shows a typical image for a fullerene-containing sample based on hydroxyapatite, alginate and chitosan. There is a layered structure with interspersed cubic particles of HA.

XRD study As noted above, the experimental fullerene contained samples are based on HA hydrogel, the X-ray diffraction spectrum of which is shown in Fig. 13.2. The XRD result shows a presence of only one phase - HA (JCPDS 9-432) with average crystallites size of 20 nm in the *c* direction. A slight deviation of the crystalline lattice parameters in the obtained HA (a = 0.939 nm; c = 0.686 nm) to the reduction side in comparison with the stoichiometric HA (a = 0.9421 nm; c = 0.6881 nm) [17] indicates the formation of a small fraction of the carbonate substituted HA.

Liquid uptake and retention abilities Biomedical applications impose special requirements for synthesized material. The material should not undergo significant volumetric changes, in order not to induce stress in surrounding tissues. So, water uptake (swelling) and water retention ability of the experimental composites are important factors.

As mentioned above, the swelling degree study was carried out in a SBF solution for 1 h at 37 °C. As expected, lyophilized samples have an increased ability to liquid uptake. As evident from the results, the including of fullerene to samples leads to a decrease in the liquid absorption degree and liquid retention ability (Fig. 13.3). Obviously, this is the result of the fullerene nanoparticles hydrophobicity.

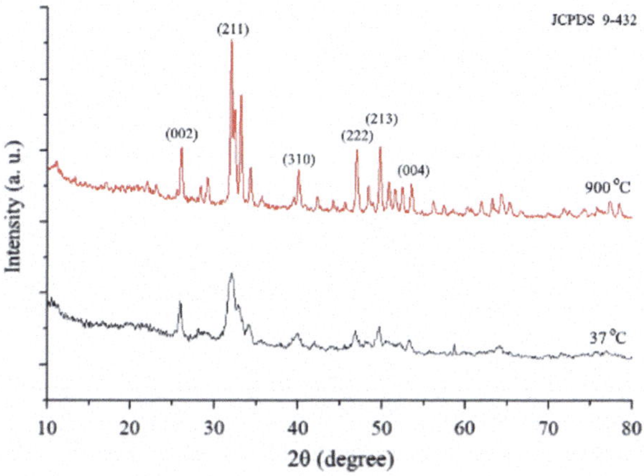

Fig. 13.2 Diffraction patterns from HA hydrogel sample, dried at 37 °C and sintered at 900 °C

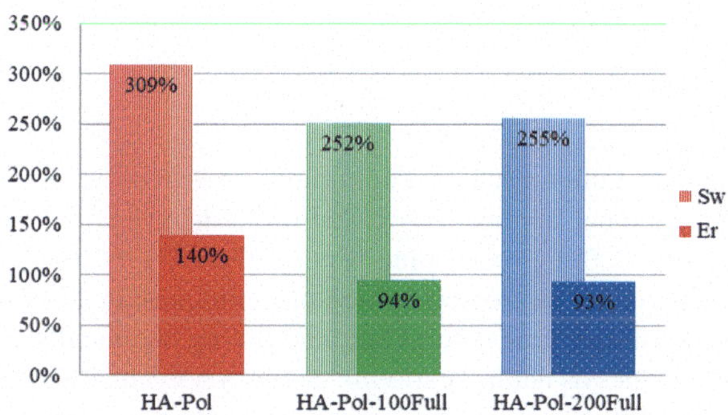

Fig. 13.3 Liquid uptakes (S_w) and liquid retention ability (E_r) of HA-Pol, HA-Pol-100Full and HA-Pol-200Full composites after saturation for 1 h

Furthermore, the fullerene-containing samples have higher shape stability after oscillations in the shaker (rpm 80) at 37 °C for 7 days during HPLC investigation.

HPLC study HPLC were carried out to test the effect of material composition on its ability to release the model drug in physiological solution. Pharmaceutical 0.05% Chlorhexidine was introduced into composites by the method of saturation. CHX is very common in dental medicine as an antibacterial agent against gram-positive and gram-negative bacteria and fungi. It has the ability to reduce the amount of harmful bacteria in the mouth and works to treat gingivitis and promote the healing of inflamed gums [4]. This is a hydrophilic molecule with a molecular weight of 505 g/mol. The results (Fig. 13.4) showed that the release of CHX from all experimental samples lasts for 48 h, after which concentrations begin to decrease due to dilution, which occurs by adding the aliquots of fresh PBS solution after each sampling.

For a control fullerene-free sample we observe unevenness and decrease the amount of released drug within the first 3 h. As it was noted at Costantini et al. [18],

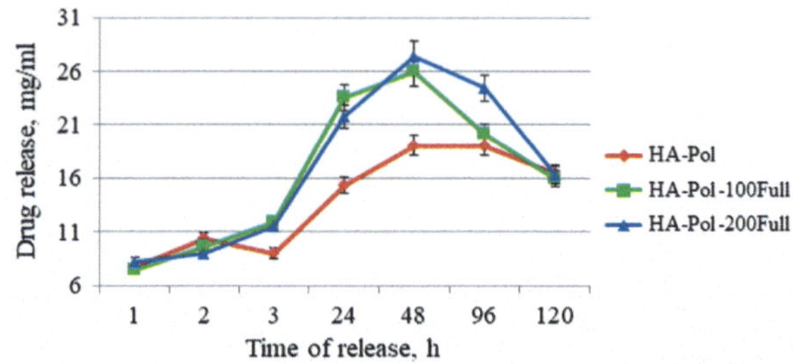

Fig. 13.4 Concentration of Chlorhexidine in PBS (μg/ml), released from the experimental samples

polymers form a diffusion barrier for the drug initial release. The addition of fullerene accelerates and equals the kinetics of CHX release, apparently because of fullerene hydrophobicity and repulsion of hydrophilic CHX molecules. For the same reason, we are seeing a faster and more complete CHX release from fullerene-containing samples.

13.4 Conclusions

The novel bioactive Fullerene C60-containing apatite/polymer polyelectrolyte composites for biomedical applications based on hydroxyapatite, sodium alginate and chitosan in form of thing membrane was synthesized by the "wet chemistry" method and investigated by various experimental techniques. In the obtained material the alginate macromolecules, as natural anionite, being chemically cross-linked by chitosan macromolecules as natural cationite, forms a polyelectrolyte scaffold in the form of a thin (up to 1 mm) membrane, containing hydroxyapatite.

XRD data indicate the formation under influence of MW irradiation of only one phase of hydroxyapatite (JCPDS 9-432) with average crystallites size of 20 nm and small carbonate content. The including of fullerene C60 to samples leads to a decrease in the liquid absorption degree for about 60% and liquid retention degree for about 50%. The fullerene-containing samples have higher shape stability after oscillations for 10 days in the shaker (rpm 80) at 37 °C.

HPLC studies have shown the ability to use the material as a system for prolonged drug release. The addition of fullerene accelerates and equals the kinetics of CHX release during 48 h. The components ratio in composite affects the thickness and mechanical properties of the membrane, its solubility and release of the antibacterial agent.

References

1. S.M. Rao, G.M. Ugale, S.B. Warad, N. Am, J. Med. Sci. **5**, 161–168 (2013). https://doi.org/10.4103/1947-2714.109175

2. R.E. Cohen, A. Mariotti, M. Rethman, S.J. Zackin, *Glossary of Periodontal Terms*, 4th edn. (American Academy of Periodontology, Chicago, 2001)
3. T. Duruel, A.S. Çakmak, A. Akman, R.M. Nohutcu, M. Gümüşderelioğlu, Int. J.Biol. Macromol. **104**, 232–241 (2017). https://doi.org/10.1016/j.ijbiomac.2017.06.029
4. L.F. Sukhodub, L.B. Sukhodub, O. Litsis, Yu. Prylutsky, Mat. Chem. Phys. **217**, 228–234 (2018). https://doi.org/10.1016/j.matchemphys.2018.06.071
5. D.Y. Lyon, L.K. Adams, J.C. Falkner, P.J.J. Alvarez, Environ. Sci. Technol. **40**(14), 4360–4366 (2006). https://doi.org/10.1021/es0603655
6. J. Safari, Z. Zarnegar, J. Saudi Chem. Soc. **18**, 85–99 (2014). https://doi.org/10.1016/j.jscs.2012.12.009
7. N.W.S. Kam, M. O'Connell, J.A. Wisdom, H. Dai, Proc. Natl. Acad. Sci. U.S.A. **102**, 11600–11605 (2005). https://doi.org/10.1073/pnas.0502680102
8. W.R. Gombotz, S.F. Wee, Adv. Drug Deliv. Rev. **64**, 194–205 (2012)
9. A. Verlee, S. Mincke, C.V. Stevens, Carb. Pol. **164**, 268–283 (2017). https://doi.org/10.1016/j.carbpol.2017.02.001
10. V. Rosa, A.D. Bona, B.N. Cavalcanti, J.E. Nör, Dent. Mater. **28**, 341–348 (2012). https://doi.org/10.1016/j.dental.2011.11.025
11. T.H. Lee, C.C. Hu, S.S. Lee, M.Y. Chou, Y.C. Chang, Int. Endod. J. **43**, 430–434 (2010). https://doi.org/10.1111/j.1365-2591.2010.01700.x
12. V.M. Kuznetsov, L.B. Sukhodub, L.F. Sukhodub, J. Nano- Electron. Phys. **6**(4), 04039–04043 (2014)
13. T. Kokubo, H. Kushitani, S. Sakka, T. Kitsugi, T. Yamamuro, J. Biomed. Mater. Res. **24**, 721–734 (1990). https://doi.org/10.1002/jbm.820240607
14. Y. Han, Q. Zeng, H. Li, J. Chang, Acta Biomater. **9**, 9107–9117 (2013). https://doi.org/10.1016/j.actbio.2013.06.022
15. J. Venkatesan, Z.J. Qian, B.M. Ryu, N. Ashok, S.K. Kim, Carbohyd. Polym. **83**, 569–577 (2011). https://doi.org/10.1016/j.carbpol.2010.08.019
16. F.A. Nejati Hafdani, N. Sadeghinia, World Acad. Sci. Eng. Technol. **74**, 257–261 (2011)
17. H.P. Klug, L.E. Alexander, *X-Ray Diffraction Procedures: For Polycrystalline and Amorphous Materials* (Wiley, New York, 1974)
18. L. Costantini, N. Bouropoulos, D.G. Fatouros, I. Kontopoulou, M. Roldo, Synthesis of carbon nanotubes loaded hydroxyapatite: Potential for controlled drug release from bone implants. J. Adv. Ceram. **5**(3), 232–243 (2016)

Chapter 14
Graphene Oxide Influences on Mechanical Properties and Drug Release Ability of Hydroxyapatite Based Composite Material

L. B. Sukhodub, L. F. Sukhodub, Yu. I. Prylutskyy, M. A. Kumeda, and U. Ritter

Abstract Bioactive composite material based on hydroxyapatite (HA), sodium alginate (Alg) with different content of graphene oxide (GO) was synthesized by the "wet chemistry" method and characterized by TEM, XRD, FTIR, HPLC analysis. Introduced the GO nanoparticles, as well as Ca^{2+} ions, as cross-linker of Alg macromolecules by the beads formation, lead to enhancement of the composites mechanical properties. HA–Alg–GO10 sample with GO content of 0.004% in relation to the HA powder has a much higher Young's modulus (1325 MPa) in comparison with GO-free HA–Alg composite (793 MPa), as well as steel sample of the same size (~706 MPa). The addition of GO reduces the degree of the composites swelling in a phosphate buffered saline for 43% and enhances the beads shape stability. Chlorhexidine bigluconate release from GO containing samples lasts for 48 h longer according to HPLC study. The findings clear demonstrate the potential possibility of applications of the HA–Alg–GO composite material in bioengineering of bone tissue to fill bone defects of various geometries with the function of prolonged release of the drug. It is assumed that HA–Alg–GO composite material can be used in 3D modeling of areas of bone tissue that have to bear a mechanical load.

L. B. Sukhodub (✉) · L. F. Sukhodub · M. A. Kumeda
Sumy State University, Sumy, Ukraine
e-mail: l.sukhodub@gmail.com

Yu. I. Prylutskyy
Taras Shevchenko National University of Kyiv, Kiev, Ukraine

U. Ritter
Technical University of Ilmenau, Ilmenau 98693, Germany

A. D. Pogrebnjak et al. (eds.), *Nanomaterials in Biomedical Application and Biosensors (NAP-2019)*, Springer Proceedings in Physics 244,
https://doi.org/10.1007/978-981-15-3996-1_14

14.1 Introduction

Bone is a composite material, consisting of about 30% of organic material (predominantly type-I collagen), 60% of mineral component, 9% of water, 1% of proteins and other biomolecules. The inorganic mineral component is a calcium-deficient carbonate-substituted apatite, similar in structure and composition to hydroxyapatite (HA)–$Ca_{10-x}(HPO_4)_x(PO_4)_{6-x}(OH)_{2-x}$. HA is used as bone graft materials in medicine for more than two decades because of its bioactivity, that means a significant ability to promote new bone growth. Ideal synthetic material must be able to form a stable interface with the surrounding natural tissue and its mechanical properties must correspond to the substitute bone [1]. But mechanical properties of pure HA are insufficient for major loadbearing applications. Alginate (Alg) is a linear unbranched anionic polysaccharide, obtained from brown seaweed and containing varying proportions of beta-D-mannuronate (M) and alpha-L-guluronate (G) residues. It is extensively used for many biomedical applications, due to its biocompatibility, low toxicity and gelation properties. Alg gels are typically nanoporous (pore size ~ 5 nm) [2], that leads to rapid diffusion of small drug molecules through the gel. So, Alg plays a significant role in controlled drug release [3]. Alg is often combined with inorganic materials, including carbon nanoparticles (CNPs), in particulary graphene. Graphene is regarded as the thinnest material in the [4], it is presented as one-atom-thick material consisting of carbon atoms arranged in a honeycomb-lattice structure [5]. Numerous oxygen functional groups in the basal planes and the borders of graphene oxide (GO) can generate the interfacial bonding with hydroxyl groups of Alg, allowing homogeneous distribution of GO in alginate matrix [6].

In this work a composite material based on HA and Alg with GO content was obtained in the form of beads and investigated for its bioactivity, physico-chemical properties and the ability to prolong the release of the drug into an environment.

14.2 Experiment Details

14.2.1 Materials

GO powder (Sigma-Aldrich, USA), calcium nitrate tetrahydrate $Ca(NO_3)_2{\cdot}4H_2O$, diammonium dihydrophosphate ($(NH_4)_2HPO_4$), ammonium hydroxide NH_4OH, calcium chloride $CaCl_2$ (Sinopharm Chemical Reagent Co., Ltd.), sodium Alg ((low viscosity, E407, China), pharmaceutical 0.05 wt% Chlorhexidine bigluconate (CH). All other reagents were analytically grade, commercially available and used as received.

14.2.2 Material Preparation

GO powder was dispersed and sonicated for 2 h (40 kHz) in ultrapure water to prepare 0.9 mg/ml stock colloidal suspension. HA hydrogel was obtained according to the "wet chemistry" method [7] under microwave irradiation (MW) in consumer microwave oven Samsung M1712NR for 3 min. After washing to achieve neutral pH, the solid precipitate was separated by centrifugation. The moisture content of the resulting HA hydrogel was about 85% by weight.

Alg powder was dissolved in deionized water at 37 °C for 5 h to form 2.0 wt% Alg suspension. Each of 0.083 and 0.83 ml of initial GO colloidal solution were added to the 2% Alg solution to obtain Alg-GO colloidal suspensions with GO concentration of 15 and 150 μg/ml, respectively.

HA hydrogel gradually added to Alg–GO suspensions with weight ratio 1:1 and treated with ultrasound for 1 min, followed by mixtures dripping into 0.25 M calcium chloride solution for 2 h. The formed beads separated from the solution by filtration, thoroughly washed with deionized water and dried at room temperature. In the finished HA–Alg–GO products the GO content was 0.0004 and 0.004 wt% in relation to the HA powder. Subsequently, the samples called as HA-Alg-GO and HA–Alg–GO10, respectively. GO-free sample called as HA–Alg were used as control.

14.2.3 Analytical Methods

The X-ray diffraction (XRD) studies were performed using the Shimadzu XRD-6000 diffractometer with Cu-Kα radiation. The average crystallite sizes (L, nm) along (002) directions were estimated by the Scherrer equation. The hexagonal lattice parameters a and c was calculated using the Miller indexes [8, 9]. Fourier transform infrared spectra (FTIR) were obtained using PerkinElmer Spectrum BX spectrometer in the range 400–4000 cm^{-1} (at 1 cm^{-1} resolution) for the samples pressed into the pellets of KBr. The Ca/P ratio was determined using the energy-dispersive X-ray fluorescence spectrometer ElvaX Light SDD.

Swelling behavior The swelling was quantified by measuring the changes in sample weight as a function of sample immersion time in phosphate buffered saline (PBS) [10, 11] at pH = 7.33. Each sample of lyophilized experimental beads had mass of 0.1 g. The swollen samples were weighed after blotting with a filter paper to remove the surface liquid. The swelling ratio (S_w) was calculated using the appropriate equation [12].

Determination of drug release kinetics The GO content influence on the drug release was investigated, using the pharmaceutical 0.05 w % CH solution which was used as a model drug. The 0.19 g of each HA–Alg, HA–Alg–GO and HA–Alg–GO10 sample placed into 6.0 ml of PBS (pH = 7.4) and incubated at 36.5 °C with continues shaking. The rate of drug release from composites was determined by taking of 600 μl aliquots of PBS from each experimental tube after 1, 2, 24, 48, 72, 96 and 120 h. An equal volume of the fresh medium then was added back to maintain a constant initial volume in the tubes. The drug release kinetics studied using high-performance liquid chromatography (HPLC; Angilent Technologies 1200, detector with UV-Vis Abs, detection at $\lambda = 280$ nm, column C18 (Zorbax SB-C18 4.6 × 150 mm, 5 μm)). As mobile phase was used 68% of 0.05 M potassium hydrogen phosphate buffer with 0.2% triethylamine (pH = 3.0 at 21 °C); 32% of acetonitrile. Isostatic treatment applied at a rate of eluent elution 2 ml/min and temperature of analytical column 40 °C.

Compressive strength study To measure the strength under the uniaxial compression, the specimens of HA–Alg, HA–Alg–GO and HA–Alg–GO10 were prepared in the form of tablets with diameter 5 mm and thickness 2.4–2.6 mm by cold pressing of granulated powder in molds using the hydraulic press (at ~100 MPa). Mechanical properties of the specimens under loading were performed using the original testing machine to determine the elastic modulus and tensile strength [13].

14.3 Results and Discussion

TEM investigation TEM images and electron diffraction (ED) patterns of the composites (Fig. 14.1) indicate the formation of needle-like HA crystals. Agglomeraates of crystallites have a size of 100 nm and more. ED confirmed that the phase composition of the studied material consists of the single phase—HA. An increase in the content of GO reduces the degree of crystallinity of the material, giving the amorphous structure. This fact confirms by ED—the contrast of the diffraction ring (002) decreases. Figure 14.1b, c show uniformly distributed GO particles in the composite material structure. If, according to previous DLS result, the average GO particles size in aqueous solution is about 1 μm, then in Alg–GO suspension the GO nanoparticles are of 20–30 nm in size. It is obvious that Alg solution in a complex with ultrasound treatment is a favorable environment for uniform immobilization and holding of GO nanoparticles.

XRD study As noted above, the HA hydrogel as a basis for the formation of composite materials, was synthesized under influence of MW irradiation. The energy-dispersive X-ray fluorescence measurements have shown that the Ca/P atomic ratio in the obtained product is 1.65, while for the stoichiometric HA this ratio is 1.67. This is evidence of the formation of calcium deficient HA with a small amount of carbonate component. This fact is confirmed by further FTIR studies that demonstrate the presence of the carbonate absorption bands at 870 and 1450 cm^{-1} [14] and indicate the formation of B-type carbonate apatite. Figure 14.2 shows the diffractograms from the investigated composite materials.

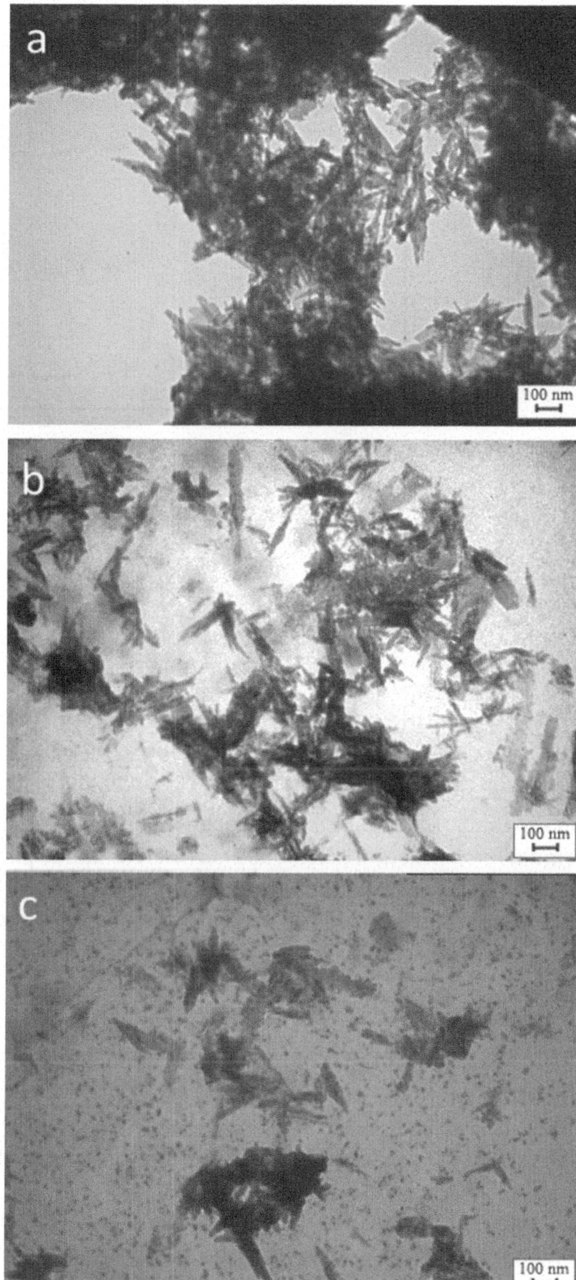

Fig. 14.1 TEM images and ED spectra of the samples: **a** HA–Alg; **b** HA–Alg–GO; **c** HA–Alg–GO10

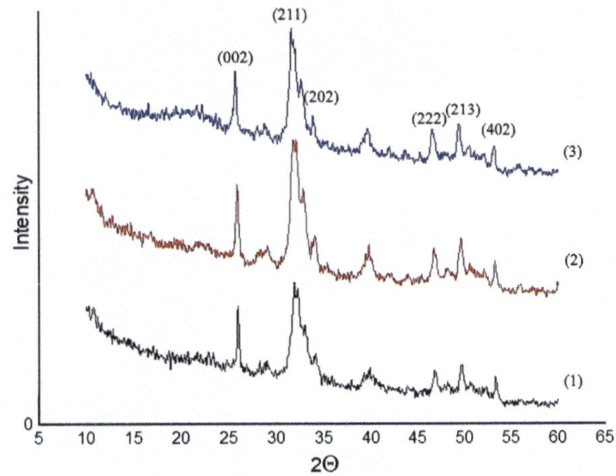

Fig. 14.2 XRD patterns of experimental samples: HA–Alg (1), HA-Alg-GO (2) and HA-Alg-GO10 (3)

Table 14.1 HA crystal structure parameters of the experimental composites

Sample	Crystallite sizes for (002) peak (nm)	HA crystal lattice parameters (nm)	
		a	c
HA–Alg	30.76	0.935	0.684
HA–Alg–GO	28.80	0.938	0.687
HA–Alg–GO10	30.19	0.943	0.690

The average crystallite sizes for all experimental samples, calculated according to the Scherer formula for diffraction peaks (002), as well as lattice parameters *a* and *c*, calculated in planes (002) and (211), are presented in Table 14.1. According to the obtained data, increasing the content of GO increases the parameter *a* of the crystal lattice.

FTIR study The characteristic IR bands for the HA–Alg sample are shown in Fig. 14.3.

Also for the comparison the additional data on the vibrations in pure HA and Alg (Table 14.2), obtained in the frame of this work are given.

The data for HA–Alg sample shows that the main bands of vibrations in the spectra belong to HA and Alg and are not significantly differ for those ones of pure HA and Alg (Table 14.2).

Finally, after adding GO in different content all characteristic peaks of the HA–Alg composite material take place in the FTIR absorption spectra (data not shown). This confirms the absence of chemical interaction between the GO and HA–Alg molecular structures and, thus, one can assume the effect of physical adsorption between them.

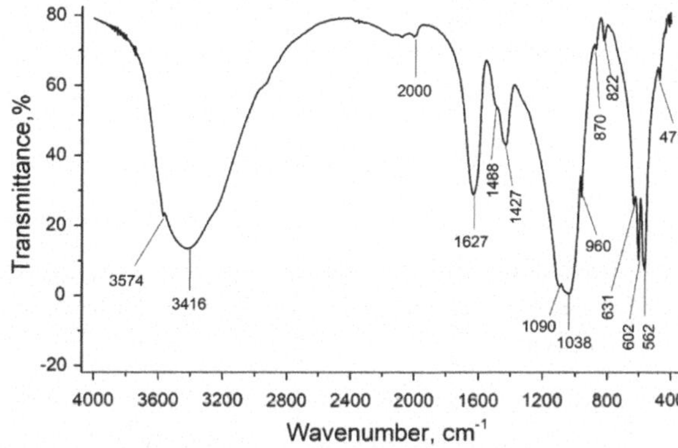

Fig. 14.3 FTIR spectra of HA–Alg sample

Table 14.2 Characteristic IR bands of HA–Alg sample

Sample/vibrational frequency (cm^{-1})			Assignments
HA–Alg	Alg	HA	
471			ν2 O–P–O in PO_4
562	572	564	ν4 PO_4 [15]
602		602	ν4 PO_4
631		630	OH librational (νL) mode [16]
870			ν2 CO_3^{2-}
960		962	ν1 P–O in PO4 [15]
1038		1044	ν3 P–O in PO_4 [15]
1090		1092	ν3 PO_4
1427	1422		COO stretching mode [17]
1488		1450	ν1 CO_3
1627	1626		ν2 H–O–H; –C=O [17]
3574	3438	3428	OH stretching (νs) mode [16]

Swelling and relative humidity studies The swelling and relative humidity ratio were determined to evaluate the ability of the experimental samples to retain their beads shape and to adsorb water. The relative humidity of the samples, dried over 2 days at 37 °C, was calculated. Relative humidity was determined as the degree of weight loss of samples after lyophilization, compared with their weight after drying at 37 °C (Table 14.3). The studies were conducted triplicates.

The results show that the addition of GO reduces the degree of the composites swelling. This fact can be explained by an increase in their density and hardness, as shown below by the results of the mechanical properties study. Comparison of HA–Alg–GO and HA–Alg–GO10 shows that an increase in the content of GO by 10 times leads to an increase in the swelling rate for about 8% and about 43% compared to GO-free sample. With the GO addition also increases the relative humidity, caused

Table 14.3 Swelling ratio and relative humidity of the experimental composites

Sample	Swelling (%)	Relative humidity (%)
HA–Alg	103.6	6.5
HA–Alg–GO	53.4	6.7
HA–Alg–GO10	61.0	6.8

by the presence of water molecules in the structure of the composite. These facts can be explained by the ability of the GO itself to adsorb and retain fluid through hydrogen bonds.

HPLC study As a reference drug for HPLC drug release studies a CH-active against Gram-positive and Gram-negative organisms [18, 19] was used. The CH was introduced into composites by the method of saturation. For this, 0.2 g of each sample in form of lyophilized beads (HA–Alg, HA–Alg–GO and HA–Alg–GO10) were immersed in 2 ml of 0.05% CH solution for 120 min with following drying at 37 °C. To evaluate the CH release kinetics, each drug contained sample was placed to the tube with 10 ml of PBS. The tubes with samples were agitated (rpm 115, temperature 37 °C) for 120 h. The quantity of released CH was plotted against the incubation time (Fig. 14.4).

The results showed that the release of CH from the HA–Alg sample occurs almost uniformly throughout the study period and lasts for 72 h. After 72 h, the concentration of CH in SBF begins to decrease due to dilution, which occurs by adding the aliqoutes of fresh PBS solution after each sampling. In Fig. 14.4 it is possible to distinguish three main stages of release of the drug from GO-containing specimens: up to 4 h, from 4 to 24 h, from 24 to 120 h. At the first stage there is a release of superficially adsorbed CH. The sample with higher GO content (HA–Alg–GO10) shows a higher

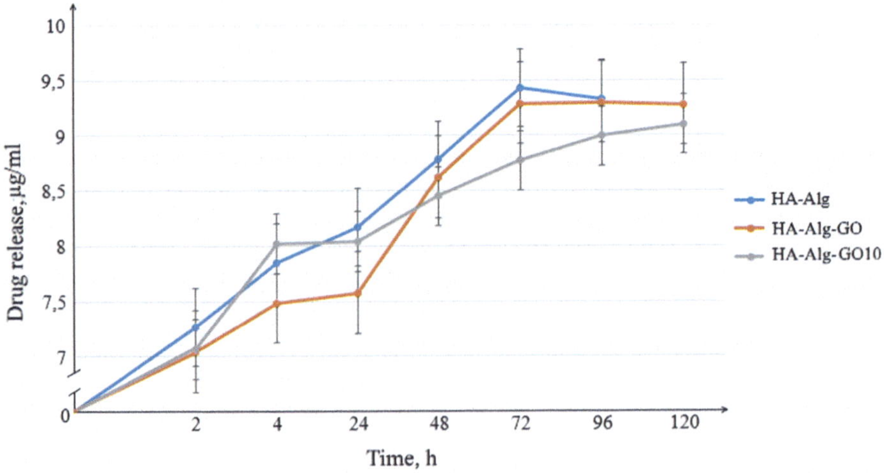

Fig. 14.4 Concentration of CH in PBS (μg/ml), released from the experimental samples for 120 h

rate and the amount of drug marketed in the solution. The fact of the largest CH release at this stage can be attributed to the ability of GO to keep fluid due to the hydrogen bonds, which are formed with the participation of functional groups of GO. This sample has a high concentration of GO, both throughout the sample volume and on the surface. During the second period (4–24 h), the swelling of the composites occurs due to their hydration, which creates conditions for further diffusion of the drug from the samples. In the third stage, the CH release from HA–Alg–GO and HA–Alg–GO10 lasts 96 and 120 h, respectively. The higher drug release speed and, at the same time, the shorter drug release duration for the HA–Alg–GO compared to HA–Alg–GO10 can be explained by the structure of the samples. Due to the low concentration of GO, the HA–Alg–GO is less dense and therefore contributes to diffusion. In the third stage, HA–Alg–GO is close to HA–Alg by the release parameters. At the same time, the sample HA–Alg–GO10 in the third stage demonstrates a slow and prolonged release of the CH.

Compressive strength study The results of the studies of compressive strength of different types of HA-based composites are presented in Table 14.4 and Fig. 14.5.

As one can see from Fig. 14.5, the addition of GO into HA–Alg slightly increases the compressive strength of the sample. The maximal relative deformations ε_{destr} before destruction lies within range 0.10–0.14 for three types of samples. In general, such increase of strength for HA–Alg–GO sample is related to high aspect ratio and

Table 14.4 Strength properties of HA-based composites

Sample	Density (g/cm^3)	Maximal relative deformation (εdestr)	Strength σ_c (MPa)	Young's modulus E (MPa)
HA–Alg	2.03	0.10	166	793
HA–Alg–GO	2.10	0.14	179	840
HA–Alg–GO10	2.11	0.12	177	1325

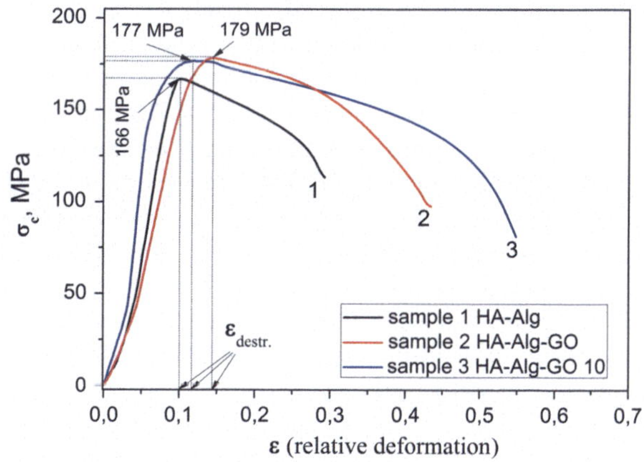

Fig. 14.5 Strength σ_c of the HA–Alg and HA–Alg–GO beads measured at uniaxial compression

excellent mechanical properties of graphene (a specific surface area is 2630 m^2/g, a tensile strength is 130.5 GPa and a Young modulus is about 1.0 TPa) [20–22]. On the other hand, it is known [23] that the interfacial adhesion between the composite components determines the load, which will be transferred between them. Thus, the increase of the strength for HA–Alg–GO may be also explained by well bonding of GO nanoparticles to the HA–Alg matrix, that leads to efficient stress transfer at the GO. The Young's modulus (*E*) estimation has shown that HA–Alg–GO10 sample had a fairly high Young's modulus (see Table 14.4) in comparison with initial HA-Alg sample.

The improvement of the mechanical properties of the composites is mainly attributed to compatibility and specific interactions of Alg hydroxyl groups with GO functional groups (hydroxyl, epoxide). The improved tensile properties are ascribed to GO and Alg hydrogen bonding. Structure of the Alg–GO composite leads to more uniform stress distribution and minimizes the presence of stress concentration center, thus significantly increase the mechanical properties of the nanocomposite [6]. In addition, the chemical cross linking between GO sheets could be elevated by adding divalent ions Ca^{2+} [24] during the beads formation, leading to enhancement in mechanical properties of the composites.

14.4 Conclusions

Bioactive composite material based on hydroxyapatite (HA), sodium alginate (Alg) with different content of graphene oxide (GO) was synthesized by the "wet chemistry" method in form of beads and characterized in details by various experimental techniques. This form of material may provide more effective filling of defects with complex geometry with minimal gap between the bone and the implant, and the space between the granules contributes to the formation of new bone tissues throughout the volume of the implant.

The obtained XRD and energy-dispersive X-ray fluorescence data indicate the formation under influence of MW irradiation of calcium deficient (Ca/P = 1.65) HA with a small carbonate content. TEM evidenced an uniform distribution of GO particles within Alg matrix after ultrasound treatment. Introduced the GO nanoparticles, as well as Ca^{2+} ions, as cross-linker of Alg macromolecules by the beads formation, lead to enhancement of the composites mechanical properties. HA–Alg–GO10 sample with GO content of 0.004% in relation to the HA powder has a much higher Young's modulus (1325 MPa) in comparison with GO-free HA–Alg (793 MPa), as well as steel sample of the same size (~706 MPa). The addition of GO leads to an increase in the density of the material from 2.03 to 2.11 g/cm^3. Degree of swelling in the PBS decreases from 103 to 61% for HA–Alg and HA–Alg–GO10, respectively. The ability of the dried samples to absorb moisture from the environment (relative humidity) increases after adding GO by 0.3%. Chlorhexidine bigluconate release from GO containing samples lasts for 48 h longer according to HPLC studies.

References

1. A. White, S. Best, I. Kinloch, Int. J. Appl. Ceram. Technol. **4**(1), 1–13 (2007)
2. T. Boontheekul, H.J. Kong, D.J. Mooney, Biomaterials **26**, 2455–2465 (2005)
3. K.Y. Lee, D.J. Mooney, Prog. Polym. Sci. **37**, 106–126 (2012)
4. Y.X. Huang, X.C. Dong, Y.X. Liu, L.J. Li, P. Chen, J. Mater. Chem. **21**, 12358–12362 (2011)
5. A.K. Geim, K.S. Novoselov, Nat. Mater. **6**, 183–191 (2007)
6. M. Ionita, M.A. Pandele, H. Iovu, Carbohyd. Polym. **94**(1), 339–344 (2013). https://doi.org/10.1016/j.carbpol.2013.01.065
7. A.S. Stanislavov, L.F. Sukhodub, L.B. Sukhodub, V.N. Kuznetsov, K.L. Bychkov, M.I. Kravchenko, Ultrasonics—Sonochemistry **42**, 84–96 (2018)
8. H.P. Klug, L.E. Alexander, *X-Ray Diffraction Procedures: For Polycrystalline and Amorphous Materials* (Wiley, New York, 1974)
9. V.M. Kuznetsov, L.B. Sukhodub, L.F. Sukhodub, Journal of nano- and electronic physics **6**(4), 04039–04043 (2014)
10. A.T. Palasz, P.B. Breña, J. De-la-Fuente, A. Gutiérrez-Adán, Theriogenology **9**, 1461–1470 (2008)
11. L.F. Sukhodub, L.B. Sukhodub, O. Litsis, Yu. Prylutsky, Mat. Chem. Phys. **217**, 228–234 (2018). https://doi.org/10.1016/j.matchemphys.2018.06.071
12. Y. Han, Q. Zeng, H. Li, J. Chang, Acta Biomater. **9**, 9107–9117 (2013). https://doi.org/10.1016/j.actbio.2013.06.022
13. L. Vovchenko, O. Lazarenko, L. Matzui, Yu. Perets, A. Zhuravkov, V. Fedorets, F. LE Normand, Phys. Status Solidi A **211**(2), 336–341 (2014)
14. C.W. Chen, C.S. Oakes, K. Byrappa, R.E. Riman, K. Brown, K.S. TenHuisen, V.F. Janas, J. Mater. Chem. **14**, 2425–2432 (2004)
15. B.O. Fowler, Inorg. Chem. **13**, 194–207 (1974)
16. J. Arends, J. Christoffersen, M.R. Christoffersen, H. Eckert, B.O. Fowler, J.C. Heughebaert, G.H. Nancollas, J.P. Yesinowski, S.J. Zawacki, J. Cryst. Growth **3**, 512–532 (1987)
17. V.K. Malesu, D. Sahoo, P.L. Nayak, Int. J. Appl. Biol. Pharmaceut. Technol. **2**(3), 402–411 (2011)
18. A. Kovtun, D. Kozlova, K. Ganesan, C. Biewald, N. Seipold, P. Gaengler, W.H. Arnold, M. Epple RSC Adv. **2**, 870–875 (2012)
19. S. Garner, M. Barbour, Oral Dis. **21**(5), 641–644 (2015). https://doi.org/10.1111/odi.12328
20. C. Lee, X. Wei, J.W. Kysar, J. Hone, Science **321**, 385–388 (2008)
21. J. Tucek, P. Błonski, J. Ugolotti, A.K. Swain, T. Enoki, R. Zboril, Chem. Soc. Rev. **47**(11), 3899–3990 (2017)
22. X. Wang, G. Shi, Phys. Chem. Chem. Phys. **17**, 28484–28504 (2015)
23. S. Fu, X. Feng, B. Lauke, Y. Mai, Composites B **39**, 933 (2008)
24. X. Jiang, Y. Ma, J. Li, Q. Fan, W. Huang, J. Phys. Chem. C **114**, 22462–22465 (2010)

Chapter 15
Effect of Surface Modification of Sputtered Ta_2O_5 Magnetron Ceramic Coatings on the Functional Properties of Antigen-Presenting Cells In Vitro Tests

S. Yakovin, S. Dudin, A. Zykova, V. Safonov, A. Goltcev, T. Dubrava, and I. Rassokha

Abstract The effect of surface treatment of Ta_2O_5 nanostructured coatings by argon ions and electron beam on the functional potential of antigen-presenting cells of the monocyte-phagocytic system has been studied. The adhesive potential, indicators of phagocytic and metabolic activity of the studied cells depending on the surface properties of magnetron sputtered tantalum pentoxide coatings were analyzed. Electron irradiation process led to the stimulation of adhesive potential, phagocytic and metabolic activity of cells on the Ta_2O_5 coated surfaces. On the contrary, the surface treatment by argon ions significantly reduced the functional activity of the studied cells.

15.1 Introduction

The biological response to the artificial material is determined by a complex of factors. The important role is played not only by the physico-chemical characteristics of the biomaterial surface, but also by the total cells response on the cell/biomaterial interface [1, 2]. It is known, that first immune response on the implantation was recorded by cells of monocytic-phagocytic system of organism [3–5]. Macrophages play important role in the immune organism reactions on the implanted materials—catheters, stents, femoral and oral implants. Immune response on the artificial implants leads to the postoperative complications, inflammatory processes and the risk of repeated surgery operations. Inflammatory and anti-inflammatory reactions

S. Yakovin · S. Dudin (✉) · A. Zykova · V. Safonov
V. N. Karazin Kharkiv National University, Kharkiv, Ukraine
e-mail: dudin@karazin.ua

A. Zykova · V. Safonov
National Science Center "Kharkov Institute of Physics and Technology", Kharkiv, Ukraine

A. Goltcev · T. Dubrava · I. Rassokha
Institute for Problems of Cryobiology and Cryomedicine NASU, Kharkiv, Ukraine

A. D. Pogrebnjak et al. (eds.), *Nanomaterials in Biomedical Application and Biosensors (NAP-2019)*, Springer Proceedings in Physics 244,
https://doi.org/10.1007/978-981-15-3996-1_15

can be regulated by biomaterial properties and surface modifications [6, 7]. The aim of present study was to investigate the effect of surface treatment by argon ions and electron beam on the structure and surface properties of tantalum pentoxide (Ta_2O_5) coatings deposited by reactive magnetron sputtering method and further correlation of the surface characteristics with immune cells response.

15.2 Materials and Methods

Magnetron, inductively coupled plasma (ICP) source, and ion source were included in the technological system for synthesis of coatings [8]. The ICP source was designed to create activated flow of reactive gas molecules, as well as flow of low-energy ions and electrons. The ICP source was used to clean the surface of the samples before deposition and, in combination with the magnetron, for reactive deposition of metal oxides and nitrides. The ICP source was located inside the vacuum chamber that allows to choose the optimal ratio between the distances from the magnetron and the plasma source to the samples.

Volt-ampere characteristics of magnetron discharge with tantalum target in argon for different values of reactive gas flow were previously presented [9]. Figure 15.1 shows the dependence of the magnetron discharge voltage and current on the oxygen flow. With the oxygen flow rise, the voltage increases, and the discharge current falls in the case of the tantalum target.

With the oxygen flow increasing, the discharge shifts to the poisoning mode with reduced deposition rate. On the other hand, non-stoichiometric coatings are formed in the case of insufficient oxygen flow. The technological process allows deposition of stoichiometric transparent coatings in the metallic mode of the magnetron target far from the poisoning mode.

The ion source "Radical M" of medium energies (0.5–3 keV) was used to clean and activate the surface of the samples before the coatings deposition, as well as

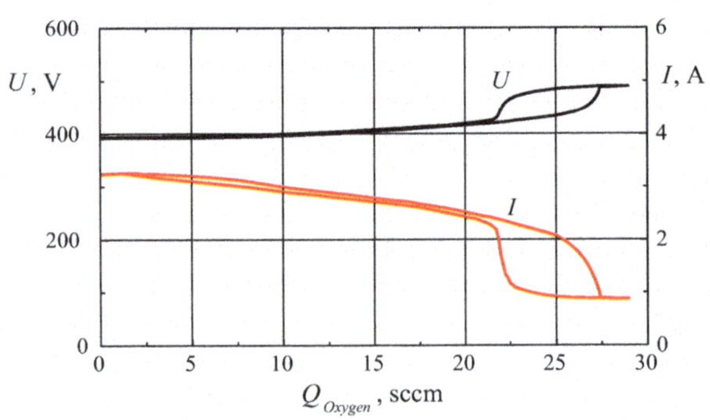

Fig. 15.1 Dependencies of the magnetron discharge voltage and current on the oxygen flow Q_{Oxygen} in the case of the tantalum target

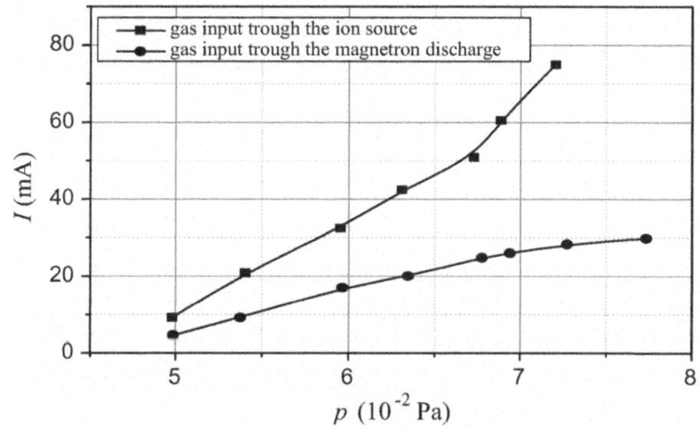

Fig. 15.2 Dependence of the discharge current of the ion source on the gas pressure in the process chamber upon gas input into the magnetron discharge or to the ion source

ion assistance in the process of film synthesis. There is a possibility to change the density, morphology and stoichiometrical composition of the coatings during deposition process by controlling the energy, current density and composition of the ion beam. The influence of gas pressure and flow rate on the current-voltage characteristics was analyzed.

The plots shown in Fig. 15.2 correspond to the technological regime of coatings deposition starting from the minimum gas pressure of 5×10^{-2} Pa for the gas input to the magnetron discharge. As can be seen from the figure, it is possible to control the current density of ions in wide range regulating the amount of gas passing through the ion source or through the magnetron discharge. The allowable range of the current density regulation reaches from 0.1 to 10 mA/cm^2 at a constant average energy of ions.

Ta_2O_5 coatings were deposited on Petri dishes glass substrates in a high vacuum pumping system with a base pressure of about 10^{-4} Pa by ion-assisted magnetron sputtering. Oxygen for the reactive deposition was delivered through the ICP plasma source $Q = 60$ sccm, magnetron voltage was $U = 700$ V, magnetron current was about $I = 5.7$ A.

For other substrates, the deposition process was carried out with simultaneous bombardment of the growing film by argon ions using the ion source. The ion source parameters were as follows: magnetic coil current 1.5 A, ion acceleration voltage 2.5 keV, ion source current 30 mA.

In electron bombardment research the electron beam was created by electron gun of type UL-119. The electrons energy was 20 keV, current density on the sample surface 14 μA/cm^2, the irradiation time 1500 s.

The coating thickness was measured by Calotest. The roughness parameters of the oxide ceramic coatings were evaluated by profilometer Hommel. The surface morphology and topography were observed by scanning electron microscope JSM 5500 LV. The chemical composition of the coatings was analyzed by energy dispersive X-ray (EDX) spectroscopy (Oxford Link ISIS 300). Advancing contact angles and wettability of the coated surfaces were evaluated by tensiometric measurements (Kruss 12). The total surface free energy, polar and dispersion parts were estimated by Owens-Wendt-Rabel-Kaeble methods.

Adhesive and proliferative activity of immune cells was evaluated by standard protocols. Phagocytes of peritoneal cavity of mice CBA/H line were isolated and cultivated on the coated substrates. The phagocytes cells were cultured with density 1×10^7 cells per dish on the control and oxide coated glass Petri dishes in an CO_2 incubator (5% CO_2) at 37 °C and 95% humidity. Cell adhesive potential was evaluated after 30 min of cultivation on coated/uncoated substrates. After the indicated time, non adhesive cells were removed by washing the tested substrates twice with Hanks solution. Adherent cells were fixed with methanol for 5 min and counted. Visual control of the cell cultures was carried out using a light phase contrast microscope «Primo Star» («Zeiss», Germany) and inverted microscope «Axiovert 40C» («Zeiss», Germany). Statistical analysis of test results was performed using the nonparametric Mann-Whitney U test. The difference was considered statistically significant at $P < 0.05$.

15.3 Results and Discussion

The deposition process carried out with simultaneous bombardment by argon ions was resulted in the structural changes of the growing films.

Also, the changing of surface topography and roughness parameters of the oxide ceramic coatings was evaluated by scanning electron microscopy. The coating thickness and cross-section structure were obtained from the SEM cross-section measurements (Fig. 15.3).

a) b)

Fig. 15.3 SEM cross-section images of as-deposited oxide ceramic coatings: **a** Ta_2O_5 coatings, **b** Ta_2O_5 coatings deposited with simultaneous bombardment by argon ions

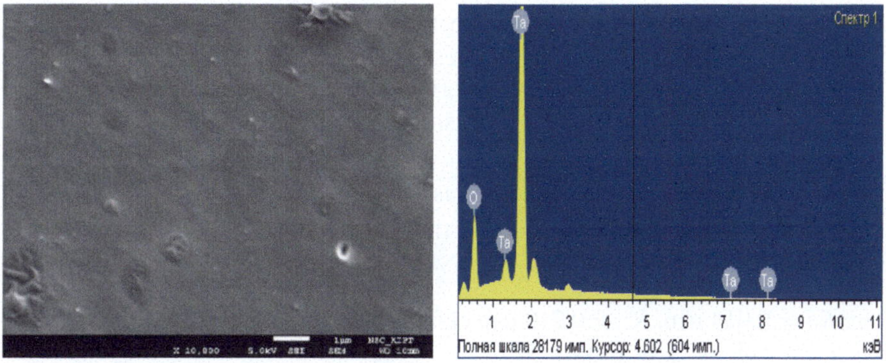

a) b)

Fig. 15.4 SEM surface images and EDX spectra of Ta_2O_5 coatings: **a** Ta_2O_5 coatings surface, **b** EDX spectra

Scanning electron microscopy images of the magnetron sputtered Ta_2O_5 ceramic coatings have revealed a relatively flat surface with no cracks. EDX spectra demonstrate the stoichiometric composition of magnetron sputtered tantalum pentoxide coatings (Fig. 15.4).

The difference of structure parameters of Ta_2O_5 coatings and Ta_2O_5 coatings deposited with simultaneous bombardment by argon ions was described in [10]. The coatings had a nano-globular structure with characteristic sizes of structural elements of about 15 nm in the case of tantalum pentoxide films without additional treatment and 20 nm with argon ion bombardment of the growing films.

The thickness and roughness parameters of as-deposited coatings, after simultaneous bombardment by argon ions and electron beam post treatment are presented (Table 15.1).

The roughness parameters were principally changed during argon ions bombardment: roughness $Ra = 0.017$ μm, $Rz = 0.240$ μm in the case of Ta_2O_5 coatings, $Ra = 0.035$ μm, $Rz = 0.490$ μm in the case of Ta_2O_5 coatings deposited with simultaneous bombardment by argon ions. In contrast, the post-treatment processing by electron beam created by electron gun was not resulted in the principal changing of surface and structural parameters of oxide ceramic coatings. The structure of electron beam irradiated coatings was similar to as-deposited coatings, roughness parameters

Table 15.1 Thickness and roughness parameters

Substrate/coating	Thickness (μm)	Roughness *Ra* (μm)	Roughness *Rz* (μm)
Glass/Ta_2O_5	2.09	0.017	0.240
Glass/Ta_2O_5 (electron beam irradiation)	2.09	0.014	0.210
Glass/Ta_2O_5 (argon ions bombardment)	1.62	0.035	0.490

Table 15.2 The surface free energy SFE and its polar and dispersion parts by Owens-Wendt-Rabel-Kaeble method

Substrate/coating	γ (mN/m)	γ^d (mN/m)	γ^p (mN/m)
Glass	56.28	30.31	25.97
Glass/Ta_2O_5	43.96	31.06	12.90
Glass/Ta_2O_5 (electron beam irradiation)	48.37	32.16	16.21
Glass/Ta_2O_5 (argon ions bombardment)	41.11	30.03	11.08

were $Ra = 0.014$ µm, $Rz = 0.210$ µm close to the case of Ta_2O_5 coatings in good agreement with previous results [9].

Variation of structural parameters may lead to a change of surface characteristics and functional properties of deposited coatings. X-ray diffraction profiles of as-deposited Ta_2O_5coatings demonstrated an amorphous nature for as-deposited magnetron sputtered Ta_2O_5 coatings, no peaks were observed.

XPS survey spectra of the ceramic Ta_2O_5 coatings deposited by magnetron sputtering method were previously analyzed [11]. All spectra consist of well-defined XPS lines of Ta 4f, 4d, 4p and 4s; O 1s; C 1s. Ta 4f doublets are typical for Ta_2O_5 coatings with two main peaks. Ta 4f doublets are typical for Ta_2O_5 coatings and have two peaks: Ta 4f7/2 at ~26.5 eV and Ta 4f5/2 whose binding energy is higher by 1.9 eV. The Ta 4f lines of the deposited films are in a good agreement with the Ta 4f doublet representative of the Ta-O bond in Ta_2O_5. The Ta/O ratio estimated from the spectra was about 0.4 for all investigated coatings. The O 1s peaks are centred at binding energies 530.6 eV for the deposited Ta_2O_5 coatings.

The surface free energy (SFE) plays an important role in the mechanism of cell/biomaterial response [12]. The SFE and its polar and dispersion parts were estimated by Owens-Wendt-Rabel-Kaeble method for liquid system: α-bromonaphthalene–formamide–ethylene glycol–diidomethane–glycerol–water (Table 15.2) at temperature 20 °C.

The data demonstrate that the surface free energy of Ta_2O_5 coatings was in the range 40-50 mN/m and SFE polar parts were in the range 11–16 mN/m. Tantalum pentoxide ceramic coatings deposited with simultaneous bombardment by argon ions possess the minimal values of SFE.

The data demonstrate (Table 15.2) that the properties of oxide ceramic coatings shift in the more hydrophilic region and values of the surface free energy increase for the oxide coated substrates after electron beam irradiation post-treatment.

The topographic and physico-chemical characteristics of the surface act as factors that modulate the adhesive potential of cells [13]. The adhesive potential of phagocytes cells of the peritoneal cavity of CBA/H mice depending on the surface properties of tantalum pentoxide coatings was analysed. Table 15.3 shows the results of studying the adhesive potential of the phagocytes of the peritoneal cavity of CBA/H mice on glass (control) and Ta_2O_5 coatings before and after surface treatment by electrons and argon ions.

Table 15.3 Adhesive potential of the phagocytes cells on glass and Ta_2O_5 coated substrates

Substrate/coatings	Adhesive potential of phagocytes cells, %	Percent of adhesive cells, %
Glass (control)	3.76 ± 0.34	100
Glass/Ta_2O_5	$2.66 \pm 0.20^*$	70.74^*
Glass/Ta_2O_5 (electron beam irradiation)	$4.99 \pm 0.43^*$	132.70^*
Glass/Ta_2O_5 (argon ions bombardment)	$2.61 \pm 0.14^*$	69.41^*

*The difference statistically significant to control, $P < 0.05$

The minimal values of cell adhesive potential in the case of Ta_2O_5 coatings deposited with simultaneous bombardment by argon ions were observed (Table 15.3). Adhesive potential of phagocyte cells on the Ta_2O_5 coated surface after electron irradiation was 30% higher than on the uncoated surface (glass substrates). Thus, the electron irradiation process led to the stimulation of adhesive potential, phagocytic and metabolic activity of cells on the Ta_2O_5 coated surfaces. On the contrary, the surface treatment by argon ions significantly reduced the functional activity of the antigen-presenting cells. The effect of material surface modification on the several inflammatory events in vivo was previously observed [14, 15]. Macrophages preferentially accumulate on rough surface in vitro [16]. A correlation between a decreasing number of interfacial ED1 positive macrophages and increasing surface roughness was found after one week, but not after 6 and 12 weeks. In present study, the surface roughness increasing after argon ions bombardment did not lead to the significant increasing of adhesive potential of phagocyte cells in vitro tests. On the contrary, the changing of surface free energy, polar part parameters, and shift to more hydrophilic region can effect on the increasing of adhesive potential, phagocytic and metabolic activity of cells after electron irradiation process.

15.4 Conclusions

The deposition process carried out with simultaneous bombardment by argon ions and post-treatment by electron irradiation were resulted in structural changes of the growing films. The change of structural parameters leads to changing of coatings surface characteristics.

The roughness parameters were significantly increased in the case of Ta_2O_5 coatings deposited with simultaneous bombardment by argon ions. In contrast, the post-treatment by electron beam was not resulted in the principal change of roughness parameters. The structure of electron beam irradiated coatings was similar to as-deposited Ta_2O_5 coatings.

Tantalum pentoxide ceramic coatings deposited with simultaneous bombardment possess the minimal values of SFE. The Ta_2O_5 coatings demonstrate more hydrophobic properties after surface treatment by argon ions. On the contrary, properties of tantalum pentoxide ceramic coatings shift in the more hydrophilic region and values of the surface free energy increase for the coated substrates after electron beam irradiation post-treatment.

It was shown that the argon ions bombardment during tantalum pentoxide coating deposition process reduces adhesive and proliferative potential of antigen-presenting cells. On the contrary, electron irradiation process led to the stimulation of adhesive potential, being a primary link in providing functional (phagocytic and metabolic) activity of these cells on the Ta_2O_5 coated surfaces. The change of surface free energy, polar part parameters, and shift to more hydrophilic region can lead to the increasing of functional activity of cells of monocytic-phagocytic system after electron irradiation process.

Results demonstrate that surface modification can affect the adhesive potential of immune cells that is important for further clinical applications.

Acknowledgements The research was supported by the international scientific cooperation program between National Academy of Science of Ukraine and National Academy of Science of Belorussia # 17-03-18.

References

1. T.A. Blaine, J. Bone Jt. Surg. **78-A**(8), 1181 (1996)
2. C.M. Takebe, A. Champagne, A. Diener, B. Nebe, F. Luthen, P. Becker, U. Beck, H.G. Neumann, J. Rychly, Biomaterials **26**(4), 383 (2005)
3. M. Bartneck, K.H. Heffels, Y. Pan, M. Bovi, G. Zwadlo-Klarwasser, J. Groll, Biomaterials **33**, 4136 (2012)
4. C.A. Janeway, R. Medzhitov, Annu. Rev. Immunol. **20**(1), 197 (2002)
5. P. Thomsen, C. Gretzer, Curr. Opin. Solid State Mater. Sci. **5**, 163 (2001)
6. C.R. Jenney, J.M. Anderson, J. Biomed. Mater. Res. **49**, 435 (2000)
7. S. MacLauchlan, E.A. Skokos, N. Meznarich, D.H. Zhu, S. Raoof, J.M. Shipley, R.M. Senior, S. Offenbacher, K. Ishibashi, L.F. Cooper, J. Biomed. Mater. Res. **64 A**(2), 207 (2003)
8. S. Yakovin, S. Dudin, A. Zykov, V. Farenik, Problems At. Sci. Technol. Ser. "Plasma Physics" **71**(1), 152 (2011)
9. A. Zykova, V. Safonov, A. Goltsev, T. Dubrava, I. Rossokha, N. Donkov, S. Yakovin, D. Kolesnikov, I. Goncharov, J. Phys: Conf. Ser. **700**, 012027 (2016)
10. A. Zykova, V. Safonov, A. Goltsev, T. Dubrava, I. Rossokha, J. Smolik, R. Rogovska, S. Yakovin, D. Kolesnikov, I. Sudzhanskaya, I. Goncharov, Surf. Coat. Technol. **301**, 114 (2016)
11. N. Donkov, E. Mateev, A. Zykova, V. Safonov, D. Kolesnikov, I. Goncharov, I. Sudzhanskaya, S. Yakovin, J. Phys: Conf. Ser. **558**, 012036 (2014)
12. N.J. Hallab, K.J. Bundy, K. O'Connor, R.L. Moses, J.J. Jacobs, Tissue Eng. **7**(1), 55 (2001)
13. K. Anselme, P. Linez, M. Bigerelle, D. Le Maguer, A. Le Maguer, P. Hardouin, H.F. Hildebrand, I. Alain, J.M. Leroy, Biomaterials **21**, 1567 (2000)

14. A.S. Shanbhag, J.J. Jacobs, J. Black, J.O. Galante, T.T. Glant, J. Biomed. Mater. Res. **28**, 81 (1994)
15. M. Werthen, A. Sellborn, M. Kalltorp, H. Elwing, P. Thomsen, Biomaterials **8**, 827 (2001)
16. A. Rich, A.K. Harris, J. Cell Sci. **50**, 1 (1981)

Chapter 16
Features of Bacterial Cellulose Hydroxyapatite Nanocomposites Obtained by Two Different Techniques

A. Talipova, A. Kistaubayeva, A. Pogrebnjak, A. Turlybekuly, I. Savitskaya, and S. Saidildina

Abstract Nanocomposite material based on bacterial cellulose and hydroxyapatite was obtained using the biomimetic mineralization method (imitation of the formation of the mineral component in natural bone tissue) and the synthesis of hydroxyapatite nanoparticles in the presence of fibrillar fragments of BC. The biocomposite BC/HA-1 obtained by biomineralization showed high physicochemical parameters compared with the composite obtained by synthesis of hydroxyapatite nanoparticles in the presence of fibrillar fragments of BC (BC/HA-2). The ultimate tensile strength of the composite BC/HA-1 was 92.07 + 1.4 MPa, the composite BC/HA-2 was 76.09 + 1.1 MPa. Young's modulus (the coefficient of proportionality between stress and strain) of the BC/HA-1 composite was 59.04 + 0.6 MPa (BC/HA-1), which is 18.6% higher than that of the BC/HA-2 composite (48.03 + 0.8 MPa).

16.1 Introduction

Attention to the creation of artificial composite materials, close to the natural bone tissue, is growing rapidly and an important problem is to obtain a composite that would be as close as possible in its structure and properties to its natural counterpart. The primary role is given to composite materials based on natural polymers (cellulose, collagen, chitosan, chitin etc.) as a substitute for collagen and hydroxyapatite (HA) nanocrystals as the main mineral component of natural bone tissue. As one of the most common biodegradable and biocompatible natural polymers, cellulose has attracted considerable attention for the application of bone scaffolds.

A. Talipova (✉) · A. Kistaubayeva · I. Savitskaya · S. Saidildina
Al-Farabi Kazakh National University, Almaty, Kazakhstan
e-mail: talipova.aizhan@gmail.com

A. Pogrebnjak
Department of Nanoelectronics, Sumy State University, Sumy, Ukraine

A. Turlybekuly
East Kazakhstan State Technical University named after D. Serikbayev, Oskemen, Kazakhstan

A. D. Pogrebnjak et al. (eds.), *Nanomaterials in Biomedical Application and Biosensors (NAP-2019)*, Springer Proceedings in Physics 244,
https://doi.org/10.1007/978-981-15-3996-1_16

Bacterial cellulose (BC) is a unique natural polymer consisting of fibers with a diameter of 20–175 nm, forming a nano-gel film that has a specific inner surface area of at least 500 m^2/g [1]. BC has unique properties that are absent in plant [2]. Unlike plant cellulose, BC is a chemically pure extracellular product, as it does not contain lignin, hemicellulose, pectin and wax. BC has a high degree of crystallinity, its density is 300–900 kg/m^3, has high mechanical strength (up to 20 MPa), absorbs and holds up to 20 g of water per 1 g of dry polymer [3–6].

Along with other natural polymers, bacterial cellulose has unique physico-chemical and sorption characteristics, which makes it a promising material as the basis for a bone implant.

As a precursor of bone tissue, a composite based on BC and hydroxyapatite nanocrystals (HA), which is biocompatible with living organisms, is considered promising. The possibility of directed synthesis of a BC and the inclusion of various additives in it allows us to consider such materials as a universal biomaterial for bone and tissue engineering in general [7–9].

Due to its composition, structure, and unique physico-biological characteristics, materials based on hydroxyapatite are widely used in medicine in the form of ceramics, cements, and composites. They are used in the restoration of bone and dental tissue defects and spinal surgeries, as eye and ear implants, in bone grafting in maxillofacial operations, in the form of coatings for metal implants, etc.

However, prior to clinical trials, it is necessary to conduct detailed studies of the structure of both individual components and composites as a whole, which will help in explaining the behavior of the material in vitro and in vivo studies.

16.2 Materials and Methods

The bacterial strain *Komagataeibacter xylinus* C-3 was obtained from the collection cultures of the Department of Biotechnology of the Al-Farabi Kazakh National University.

16.2.1 Synthesis of Gel Film BC

The synthesis of cellulose strains of acetic acid bacteria was carried out on nutrient media containing aqueous solutions of yeast extract, glucose, peptone, ethanol and beer wort in concentrations established by optimizing the nutrient medium with a pH of 5.9–6.0. A 48-h culture of acetic acid bacteria grown on a medium containing yeast extract and beer wort in a 1: 1 ratio with 2 wt% glucose 1 vol. % ethanol. Cultivation was carried out at 30 °C for 5 days, after which the cellulose was separated and periodically washed with 0.5–1% aqueous NaOH solution while boiling until the cells were removed. Then, the cellulosic film was washed from the NaOH solution with distilled water, 0.5% acetic acid solution and again distilled water until neutral.

The resulting cellulose was stored as a gel film in distilled water at 5 °C [10]. The biomass of the BC films was determined after preliminary drying in a thermostat with a dry heat at 80 °C to a constant mass of the sample.

16.2.2 Development of Composite BC/HA

Nanocomposite material based on bacterial cellulose and hydroxyapatite was obtained using the biomimetic mineralization method (imitation of the formation of the mineral component in natural bone tissue) and the synthesis of hydroxyapatite nanoparticles in the presence of fibrillar fragments of BC.

Biomimetic mineralization method. For this, a polyvinylpyrrolidone (PVP) solution was prepared by adding 0.1 g of PVP to 500 ml of distilled water. Pre-obtained gel films of BC, were treated in PVP solution at 28 °C for 2 days. Then the PVP-treated BC-membrane was washed several times in distilled water. Next, these films were immersed in 0.1 mol/l $CaCl_2$ at 37 °C for 3 days to activate the hydroxyl group. To induce the formation of HA, modified PVP and activated Ca^{2+} BC were soaked in simulated body fluid (SBF) solution at 37 °C for 5 days. The SBF solution was prepared according to the protocol developed by WAN et al. The resulting HA/BC nanocomposites were washed with distilled water and dried [9].

Synthesis of hydroxyapatite nanoparticles in the presence of fibrillary fragments of BC suspension was carried out by co-disintegrating aqueous suspensions of BC and hydroxyapatite nanoparticles. Disintegration was performed in a laboratory blender for 15 min at a rotation speed of 2000 rpm. Then the volume of the reaction mixture was adjusted to 200 ml with distilled water and a portion of calcium oxide was added with vigorous stirring. The resulting mixture was homogenized for 5 min, and then the synthesis of HA nanoparticles was carried out by adding phosphoric acid (at a given rate of ~1 ml/min) to the resulting aqueous suspension of BC and $Ca(OH)_2$. The reaction was considered complete when reaching pH = 6.8–7.0.

16.2.3 Scanning Electron Microscopy (SEM) Studies

The structure of samples of films of bacterial cellulose and composite was examined by field emission scanning electron microscope JSM-7800F (Jeol, Japan). Prior to the SEM observation, the films were sputter coated with a platinum-palladium alloy (Pt/Pd 80/20).

16.2.4 The Study of the Strength of Films of Bacterial Cellulose and Composite BC/HA

Strength was determined on an "Instron" machine with a uniaxial mode in terms of maximum tearing load (kg), breakage stress (MPa), elongation (%). The young's modulus of the films was determined by tensile strength with the standard test ASTM D-882-97.

16.2.5 Statistical Analysis

Statistical comparison was performed using unpaired t-test and one-way analysis of variance (ANOVA) followed by Dunnett's test for multiple comparisons. All statistical analyses were performed using SPSS 16.0 software package (SPSS Inc., USA).

16.3 Results and Discussion

The structural properties, namely, the morphology of the surface of the gel film of the BC and the BC/HA composite, the diameter and location of the microfibrils of the polymer relative to each other were examined by field emission scanning electron microscope JSM-7800F (Jeol, Japan). Figure 16.1 shows electron micrographs of the dehydrated film of BC and composites BC/HA-1, BC/HA-2.

As shown in Fig. 16.1a, the BC has an ultrathin three-dimensional nanoporous network structure consisting of nanoscale fibers, which provides an excellent basis for the formation of HA nanoparticles and facilitates cellular adhesion of bones when used in tissue engineering. After treatment of PVP with a solution and biomineralization, HA crystals formed on the fibers of the BC for 5 days (Fig. 16.1b). Figure 16.1b, c show of HA crystals are embedded as a rod-like shape in a BC fiber network, which facilitates cell adhesion and makes it promising for use as a biomaterial for bone replacement.

In addition, it is obvious the presence of a uniform distribution density of the fibers of the frame, which provides high strength films. Due to the correct location of the fibers, the degree of crystallinity of the films reaches more than 60%, and in order to break them, you need to apply a force of up to several kilograms per square millimeter. Mechanical strength is an important indicator of the quality of polymeric materials. The strength of the films and the composite was determined on a universal tensile testing machine "Instron" in uniaxial mode by tensile strength (MPa), elongation (%) under tension. For this purpose, samples were prepared in the form of strips with a size of 65 × 10 mm. The tests were carried out at a temperature of (25 ± 2) °C and a relative humidity of (55 ± 5) %, with a sample deformation rate

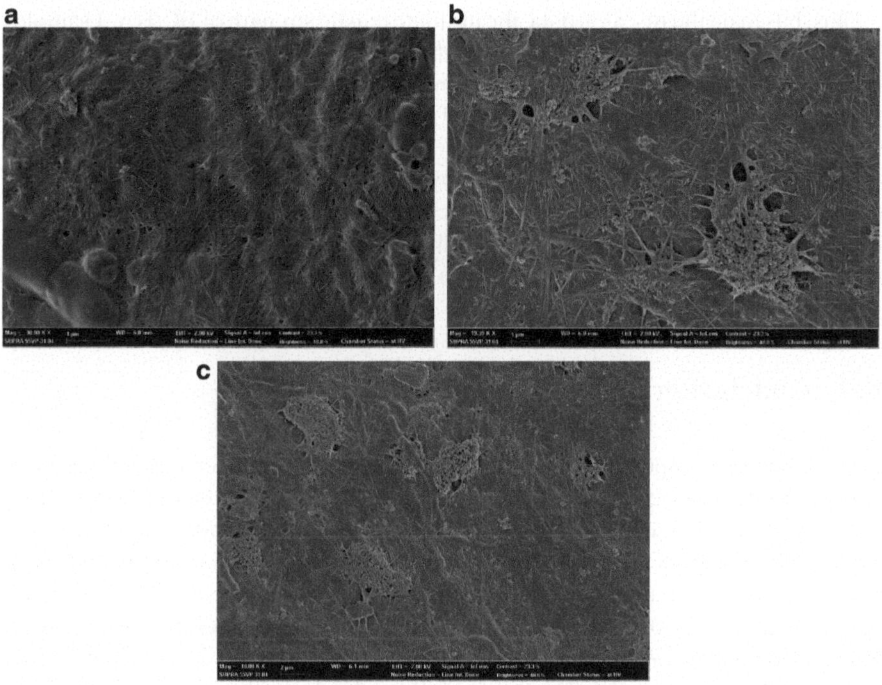

Fig. 16.1 SEM images of BC (**a**) films; BC/HA-1 composite obtained by biomineralization (**b**); BC/HA-2 composite obtained by synthesizing hydroxyapatite nanoparticles in the presence of fibrillar fragments of BC (**c**)

of 100 mm/min. Young's modulus (modulus of longitudinal elasticity) of the samples was determined by tensile strength with standard test ASTM D-882-97, designed to determine the strength of thin plastic films. The results of the study are shown in Table 16.1.

Table 16.1 Physico-chemical properties of BC/HA composite materials

Composite BC/HA	1 method	2 method
Density (g/cm^3)	61 ± 0.5	52 ± 0.3
Porosity (%)	0.55 ± 0.01	0.42 ± 0.02
Porevolume (cm^3/g)	79 ± 1.7	67.5 ± 1.9
Specific surface area (SSA) (m^2/g)	1.48 ± 0.07	2.04 ± 0.8
Ultimate tensile strength (MPa)	92.07 ± 1.4	76.09 ± 1.1
Young's modulus (MPa)	59.04 ± 0.6	48.03 ± 0.8

Thus, based on the research data, the ultimate tensile strength of BC/HA-1 samples is 92.07 ± 1.4 MPa, BC/HA-2 is 76.09 ± 1.1 MPa. This indicator of strength is associated with a high degree of crystallinity of BC, which provides resistance to high pressure (Backdahl 2006: 2141–2149). The Young's modulus of composites is 59.04 + 0.6 MPa and 48.03 ± 0.8 MPa.

The biocomposite BC/HA-1 obtained by biomineralization showed high physicochemical parameters compared with the composite obtained by synthesis.

According to the study, it can be said that the BC/HA-1 composite obtained by biomineralization is promising for use as a biomaterial for bone replacement.

16.4 Conclusion

The obtained biocomposite material based on the gel film of bacterial cellulose and hydroxyapatite. Nanocomposite material based on bacterial cellulose and hydroxyapatite was obtained using the biomimetic mineralization method (imitation of the formation of the mineral component in natural bone tissue) and the synthesis of hydroxyapatite nanoparticles in the presence of fibrillar fragments of BC. The biocomposite BC/HA-1 obtained by biomineralization showed high physicochemical parameters compared with the composite obtained by synthesis. The ultimate tensile strength of the composite BC/HA-1 was 92.07 ± 1.4 MPa, the composite BC/HA-2 was 76.09 ± 1.1 MPa. Young's modulus (the coefficient of proportionality between stress and strain) of the BC/HA-1 composite was 59.04 ± 0.6 MPa (BC/HA-1), which is 18.6% higher than that of the BC/HA-2 composite (48.03 ± 0.8 MPa). These indicators are quite high compared with the values of the Young's modulus of many flat oriented layers of organic polymers.

References

1. L.R. Lynd, P.J. Weimer, W.H.V. Zyl, I.S. Pretoriu, Microbial cellulose utilization: fundamentals and biotechnology. Microbiol. Mol. Biol. Rev. **66**, 506 (2002)
2. N. Shah, M. Ul-Islam, W.A. Khattak, J.K. Park, Overview of bacterial cellulose composites: a multipurpose advanced material. Carbohydr. Polym. **98**, 585–598 (2013)
3. A.S. Guzun, M. Stroescu, S.I. Jinga, G. Voicu, u A.M. Grumezesc, A.M. Holban, Plackett-Burman experimental design for bacterial cellulose-silica composites synthesis. Mater. SciEng. C Mater. Biol. Appl. **49**, 280 (2014)
4. W.C. Lin, C.C. Lien, H.J. Yeh, C.M. Yu, S.H. Hsu, Bacterial cellulose and bacterial cellulose-chitosan membranes for wound dressing applications. Carbohydr. Polym. **94**, 603–611 (2013)
5. R.D. Ruka, P.G. Simon, K. Dean, Altering the growth conditions of *Gluconoacetobacter xylinus* to maximize the yield of bacterial cellulose. CSIRO Mater. Sci. Eng. 1–21 (2015)
6. D.P. Romanov, YuG Baklagina, N.V. Lukasheva, A.K. Khripunov, A.A. Kachenko, V.K. Lavrentyev, Investigation of nanocomposites based on hydrated calcium phosphates and cellulose Acetobacter xylinum. J. Glas. Phys. Chem. **34**, 192–200 (2008)

7. M.J. Kim, M.J. Yeo, M. Kim, G.H. Kim, Biomimetic cellulose/calcium-deficient hydroxyapatite composite scaffolds fabricated using an electric field for bone tissue engineering. RSC Adv. **8**, 20637–20647 (2018)
8. T. Niamsapa, N.T. Lama and P. Sukyaia, Production of hydroxyapatite-bacterial nanocellulose scaffold with assist of cellulose nanocrystals. Carbohydr. Polym., 159–166 (2019)
9. N. Yin, S.Y. Chen, Y. Ouyang, L. Tang, J.X. Yang, H.P. Wang, Biomimetic mineralization synthesis of hydroxyapatite bacterial cellulose nanocomposites. Prog. Nat. Sci.: Mater. Int. **21**, 472–477 (2011)
10. Z. Cai, J. Kim, Bacterial cellulose/poly (ethylene glycol) composite: characterization and first evaluation of biocompatibility. Cellulose **17**, 83–91 (2010)

Chapter 17
Quality Parameters of Cellulose–Chitosan Based Edible Films for Probiotic Entrapment

D. H. Shokatayeva, A. Talipova, I. Savitskaya, A. Pogrebnjak, A. Kistaubayeva, and L. V. Ignatova

Abstract New carboxymethyl cellulose and chitosan edible films have been developed and characterized. Model probiotic bacteria *Lactobacillus rhamnosus* GG and *Lactobacillus acidophilus* AA-1 were incorporated after the film synthesis via bacteria diffusion and adsorption. The film formation and mechanical properties were characterized, as well as the viability of the entrapped bacteria. Inclusion of Ch in CMC film leads to a decrease in swelling ratio but increased rigidity of the films. Presence of Ch in such films inhibits bacteria incorporated in them. Overall, these novel CMC films are regarded as promising inexpensive and friendly matrices for food protection and packaging applications.

17.1 Introduction

Maintaining or improving quality and safety of food is one of the most important roles of its packaging. In the last decade, much attention has been paid to edible films and coatings. Edible films were performed thin layer materials that can be used as a coating to separated food components; food packaging; carrier for the delivery of active compounds or as packaging material for storing and protecting food [1]. In addition to barrier function, i.e. extending the shelf life and reducing the risk of pathogenic microorganisms growth on food products, such films can be considered as carriers of biologically active compounds [2]. It may be antioxidants, nutraceuticals, antimicrobials, flavors [3, 4]. Among them, probiotic bacteria are particularly relevant, since it is believed that these living microorganisms have a beneficial effect on the health of host, replenishing the natural gastrointestinal microbiota [5]. The inclusion of probiotics in polymer films can protect them from premature degradation and improve controlled release [6]. Although in the case of

D. H. Shokatayeva · A. Talipova (✉) · I. Savitskaya · A. Kistaubayeva · L. V. Ignatova
Al-Farabi Kazakh National University, Almaty, Kazakhstan
e-mail: talipova.aizhan@gmail.com

A. Pogrebnjak
Department of Nanoelectronics, Sumy State University, Sumy, Ukraine

A. D. Pogrebnjak et al. (eds.), *Nanomaterials in Biomedical Application and Biosensors (NAP-2019)*, Springer Proceedings in Physics 244,
https://doi.org/10.1007/978-981-15-3996-1_17

edible bioactive films and coatings, this is not even required, since it is assumed that the film/coating itself is consumed with food.

Edible films and coats are classified into three categories based on the components: hydrocolloids (polysaccharides, proteins, alginates); lipids and waxes; and composite films. Polysaccharides used for edible films or coatings are cellulose derivatives, dextrans, inulin, alginate, chitosan, carrageenan, starch derivatives, pectin derivatives [7, 8]. In recent years cellulose-based edible films are used for probiotic entrapment [9–11]. Carboxymethyl cellulose (CMC) is successfully used for these purposes most often [12, 13]. Films and coatings based on cellulose and its derivatives are transparent, flexible, odourless and tasteless. CMC is more resistant to water, however, the water vapor permeability is quite high. It can be reduced by creating a composite film with another biopolymer incorporated into it having high moisture barrier [14].

From this point of view, edible films and coatings obtained on the basis of chitosan (N-deacetylated derivative of chitin) deserve attention. Chitosan (Ch) is a biodegradable biopolymer having excellent film-forming and sorption properties. Therefore, it is often used for the manufacture of film packaging and food coatings [15]. The interesting fact that Ch was successfully used to create microcapsules with probiotics [16–19]. This is on the one hand. However, on the other hand, it has fungicidal and bactericidal properties [20, 21]. Thereby in this study were investigated physicochemical and biological properties of films with probiotic cells immobilized in edible films based on CMC and CMC + Ch. To the best of our knowledge, the potential use of such systems to entrap probiotic bacteria has never been explored before.

17.2 Materials and Methods

Preparation of bacteria *Lactobacillus rhamnosus* GG (LGG) was bought from the Belgian coordinated collection of microorganisms. Strain *Lactobacillus acidophilus* AA-1 (LA) isolated by our group from traditional yogurt was used. The strain was stored in glycerol (50%) at −20 °C. The stored cells were activated twice on MRS agar plates before use. After 48 h of growth, one colony was selected, inoculated into 20 ml MRS broth and incubated for 24 h at 37 °C. After this, the culture was transferred to fresh MRS broth and incubated at 37 °C for 18 h under anaerobic conditions. The cells were collected in the stationary growth phase by centrifugation at 6000 g for 15 min at 4 °C. The supernatant was discarded and cell pellet was washed twice with sterilized distilled water. The precipitate was resuspended in 3 ml of sterilized 0.85% sodium chloride solution. The concentration of cells after this treatment was around 2×10^{10} CFU/ml. Freshly prepared concentrated cell suspension was quantitatively evaluated by plating on MRS agar and used immediately.

Film formation Carboxymethyl cellulose (CMC) molecular weight 250 kDa with a degree of substitution 0.80–0.85 and Chitosan (Ch) of medium molecular weight (100–300 kDa) and a degree of deacetylation of 75–85% were purchased from Sigma-Aldrich Chemie GmbH.

CMC was dissolved in a full dissolution water. The total polymer concentration was set to 2 wt%. The films were plasticized by incorporating 1 wt% glycerol into the polymer solution. This level of glycerol has been chosen to produce flexible films without excessive surface stickiness. After obtaining a clear solution, 15 g of it were poured into Petri dishes and allowed to cure in an oven at 50 °C for 20 h. The obtained films were removed from the glass plate and conditioned at 20 ± 1 °C and relative humidity 55 ± 1%.

Preparation of CMC/Ch film carried out by the "submerging" ex situ method. Ch powder was dissolved in an aqueous solution of 1% acetic acid to a final concentration of 0.6%. CMC films were placed in Ch solution and incubated for 6 h at room temperature. The excess of Ch solution was removed by placing obtained film between two sheets of filter paper. The whole procedure was performed using autoclaved (121 °C, 15 min) materials and under sterile conditions in a laminar air flow box. The films were stored in sterile vials with lids at 4 °C and used in further experiments.

Model probiotic bacteria (*L*GG and LA) were incorporated in the films after the film synthesis, via bacteria diffusion and adsorption. For this the films were soaked in a bacteria medium for 30 min at room temperature, then the excess of medium was removed to allow a moderate drying at 37 °C for 30 min.

To estimate the viable counts and matrix effect, the included in the film bacteria were released by resuspending 10 mg of film either in a PBS buffer solution (pH 7.4) at 37 °C for 30 min, under stirring. The viability of released cells was determined by plating serial dilutions of resulting suspension on MRS agar. Colony-forming units were counted after 48 h of anaerobic incubation at 37 °C. The total count of viable bacteria was expressed as log colony forming units per gram (log CFU/g, CFU/g = CFU/plate × dilution factor).

Mechanical tests The universal uniaxial tensile tester machine (Instron, USA) was used to access the mechanical properties of the films. Tensile grips with 35 mm were used to hold the specimens which consisted of 40 mm × 40 mm films. Up to four repetitions were made for each film formulation using a grip speed of 1.0 mm/s.

Moisture content The moisture content was assessed according to a gravimetric method. Pre-weighed aluminum pans containing edible films (approximately 0.5 g) were dried at 105 °C in hot air oven until they reached to constant weight. The moisture content was calculated using the following equation: Percentage of residual water content = $(W_i \times W_f)/W_i \times 100$, where w_i and w_f are the initial and final weight of the edible films, respectively.

Scanning electron microscopy (SEM) studies The surface morphology of lyophilized film samples was observed by field emission scanning electron microscope JSM-7800F (Jeol, Japan). Prior to the SEM observation, the films were sputter coated with a platinum-palladium alloy (Pt/Pd 80/20).

Statistical analysis Statistical comparison was performed using unpaired t-test and one-way analysis of variance (ANOVA) followed by Dunnett's test for multiple comparisons. All statistical analyses were performed using SPSS 16.0 software package (SPSS Inc., USA).

17.3 Results and Discussion

A photograph of the films obtained by the procedure described in experimental section is presented in Fig. 17.1.

The films made from CMC alone or composite CMC/Ch film are reasonably flexible and transparent.

The inclusion of probiotic bacteria in a film is carried out in two ways: the incubation of finished films in bacterial suspension for a certain time or mixing of bacteria inoculum in a mixture of solutions of individual components (in our case CMC/Ch/glycerol before casting and hardening).

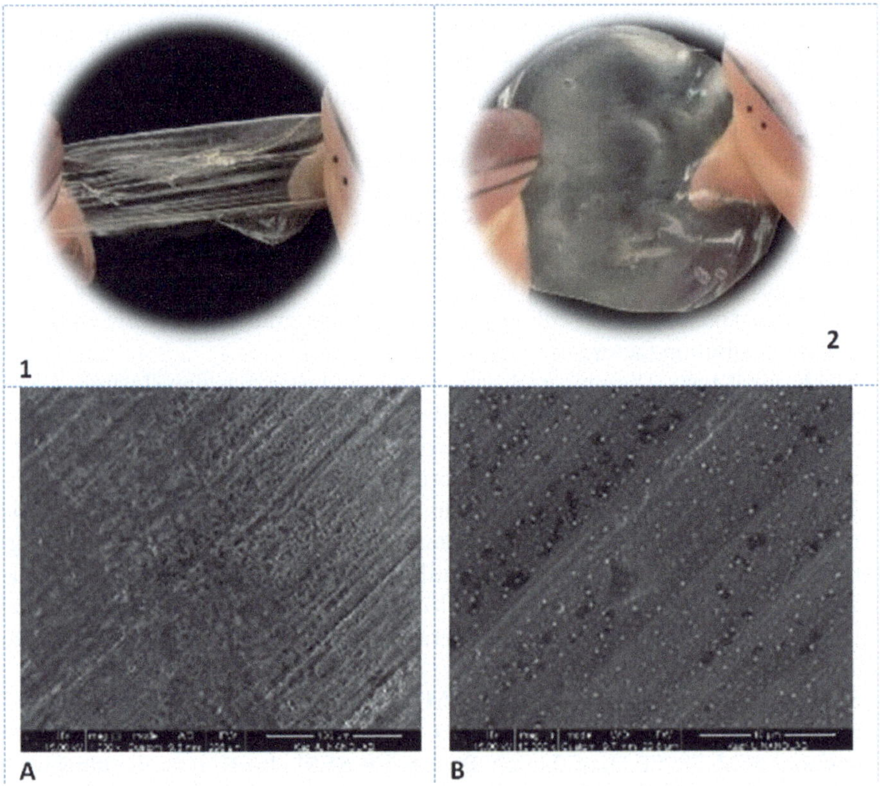

Fig. 17.1 **a** Photograph of CMC (1) and CMC/Ch (2) based films (1—CMC; 2—CMC/Ch) after being casted and dried in Petri dishes. **b** SEM images of different cross-linked films (1—CMC; 2—CMC/Ch)

In the first approach, bacteria should diffuse and adsorb to the film, in the second approach, it is expected that bacteria will be captured in a matrix of cellulosic polymer before casting and hardening [6]. However, since the films are not formed at temperatures below 50 °C, the viability of bacteria included by the second method can be significantly reduced [11]. In this connection, the first method was used in work, i.e. probiotic bacteria were immobilized in already finished films. *Lactobacillus rhamnosus* GG (LGG) and *L. acidophilus* AA-1 (LA) were selected as model bacteria.

The SEM results showed that the structure of control CMC and CMC/Ch films was homogeneous, uniform and compact with no micropores (Fig. 17.1). However, probiotic films showed a higher number of holes than control films. The bacterial cells were embedded in the film matrix (tiny rod-like shapes) that could result in increasing cell-protective effects of films (Fig. 17.2).

Edible films should be adequately resistant to external stresses to use as food packaging materials. Furthermore, films must be flexible and strong during packaging and storage. The obtained films were subjected to tensile tests in order to study their mechanical properties. Tensile strength (TS) and elongation at break (EB) are two key indicators of edible films used for packaging. The TS and EB of the control and probiotic films are shown in Table 17.1.

The CMC/Ch film presents the higher TS (maximum load at break) while the CMC is the weaker one. On the other hand, a significant difference in the TS ($p < 0.05$) is observed in films with probiotics. It is also interesting to note that the

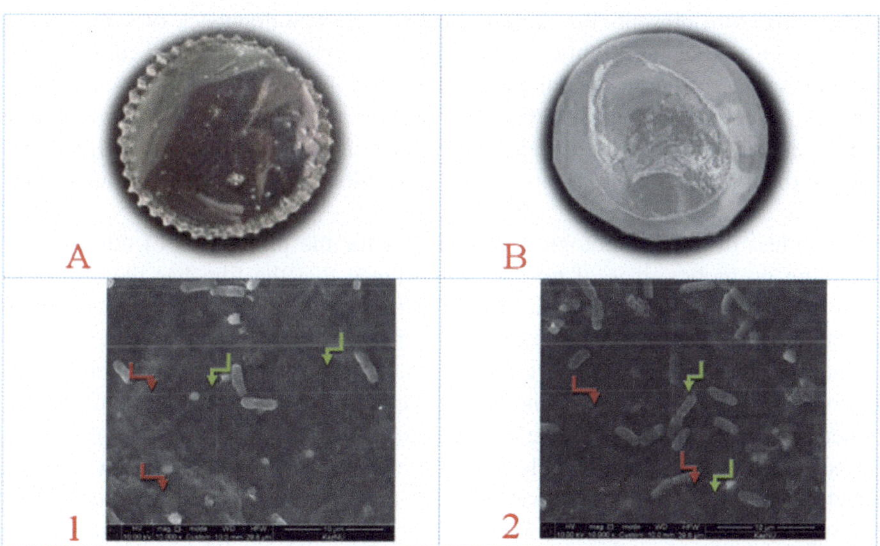

Fig. 17.2 Photograph (**a**) and SEM cross-section image (**b**) of CMC (1) and CMC/Ch (2) films loaded with probiotic *Lactobacillus*. Cells included in (cells included in film structure are shown with red arrows, green arrows show cells attached on surface of material)

Table 17.1 Physicochemical and optical properties of carboxymethyl cellulose films

Type of film	Parameters		
	TS (MPa)	EB (%)	MC (mg/g) of film
CMC	20.3 ± 1.2^a	15.6 ± 0.2^a	340.3 ± 8.2^a
CMC/Ch	28.4 ± 0.8^b	12.3 ± 0.1^b	228.4 ± 7.9^b
CMC/probiotic	15.4 ± 0.2^b	19.4 ± 0.3^b	360.3 ± 10.2^a
CMC/Ch/probiotic	23.7 ± 1.4^b	18.2 ± 0.08^b	237.8 ± 9.4^b

Results are represented as mean ± standard deviation. Values with different superscript letters in each column are significantly different ($p < 0.05$)
a, *b* show statistical difference of values

tensile strength and elongation to break follow an opposite trend. Inclusion of Ch in CMC film improves the TS but decreases the EB. In the present study, addition of probiotic cells into CMC and CMC/Ch films significantly reduced TS as well as EB. This indicates a lower strength and flexibility of the obtained probiotic films. Probably this phenomenon is a consequence of the increase in molecular mobility and free volume in polymer chains in films during immobilization of bacteria in them. Overall, the mechanical properties obtained films are promising.

The moisture content (MC) after drying not only affects the rate of viability decrease during long storage periods but it facilitates melting of edible films in the mouth. Therefore, the quantitative determination of MC in films is important. The MC of the films is shown in Table 17.1.

The CMC film presents the highest swellability. The drawback of such a huge water uptake is that the highly swollen film behaves as a weak gel and thus its handling is rather difficult in contrast to other formulations.

It was observed that the CMC/Ch films present a rather similar trend with a low swelling capacity. In the present study, the addition of probiotics caused no significant effect on the film moisture ($p > 0.05$). Similar results were found in previous studies [13].

Regardless of the technological mechanism for the production of edible films containing probiotics, it is of primary importance that the film has the ability to provide of the viability of probiotic strains in films in order to assess their suitability as probiotic carriers.

One of the most important effective factors on the viability of probiotics is temperature of distribution and storage conditions [5]. The optimal storage temperature, ensuring safety of the population level of probiotics in the composition of functional foods as well as edible films is 4–6 °C. This is explained by the fact that in such a temperature range the level of bacterial metabolism significantly decreases ensuring high cell viability [11–13]. In this regard, the survival of probiotics in the experimental films was determined under such temperature conditions. Figure 17.3 shows the survival of probiotic strains in CMC and CMC/Ch films during storage at 4 °C.

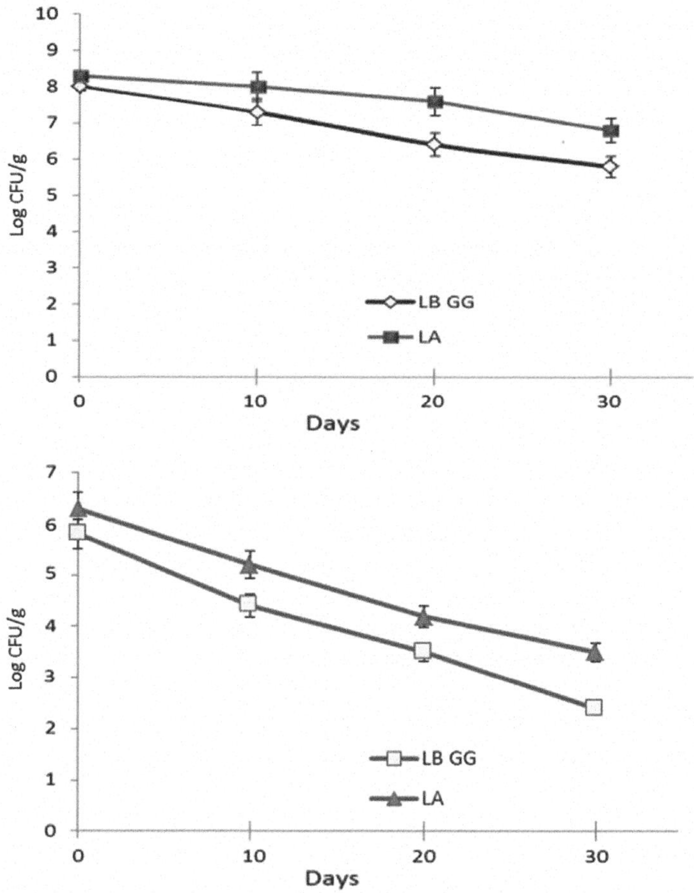

Fig. 17.3 Survivability of *L. rhamnosus* GG and *L. acidophilus* AA-1 during storage at refrigerated (4 °C) temperature. The error bars represent the standard deviation

It has been established that the bacteria trapped in the CMC film have a rather high viability around 8 Log CFU/g. The probiotics showed <2 log CFU decrease in viability at the end of storage. Such indicators are considered as good [11, 13, 22]. In general, viable LGG and LA bacteria can be effectively entrapped in the CMC based films.

In addition, MC in CMC/Ch is reduced compared to CMC films, and decrease in water activity (increase in osmotic stress) in media can affect the viability of probiotics. Furthermore, poor swelling of the film can impede the release of bacteria from it, leading to lower viability counts (thus resulting in a lower viability counting). It should also be noted that viability of LA was significantly higher than that of LGG probiotics at the end of storage. This may occur due to its higher tolerance to detrimental conditions of edible films during storage, what has been observed in other studies [13, 23, 24].

On the other hand, there are quite positive examples of quite tolerant "relationships" of probiotic cells with Ch. But this is only observed when probiotics are found in microcapsules with Ch. And in these capsules, Ch is used as an additional coating (layer-by-layer), but the basis of the microcapsule is either CMC [19] or alginate [16–18]. Therefore, cells of probiotics are in contact only with these polymers.

In general, viable LGG and LA bacteria can be effectively entrapped in the CMC based films and their release is strongly dependent on the film composition, physicochemical and biological properties of films.

17.4 Conclusion

The film properties can be controlled by physical and chemical features of components included in it. Incorporation of Ch в CMC film leads to a decrease in swelling ratio but increased rigidity of the films. Viable LGG and LA could be effectively entrapped in the films after soaking them in a bacteria medium. Presence of Ch in such films inhibits bacteria entrapped in them. However, these films may inspire the formation of related systems, such as micro-beads as carriers for probiotic bacteria and delivery in the gastrointestinal tract.

The systems developed represent promising advances in the search for new applications of edible cellulose based films and coatings as carriers of diverse probiotics and open new possibilities for the development of novel food probiotic products.

References

1. D.Z. Suput, V.L. Lazi´c, S.Z. Popovi´c and N.M. Hromis, Food Feed Res. **42**, 11–22 (2015)
2. J. Wyrwa, A. Barska, Eur. Food Res. Technol. **243**, 1681–1692 (2017)
3. P. Muranyi, J. Food Process Technol. **4**, 1–14 (2013)
4. A.E. Kapetanakou, P.N. Skandamis, Curr. Opin. Food Sci. **12**, 1–12 (2016)
5. P.J.P. Espitia, R.A. Batista, H.M. Azeredo, C.G. Otoni, Food Res. Int. **90**, 42–52 (2016)
6. F. Pavli, C. Tassou, G-J.E. Nychas, N. Chorianopoulos, Int. J. Mol. Sci. **19**, 150 (2018)
7. D.Z. Suput, V.L. Lazic, S.Z. Popovic, N.M. Hromis, Food Feed Res. **42**, 11–22 (2015)
8. A. Dhanapal, P. Sasikala, L. Rajamani, V. Kavitha, G. Yazhini, M. Shakila Banu, Food Sci. Qual. Manag. **3**, 9–17 (2012)
9. L. Sánchez-González, J. Iván Quintero Saavedra, A. Chiralt, Food Control **35**, 200–206 (2014)
10. N. Romano, M. José Tavera-Quiroz, N. Bertola, P. Mobili, A. Pinotti, A. Gómez-Zavaglia, Food Res. Int. **64**, 560–566 (2014)
11. P. Singh, S. Magalhaes, L. Alves, F. Antunes, M. Miguel, B. Lindman et al., Food Hydrocolloid **88**, 68–74 (2019)
12. J. Odila Pereira, J. Soares, S. Sousa, A.R. Madureira, A. Gomez, M. Pintado, LWT-Food Sci. Technol. **73**, 543–550 (2016)
13. B. Ebrahimi, R. Mohammadi, M. Rouhi, A.M. Mortazavian, S. Shojaee-Aliabadi et al., LWT—Food Sci. Technol. **87**, 54–60 (2018)

14. F. Garavand, M. Rouhi, S.H. Razavi, I. Cacciotti, R. Mohammadi, Int. J. Biol. Macromol. **104**, 687–707 (2017)
15. P. Varela, S.M. Fiszman, Food Hydrocolloid **25**, 1801–1812 (2011)
16. I. Trabelsi, W. Bejar, D. Ayadi, H. Chouayekh, R. Kammoun, S. Bejar et al., Int. J. Biol. Macromol. **61**, 36–42 (2013)
17. T.W. Yeung, E.F. Üçok, A.K. Tiani, D.J. McClements, D.A. Sela, Front Microbiol **7**, 494 (2016)
18. M. De Araújo Etchepare, G.C. Raddatz, E.M. de Moraes Flores, L.Q. Zepka, E. Jacob Lopes, J.S. Barin et al., LWT-Food Sci. Technol. **65**, 511–517 (2016)
19. P. Singh, B. Medronho, L. Alves, M. Miguel, B. Lindman, B. Medronho, Carbohyd. Polym. **175**, 87–95 (2017)
20. D. Raafat, H.-G. Sahl, Microb. Biotechnol. **2**, 186–201 (2009)
21. A.P. Martínez-Camacho, M.O. Cortez-Rocha, J.M. Ezquerra-Brauer, A.Z. Graciano-Verdugo, F. Rodríguez-Félix, M.M. Castillo-Ortega et al., Carbohyd. Polym. **82**, 305–315 (2010)
22. C. Soukoulis, L. Yonekura, H.-H. Gan, S. Behboudi-Jobbehdar, C. Parmenter, I. Fisk, Food Hydrocolloid **39**, 231–242 (2014)
23. K. Shukla, K. Mishra, A. Arotiba, B. Mamba, Int. J. Biol. Macromol. **59**, 46–58 (2013)
24. M. Rouhi, R. Mohammadi, A. Mortazavian, Z. Sarlak, Dairy Sci. Technol. **95**(2), 115–133 (2015)

Chapter 18
Synthesis of Silver Nanoparticles and Therapeutic Films for Ophthalmology Based on Them

V. Skobeeva, V. Smyntyna, V. Ulyanov, M. Makarova, V. Tkachenko, N. Malushin, and N. Molchaniuk

Abstract This article presents the results of the silver nanoparticles synthesis by a simple and ecological method of chemical reduction of silver from silver nitrate in an aqueous solution of sodium citrate and their use in the treatment of eye diseases. The technological conditions that lead to formation of silver nanoparticles with spherical shape and size of 30 nm with an intensive band of surface plasmon resonance are established. The effect of the application of therapeutic films with nanoparticles on the ultrastructure of the epithelium and stroma of the cornea of a rabbit after modeling in animals of moderately severe bacterial keratitis has been studied. It has been established that the application of therapeutic films with silver nanoparticles, as compared with antibiotics, leads to more efficient regeneration of the anterior corneal epithelium, contributes to the rapid restoration of the stroma and active formation of the delicate collagen skeleton typical of healthy corneal tissue.

18.1 Introduction

Interest in studies of Ag nanoparticles (Ag NPs) is due to a number of their unique properties that differ from the properties of nanoparticles of other metals, namely, Ag NPs have the greatest intensity of the surface plasmon resonance band (SPR), high absorption coefficient, the phenomenon of giant Raman scattering. The development of nanotechnology and modern research on the potential biological

V. Skobeeva (✉) · V. Tkachenko · N. Malushin
Research Institute of Physics, Odessa I.I. Mechnikov National University Odessa, Odessa, Ukraine
e-mail: v_skobeeva@ukr.net

V. Skobeeva · V. Smyntyna
Odessa I.I. Mechnikov National University, Odessa, Ukraine

V. Ulyanov
Odessa National Medical University, Odessa, Ukraine

M. Makarova · N. Molchaniuk
The Filatov Institute of Eye Diseases and Tissue Therapy, National Academy of Medical Sciences of Ukraine Odessa, Odessa, Ukraine

A. D. Pogrebnjak et al. (eds.), *Nanomaterials in Biomedical Application and Biosensors (NAP-2019)*, Springer Proceedings in Physics 244,
https://doi.org/10.1007/978-981-15-3996-1_18

properties of silver nanoparticles confirm their applicability in medicine. Due to their high antibacterial activity, anti-inflammatory action, low toxicity, silver ions and nanoparticles are used as an alternative to antibiotics. Silver NPs have been used as wound dressings for the treatment of burns, in the form of coatings for medical devices impregnated with silver nanoparticles [1, 2], in the development of restoration materials for dentistry [3]. In a review article [4] R.P. Allaker reports on the use of nanoparticles in combating infections of the oral cavity, on the use of nanoparticles in photodynamic therapy. Studies by the authors [5] showed that Ag NPs show the best antimicrobial activity compared to the standard antibiotic chloramphenicol. This confirms the possibility of using Ag NPs as an antimicrobial agent for suppressing gram-negative and gram-positive bacteria. Ag NPs are non-toxic to humans and do not create side effects compared to antibiotics. Compared with conventional antibiotics, the development of bacterial resistance is less likely when using nanostructured media. Such nanomaterials can attack a wide range of targets in bacteria, for example, membrane proteins, the lipid bilayer, and damage DNA when nanoparticles are absorbed [6]. An analysis of the literature data showed that, despite numerous examples of the use of Ag NPs in medical practice, there is practically no information on the use of Ag NPs in ophthalmology.

In addition, for the practical implementation of the unique properties of Ag NPs, the issue of using nanoparticles with stable and controlled physical and optical characteristics (size, shape, spectrum, and absorption intensity under local plasmon resonance conditions) is topical. This task is closely related to the study of the influence of synthesis conditions on the formation of Ag NPs with the properties that are required for a specific medical application. Note that with the development of modern technologies for producing Ag NPs with the prospect of their further practical application, simple, low-toxic and environmentally friendly synthesis methods are relevant.

In this work, on the basis of these considerations, the method of chemical reduction of silver salt in an aqueous solution of sodium citrate, in which sodium citrate ($Na_3C_6H_5O_7$) performs the functions of both a reducing agent and a particle stabilizer. Sodium citrate is the sodium salt of citric acid, which has a synthetic origin and a very low level of danger. Sodium citrate is often used as a drug. Anionic surfactant, sodium dodecyl sulfate (SDS), which also has a low level of toxicity, was used as additional stabilizers. The above information justifies the use of these substances as media in which silver NPs are formed. However, the fact that sodium citrate performs a dual role in the synthesis (silver reducing agent and NP growth stabilizer) makes the synthesis process dependent on many factors. The problem is that the processes of nucleation and growth of nanoparticles depend on the rate of chemical reaction of silver ion reduction, on the concentration of the reacting components, their ratio, the method of introducing the reagents into the solution.

Thus, the technological part of this work on the synthesis of Ag NP includes the establishment of optimal conditions that ensure the growth and stability of NP of different sizes.

The purpose of the second part of the work is to study the effect of applications of therapeutic films with silver nanoparticles of 30 nm in size, obtained by the citrate method, on the ultrastructure of the epithelium and stroma of the rabbit cornea after modeling in animals of moderate bacterial keratitis.

18.2 Material and Research Methods

The object of the study of the effect of Ag NPs on ophthalmic diseases was rabbits, in which moderately severe bacterial keratitis was modeled.

Bacterial keratitis in rabbits was modeled according to the patent scheme [7].

After receiving a model of keratitis of moderate severity, all animals were equally divided into 3 groups:

1. control group—rabbits with moderate keratitis
2. group—rabbits with application of a film with silver nanoparticles on the cornea.
3. group—rabbits with antibiotic application (decamethoxin) on the cornea.

Films were obtained by casting onto a polished surface using polyacrylamide, polyvinyl alcohol, polyvinylpyrrolidone, and sodium carboxymethylcellulose as the base. The concentration of Ag NPs in the film was 10 mg/g.

Electron microscopic studies of the ultrastructure of the rabbit cornea were carried out on an electron microscope TEM-100-01. The studied material was processed according to the described generally accepted method [8]. Ultra-thin sections were contrasted by the Reynolds method [9]. Electron microscopic studies of Ag NPs were carried out using scanning electron microscopy (SEM), atomic force microscopy (AFM). To control the synthesis of Ag NPs, we used the method of optical spectrometry in the UV and visible spectral regions.

18.3 Results and Discussion

18.3.1 Synthesis and Optical Properties of Ag NPs

Silver nanoparticles were synthesized by the method of chemical reduction of silver ions from a solution of silver salt ($AgNO_3$) by a reducing agent—sodium citrate ($Na_3C_6H_5O_7$). The reaction used solutions of silver nitrate and sodium citrate with an equimolar concentration of 1×10^{-5} M; 1×10^{-4} M; 1×10^{-3} M; 5×10^{-3} M. The synthesis reaction temperature is 100 °C. The synthesis time varied depending on the conditions of registration of the local surface resonance band (SPR) of Ag NPs (10; 15; 20; 30; 40; 50; 60; 70; 80 min).

The experimental results showed that the process of synthesizing NPs significantly depends on the concentration of the initial reaction products. It was noted that at concentrations of silver salt 10^{-5} M at 100 °C for 120 min, the SPR band in Ag NPs is not registered. It was established that with a further increase in the concentration of $AgNO_3$, (1×10^{-4} M to 5×10^{-3} M), the formation of silver NPs is determined by the kinetics of the chemical reaction.

The kinetics of the formation of Ag NPs in the reduction reaction was studied using spectrometry in the UV and visible spectral regions at different initial concentrations of $AgNO_3$ and $Na_3C_6H_5O_7$. The basis for the application of the optical method is the fact that the intensity, shape and half width of the SPR band of Ag NPs depend on the number of nanoparticles in the solution and their size dispersion, and the position of the SPR band maximum depends on the size of the NP. The appearance of this band in the absorption spectra is evidence of the formation of silver metal nanoparticles, in which surface plasmons are formed when interacting with electromagnetic irradiation.

In the absorption spectrum of the obtained silver nanoparticles, an intense SPR band is recorded, the maximum of which is localized in the wavelength range of 415 ± 420 nm. This position of the band maximum SPR corresponds to the average NP size of 30 nm. The maximum intensity of the absorption band was achieved at component concentrations in the range of 1×10^{-3} M to 5×10^{-3} M (curves 2 and 3, Fig. 18.1).

It is characteristic that the position of the band maximum does not depend on the concentrations of the components in the range (1×10^{-4} M to 5×10^{-3} M), which indicates the formation of nanoparticles of the same size (30 nm) at the end of the synthesis process. The dependence of the absorption spectra of Ag NPs on the synthesis time for a concentration of 5×10^{-3} M is shown in Fig. 18.2. An increase in the intensity and a decrease in the half-width of the SPR band with increasing synthesis time (1, 20 eV, 1.12 eV, 1.07 eV for curves 2, 3, 4, respectively) is observed. The increase in the absorption intensity and the narrowing of the LPRS

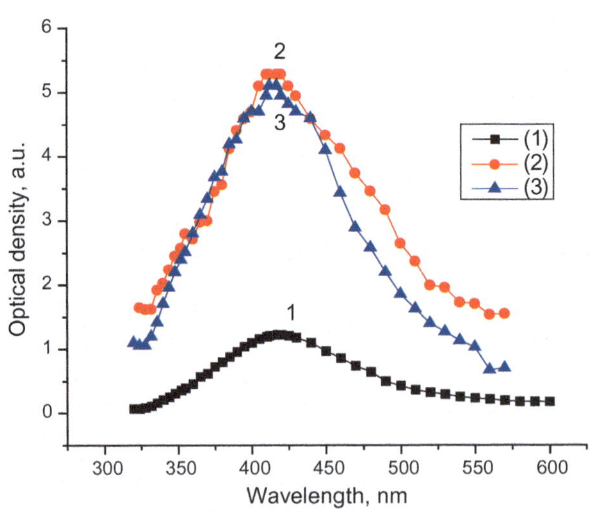

Fig. 18.1 The absorption spectra of Ag NPs obtained at different concentrations of $AgNO_3$ and $Na_3C_6H_5O_7$: 1×10^{-4} M (1), 1×10^{-3} M (2), 5×10^{-3} M (3)

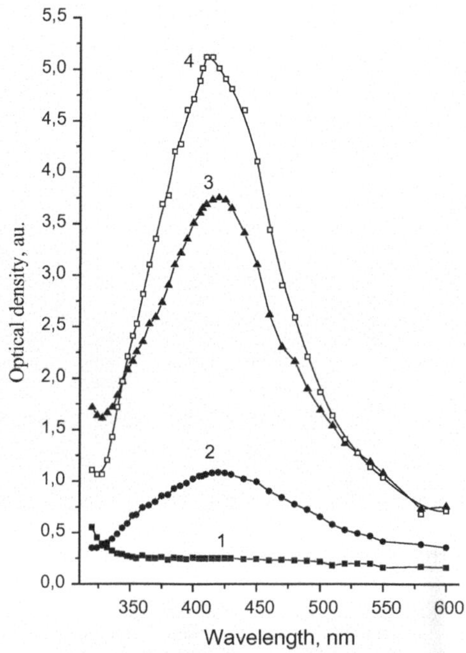

Fig. 18.2 Dependence of the absorption spectra of silver NPs on the synthesis time, min: 10 (1); 13 (2); 15 (3); 20 (4). The concentrations of $AgNO_3$ and $Na_3C_6H_5O_7 = 5 \times 10^{-3}$ M

band may be evidence in favor of the following mechanism for the formation of Ag NPs. At the first stage, particles of different sizes are formed, then smaller particles dissolve according to the mechanism of Ostwald ripening and, at the same time, the particles grow due to the recovery of silver ions on the surface of nanoparticles.

When the concentration of the components is more than 5×10^{-3} M, as a result of coagulation of the nanoparticles, the formation of silver precipitate in solution was observed. Thus, silver NPs used for research in ophthalmology were obtained at a component concentration of 5×10^{-3} M and had a spherical shape with an average size of 30 nm (Fig. 18.3.).

The effect of applications of therapeutic films with silver nanoparticles 30nm in size on the ultrastructure of the cornea of rabbits with moderate keratitis was studied.

The ultrastructure of the cornea of the control group of rabbits with moderate keratitis is shown in Fig. 18.4.

In the central region of the cornea, there is an accumulation of blood cells and histiocytic cells, as well as single collagen fibrils between them. The cells of the anterior epithelium in the subliminal region of the cornea, including the basal ones, have signs of dystrophic changes. They contain few organelles, membrane organelles are vacuolated, the nuclei are small with loosely arranged chunks of chromatin. Some of the cells have signs of cytoplasmic edema. In the intercellular fissures between the basal cells, electron-dense deposits are visible. Moreover, most of the basal cells also

Fig. 18.3 Image of silver NPs obtained in a raster electron microscope

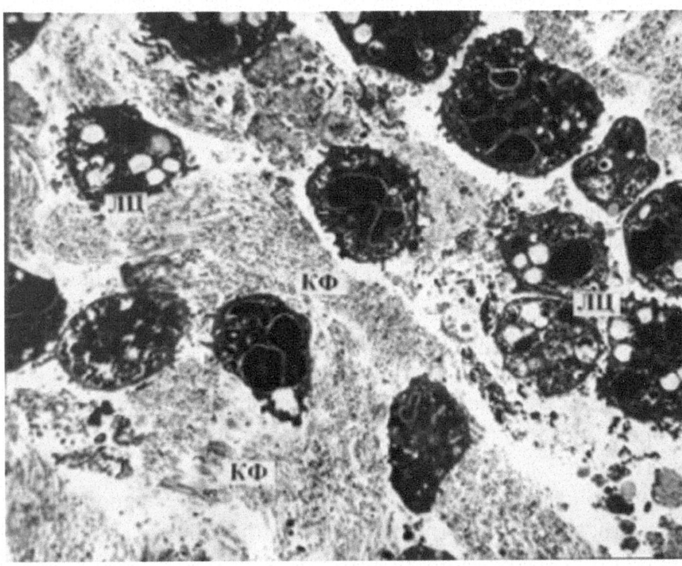

Fig. 18.4 Ultrastructure of a rabbit cornea during modeling of moderate bacterial keratitis. The accumulation of leukocytes and destroyed collagen lamellae in the stroma. Collagen fibrils in a state of destruction. Electron micrograph X 2500. Legend: ЛЦ—leukocytes, КФ—collagen fibrils

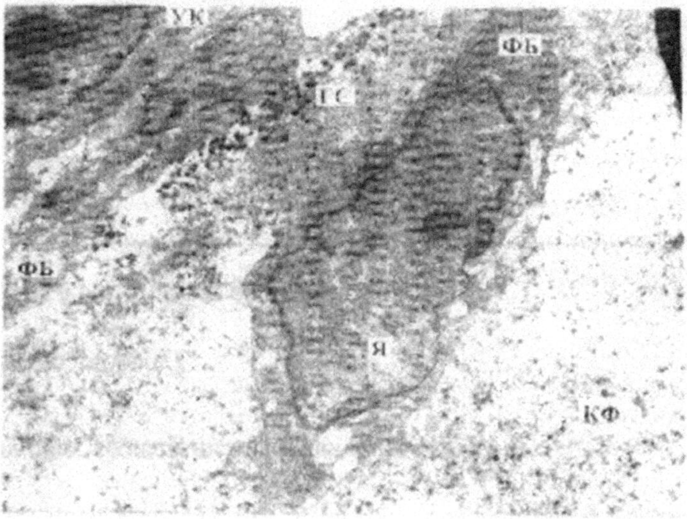

Fig. 18.5 Ultrastructure of the cornea in the area of reparative regeneration of ACE at the border with the performed stromectomy and with the subsequent application of the film with silver nanoparticles. A flattened cell in contact with fibroblast. Large fibroblast and accumulations of silver granules in the connecting skeleton. Electron micrograph X 6000. Legend: УК—a flattened cell, ГС—silver granules, КФ—collagen fibrils, FB—fibroblast, Я—nucleus

contain a homogenized electron dense cytoplasm. The basement membrane does not contain semi-desmosomes.

Ultrastructure of the cornea in the area of reparative regeneration of the anterior corneal epithelium (ACE) at the border with the performed stromectomy and with the subsequent application of films with silver nanoparticles is shown in Fig. 18.5.

In the area of the formation of the epithelial layer in the basal layer, single, slightly differentiated cells are located, located in one row, having short finger-shaped outgrowths with the help of which these cells come into contact with each other, forming interdigitations. They form large intercellular clefts in the form of cavities. Dense intercellular connections are also observed here. In the cytoplasm of these cells, a large nucleus is determined in the karyoplasm of which chromatin is located in a diffuse state, in the center of the karyoplasm there are a large number of free ribosomes and polysomes, and small cisterns of the granular endoplasmic reticulum (ЗЭС). In some places, cells of the same type are in contact with single, large, flattened cells with a fusiform nucleus containing chromatin in a diffuse state, and its margin is also determined. The cytoplasm of these cells contains polysomes, a few mitochondria, with focal destruction of cristae, vacuoles, apparently formed from mitochondria after the destruction of the inner membrane. Under the aforementioned cells are clusters of silver granules and fibroblasts in contact with these cells. Deeper, in the connective tissue skeleton, consisting of thin, short fibrils, silver granules and large fibroblasts with a well-defined ЗЭС are also diffusely located, in the cavity of the cisterns of which there is a moderate electron density, which indicates protein synthesizing activity.

18.3.2 Ultrastructure of the Cornea in the Area of Reparative Regeneration of ACE at the Border with the Performed Stromectomy and with the Subsequent Application of the Film with Decamethoxin

The anterior corneal epithelium in direct contact with the area where the stromectomy was performed consists of a well-formed layer of surface cells, under which there are large, poorly differentiated basal cells of a round or oval shape, a nucleus containing chromatin in a diffuse state and electron-transparent karyoplasma. Their cytoplasmic backbone consists of short thin fibrils, tonofilaments, rod-shaped or round mitochondria. The number of mitochondria in the cells is not the same, from single organelles in a number of cells to their accumulations in others. A small number of short ЗЭС cisterns containing ribosomes on the profiles of their membranes and slightly free ribosomes are also determined in these cells. The intercellular contacts in the forming layer are dense, sometimes narrow gaps are visible. Interdigitations are determined in places. The basement membrane is not formed. In some areas under the layer of basal cells, elements of the formation of the basement membrane are observed. Under the basal cells in the electron-transparent zone are located short, thin, lying diffusely or in the form of small clusters, collagen fibrils. Near this area, basal cells, both slightly and more differentiated, containing an increased number of organelles, are detected. Subepithelially located tender connective tissue skeleton. In the PER region, a layer of surface cells and a layer of basal cells are observed. The cells of the surface layer, especially those that border the cells of the basal layer, contain an increased number of polysomes, small mitochondria, and bundles of tonofilaments. Basal cells are more differentiated, a large number of typical organelles are revealed in their cytoplasm. Cells are mostly closely adjacent to each other. In places, slightly pronounced, interdigitations and gap junctions are visible. The basement membrane under these cells is more formed, single semi-desmosomes and anchor filaments are visible. Subepithelially visible elements of the coarse connective tissue (Fig. 18.6).

18.4 Conclusions

The kinetics of the formation Ag NPs in the reaction of silver reduction with sodium citrate was studied at various initial concentrations of $AgNO_3$ and $Na_3C_6H_5O_7$. Optimal conditions have been established that ensure the growth and stability of NPs with an average size of 30 nm. To control the synthesis of Ag NPs, the method of optical spectrometry in the UV and visible spectral regions was used.

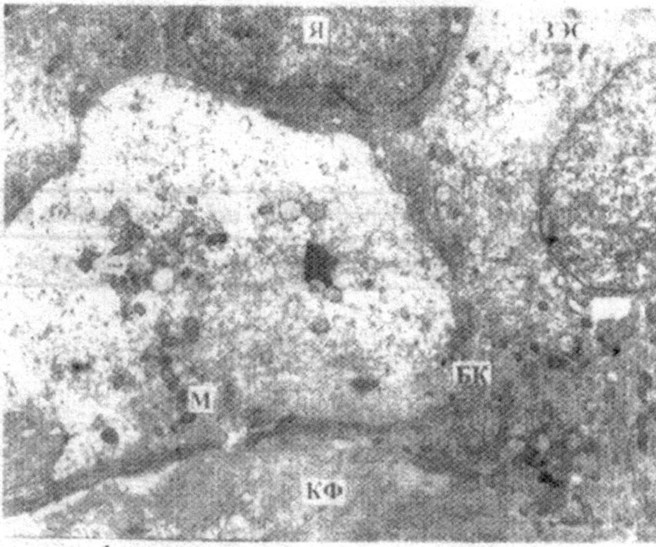

Fig. 18.6 The ultrastructure of the cornea in the area of reparative regeneration of ACE at the border with the performed stromectomy and with the subsequent application of the film with decamethoxin. Under the basal cells, which are in a state of varying degrees of differentiation, are collagen fibrils of coarse connective tissue. Electronic micrograph. X 5000. Legend: БК—basal cells, Я—the nucleus, M—mitochondria, ЗЭС—granular endoplasmic reticulum, КФ—collagen fibrils

It was shown that films with silver nanoparticles 30 nm in size obtained by the citrate method reduce the exudative phase of the inflammatory process, enhance the proliferative reaction, which leads to the regeneration of the anterior epithelium of the cornea, contribute to the rapid recovery of stroma, due to an increase in the number of fibroblasts, as well as an increase in their protein synthesizing function and the active formation of a delicate collagen skeleton characteristic of healthy corneal tissue.

After exposure to films with decamethoxin, the restoration of the anterior epithelium of the cornea and stroma is slowed down, signs of the exudative phase of the inflammatory process are expressed. In the stroma, the formation of coarse fibrous connective tissue occurs.

References

1. T. Gunasekaran, T. Nigusse, M.D. Dhanaraju, Silver nanoparticles as real topical bullets for wound healing. J. Am. Coll. Clin. Wound Spec. **3**, 82–96 (2012). https://doi.org/10.1016/j.jcws.2012.05.001
2. M. Rai, A. Yadav, A. Gade, Silver nanoparticles as a new generation of antimicrobials. Biotechnol. Adv. **27**(1), 76–83 (2009). https://doi.org/10.1016/j.biotechadv.2008.09.002
3. E. Abou Neel, L. Bozec, R. Perez, H. Kim, J.C. Knowles, Silver nanoparticles as a new generation of antimicrobials. Int. J. Nanomed. **10**, 6371–6394 (2015). https://doi.org/10.2147/IJN.S86033

4. R.P. Allaker, The use of nanoparticles to control oral biofilm formation. J. Dent. Res. **89**, 1175–1186 (2010). https://doi.org/10.1177/0022034510377794
5. A. Shukla, B.A. Makwana, Facile synthesis of silver nanoparticle and their potential application. J. Nanosci. Nanotechnol. **2**, 84–92 (2014). https://doi.org/10.11648/j.nano.20140204.14
6. R. Kummala, K.J. Brobbey, J. Haapanen, J.M. Mäkelä, M. Gunell et al., Antibacterial activity of silver and titania nanoparticles on glass surfaces. Adv. Nat. Sci. Nanosci. Nanotechnol. **10**, 015012–015017 (2019)
7. Patent for model No. UA87119U dated 1/27/2014 Drozhzhina GI, Gaydamaka TB, Vansovich E.V.
8. V.A. Ulyanov, M.B. Makarova, N.I. Movchanyuk, The effect of instillation of a colloidal solution of silver nanoparticles on the ultrastructure of the anterior epithelium and cornea strom. Ophthalmol. J. **3**(476), 63–69 (2017)
9. E.S.S. Reunoldes, The use of lead citrate at higt pH an electronopaque stain in electron microscopy cornea strom. J. Cell Biol. **17**, 208–212 (1963)

Chapter 19
A Hg (II) Fluorescent Sensor Based-on Bodipy Synthesized by Using Knorr Pyrrole

Ersin Guler, Emine Bagci, and Ahmed Nuri Kursunlu

Abstract In this study, firstly Bodipy derivative containing organic groups were prepared. Changes in the emission, absorption and excitation properties of these compounds were then examined in the presence of some metal ions. Thus, in these measurements, it was determined that the final Bodipy derivative intended to be selective of metal ions were found and the compounds obtained were tested with different metal ions. However, only mercury has shown a selectivity for the metal. The prepared compound based-on Bodipy can be used as a fluorescent sensor for the Hg (II) ions.

19.1 Introduction

Due to their fluorescent properties, Bodipy compounds generally have the potential to be used as effective drugs in cell labeling of biomolecules and in photodynamic therapy. Phthalocyanines are target compounds that can be used in photodynamic therapy because of their ability to absorb high wavelength light (about 700 nm), high triplet state quantum yields, triplet state long life, no toxic effect when light is not used, and effectively producing singlet oxygen [1–3].

One of the most important properties of Bodipy dyes is the ability to change their optical properties by changing the molecular backbone. These Bodipy dyes have sharp absorption bands, high fluorescence quantum yields, good solubility, chemical and photochemical resistance in solvent or solid state and ability to transfer energy [4, 5].

They also have very good heat and photochemical stability. Physical conditions are also very stable compounds, they are not much affected by the pH and polarity of the environment. Other advantages of Bodipy dyes are simple synthesis methods, good solubility in organic solvents and a narrow emission band. The photochemical

E. Guler (✉) · E. Bagci · A. N. Kursunlu
Faculty of Science, Department of Chemistry, University of Selcuk, Konya, Turkey
e-mail: eguler66@gmail.com

A. D. Pogrebnjak et al. (eds.), *Nanomaterials in Biomedical Application and Biosensors (NAP-2019)*, Springer Proceedings in Physics 244,
https://doi.org/10.1007/978-981-15-3996-1_19

properties of Bodipys can be readily adjusted by modification, such as emission and excitation wavelengths [6–8].

Bodipy and many derivatives make absorption and emission (500–700 nm) in the visible region. Bodipy, the basic state of singlet (S_0), is thought to result from the fluorescence of S_1 (which is the case for most organic chromophores), and the population of S_1 is assumed to be a simple HOMO → LUMO electron transition [6–8].

The light-induced electron transfer causes the transition from one group to another in the Bodipy compound of the electron excited by photon energy. This transition may be from the Bodipy core to the substituent or from the substituent to the Bodipy core.

Light-induced electron transfer is responsible for fluorescence quenching. This usually occurs in natural phenomena, which is very important in artificial systems and photosynthesis for the conversion of solar energy based on beam-effective charge separation.

The purpose of this study firstly a Bodipy derivative containing Knorr pyyroles were prepared. Changes in the emission, absorption and excitation properties of these compounds were then examined in the presence of some metal ions. Thus, in these measurements, it was determined that the final Bodipy derivative intended to be selective of metal ions were found and the compounds obtained were tested with different metal ions. However, only mercury has shown a selectivity for the metal. The prepared compound based-on Bodipy can be used as a fluorescent sensor for the Hg (II) ions.

19.2 Experiment Details

All chemicals used; isophthaloyl dichloride, 2-propin-1-amine, N1, N3-di (prop-2-yn-1-yl) isophthalamide, 1,4-bis (prop-2-yn-1-yloxy) benzene, benzene-1 2-diamine, sodium azide, lithium bromide, potassium iodide, copper sulfate, sodium ascorbate, potassium carbonate. The metals used are; chromium (Cr (III)), cobalt (Co (II)), copper (Cu (II)), zinc (Zn (II)), mercury (Hg (II)), lead (Pb (II)), cadmium (Cd (II)). Column chromatography, synthesis and characterization applications, dimethylformamide, tetrahydrofuran, dichloromethane, ethanol, ethyl acetate, petroleum ether (40–60%), n-hexane, chloroform, methanol, toluene, acetone, dimethyl sulfoxide, deutero chloroform, deutero dimethylsulfoxide, deutero methanol, acetonitrile, diethylether solvents were used.

Since the pyrrole required for the reaction medium deteriorated very quickly at light and temperature, it was distilled and used fresh each time. However, dry-dichloromethane used in Bodipy syntheses was obtained with the aid of a special apparatus and calcium hydride. Elemental analysis of the C, N and H atoms in the molecule was carried out on the Bruker brand elemental analyzer. ^{1}H-NMR, ^{13}C-NMR, ^{11}B-NMR, ^{19}F-NMR spectra were performed with a Varian 400 MHz spectrometer.

19.2.1 The Synthesis of Bodipy Including Alkly-Chloro (Compound 1)

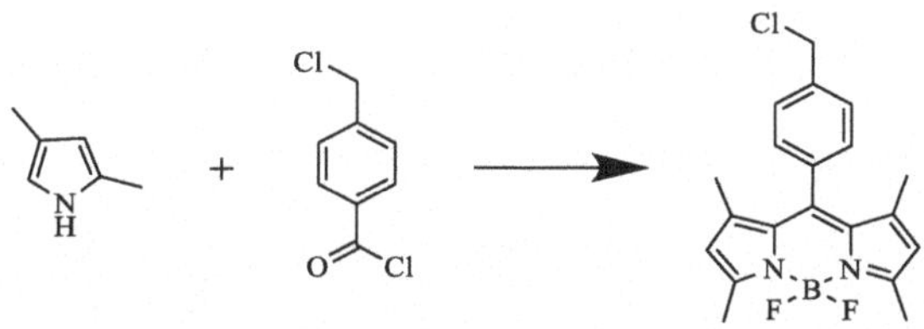

4-(chloromethyl) benzoyl chloride (1.5 g, 7.9 mmol) was added dropwise to a solution of 2,4-dimethyl pyrrole (1.65 mL) in dry dichloromethane (150 mL) at room temperature and under N_2. The solution was stirred in the dark for 2 h (at 60 °C). After cooling the solution, triethylamine (TEA) (10 mL) was added to the remaining solid, the mixture was stirred under N_2 for 30 min at room temperature, and then boron trifluoride diethyl etherate (7 eq) was added. The solution was stirred at 60 °C for 2 h and the resulting residue was purified by column chromatography (petroleum ether-EtOAc; 7: 1 ratio). An orange solid.

FT-IR values: 1640 cm^{-1} (C=N), 1600–1450 cm^{-1} (C=C), 3000–2880 cm^{-1} (C–H), (1.75 g, Yield 41%). M.P.:189 °C. ^{1}H NMR [400 MHz, $CDCl_3$]: 7.48 (d, 2H, ArH), 7.25 (d, 2H, ArH), 6.02 (s, 2H, Pyrrole-H) 5.18 (s, 2H, CH_2), 2.53 (s, 6H CH_3) 1.25 (s, 6H, CH_3).

19.2.2 The Synthesis of 8-(azidomethyl)-4,4-difluoro-1,3,5,7-tetramethyl-4-bora-3a, 4a-diaza-s-indacene (Compound 2)

A mixture of compound 1 (2.3 g, 6.2 mmol), NaN_3 (1.2 g, 18.6 mmol) and LiBr (0.54 g, 6.2 mmol) in DMF (30 mL) was stirred at room temperature for 24 h using a magnetic stirrer. H_2O (15 mL) was then added. The aqueous layer was extracted with ethylacetate. The combined organic layers were washed with water, dried with $MgSO_4$, filtered and concentrated. Column chromatography afforded a red solid, i.e., Compound 2.

FT-IR values: 1642 cm^{-1} (C=N), 1600–1450 cm^{-1} (C=C), 3000–2870 cm^{-1} (C–H), 3400 cm^{-1} (NH), M.P.: 140 °C. ^{1}H-NMR [400 MHz, $CDCl_3$]: 7.51 (d, 2H, ArH), 7.32 (d, 2H, ArH), 5.99 (s, 2H, Pyrrole-H), 4.49 (s, 2H, CH_2), 2.52 (s) 6H, CH_3), 1.27 (s, 6H, CH_3).

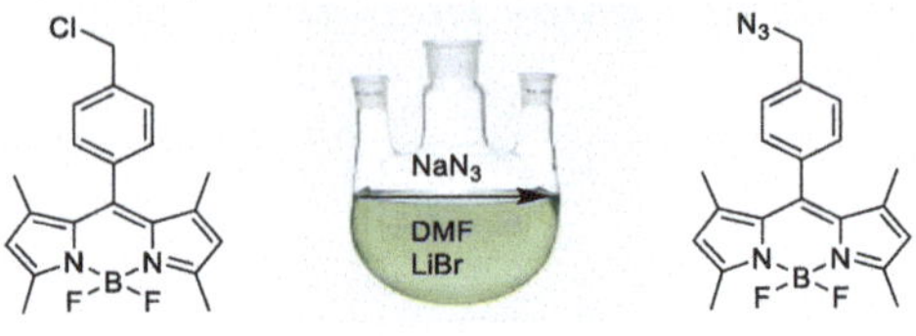

1 2

19.2.3 Synthesis of N^1, N^3-di (Prop-2-yn-1-yl) Isophthalamide (Compound 3)

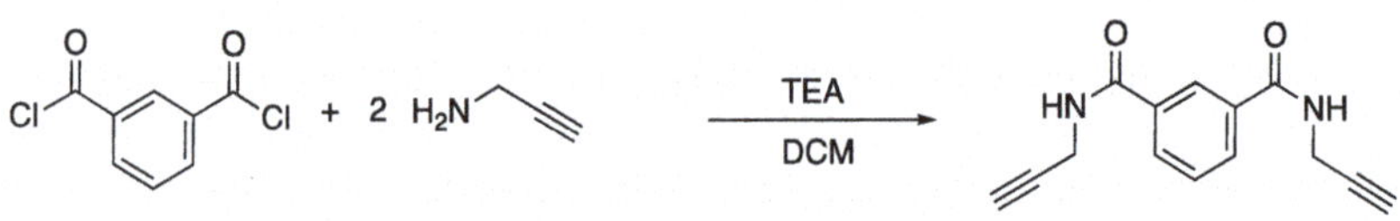

To a solution of isophthaloyl dichloride (167.0 mg, 1 mmol) in dichloromethane was added 0.5 mL of triethylamine at −5 °C (salt-ice). 0.4 mL of propargylamine was then added dropwise to this solution at room temperature and stirred for 48 h. The mixture was extracted in a DCM-water mixture and the crude product was extracted into the organic phase. The organic phase was dried with $MgSO_4$ and the solvent was evaporated in vacuo. The crude product was purified on the column (Petroleum ether 40–60%/EtOAc;1:1). A white solid was obtained.

FT-IR values: 2800–3000 cm^{-1} (CH), 1640 cm^{-1} (C=N), 1596–1448 cm^{-1} (C=C), 1682 cm^{-1} (C=O), 3370–2900 cm^{-1} (NH) Yield: 81%. ^{1}H-NMR [400 MHz, DMSO]: 9.07 (2H, t, J = 5.51 Hz), 8.34 (1H, t, J = 1.61 Hz), 8.02–7.95 (2H, m), 7.56 (1H, t, J). = 7.75 Hz), 4.10–4.08 (4H, m), 3.15 (2H, t, J = 2.49 Hz); ^{13}C-NMR [100 MHz, DMSO]: d (ppm); 165.6, 133.8, 129.9, 128.3, 126.1, 80.9, 72.6, 28.

19.2.4 The Synthesis of Bodipy

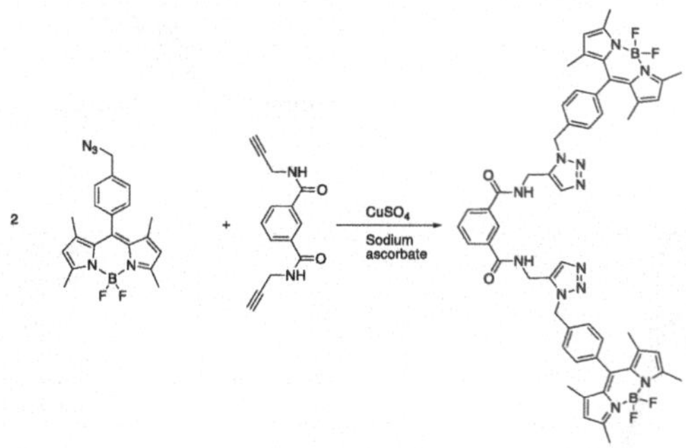

In a mixture of ethanol and water (45:5), compound 2 (0.171 g, 0.45 mmol), N^1, N^3-di (prop-2-yn-1-yl) isophthalamide (0.054 g, 0.22 mmol), $CuSO_4$ (0.034 g) (0.21 mmol) and sodium ascorbate (0.15 g, 0.76 mmol) were stirred for 3 nights with the aid of a magnetic stirrer. Purified by column chromatography (petroleum ether-EtOAc; 7:1 ratio) to give a viscous solid.

0.444 g, Yield: 40%. ^{1}H NMR [400 MHz, $CDCl_3$]: 8.25 (NH, s, 2H), 8.22 (s, 2H, ArH), 8.01 (d, 4H, ArH), 7.55 (d, 4H, ArH), 7.35 (m, 4H, ArH), 5.90 (s, 4H, Pyrrole-H), 5.19 (s, 4H, CH_2), 3.90 (s, 4H, CH_2) 2.35 (s, 12H, CH_3), 1.20 (s, 12H, CH_3). ^{13}C NMR [100 MHz, $CDCl_3$]: δ (ppm); 195.20, 161.02, 151.32, 144.22, 142.44, 141.38, 139.18, 137.22, 136.53, 135.33, 135.11, 132.31, 130.22, 129.41, 128.03, 125.21, 124.47, 123.21, 120.80, 112.44, 110.35, 59.24, 43.22, 13.44, 12.33, 10:43.

19.3 Results and Discussion

In the present study, a Bodipy based ligand was obtained as the final product. In the second stage of this study, The solutions of the Bodipy derivative and metal salts [Hg (II), Pb (II), Cd (II), Co (II), Cr (III), Cu (II), Zn (II),)] were prepared at concentrations. With these solutions, solution mixtures were prepared with a ligand/metal ratio of 1/20 and spectroscopic applications (absorption, emission and excitation, etc.) were initiated. In addition, changes in spectroscopic data, complex stoichiometry (by Job method) and determination of quenching constants due to increase in metal concentration were investigated.

As a result of this data, it was determined which metal ion is more selective for the Bodipy derivative and the degree to which it can be used as a chemosensor due to spectroscopic changes.

In the absorption study for Bodipy, ligand solution at a concentration of 1.10^{-6} M and metal salt solutions at a concentration of 1.10^{-5} M were prepared. These two solutions were mixed in a 1:1 ratio and spectroscopic applications such as absorption

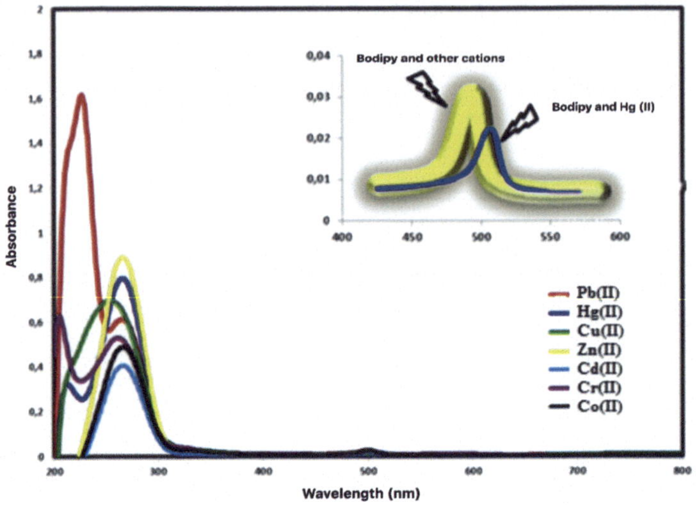

Fig. 19.1 Absorption spectra of Bodipy in presence of Pb (II), Hg (II), Cu (II), Zn (II), Cd (II), Cr (II) and Co (II)

and emission were performed. Metal salts identified as metal ion sources were used here [Cu (II), Zn (II), Hg (II), Cr (III), Co (II), Pb (II), Cd (II)]. After the addition of metal salts, other metal ions except mercury (II) ions did not cause a significant change in the absorption spectrum of Bodipy. However, a charge transfer from the Bodipy nucleus to the mercury was observed and revealed the Hg (II) sensitivity of Bodipy (Fig. 19.1).

Emission measurements were carried out to support absorption measurements. The prepared metal ion-Bodipy solutions were excited at 470 nm excitation to obtain fluorescence spectra. The emission band of Bodipy appeared around 545 nm. The emission spectra of the prepared metal ion-Bodipy solution mixtures were taken. The emission band around 545 nm showed no significant effect while other metal ions showed a significant quenching effect due to Hg (II) ions. Emission band strength decreased from 450 units to 30 units. The damping of the emission of Bodipy with Hg (II) ion sensitivity can be attributed to the paramagnetic character of the mercury complexes and it is possible to talk about a PET type charge transfer event due to the chelate effect with triazole nitrogens (Fig. 19.2).

Complex stoichiometry experiments for Bodipy with proven mercury (II) sensitivity were performed with the Job method. The Hg (II) molar fraction graph corresponding to the emission intensity gave a maximum at 0.5. Thus, the metal: ligand ratio was determined to be 1:1 (Fig. 19.3).

Here, the binding constants were calculated using the Stern-Volmer equation. This equation;

$$I_0/I = 1 + K_{sv}[M]$$

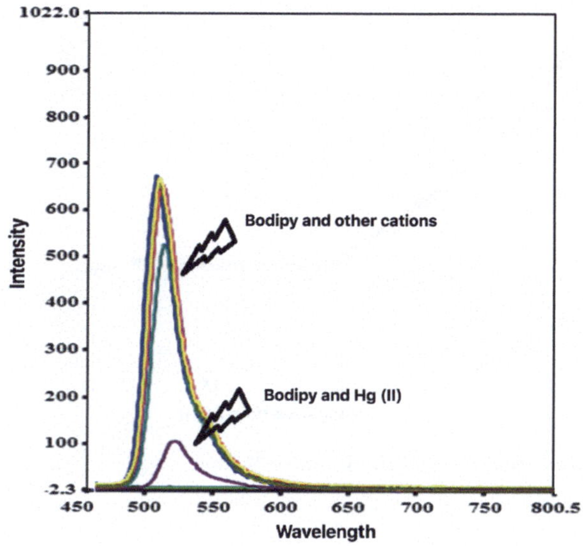

Fig. 19.2 Emission spectra of Bodipy in presence of Pb (II), Hg (II), Cu (II), Zn (II), Cd (II), Cr (II) and Co (II)

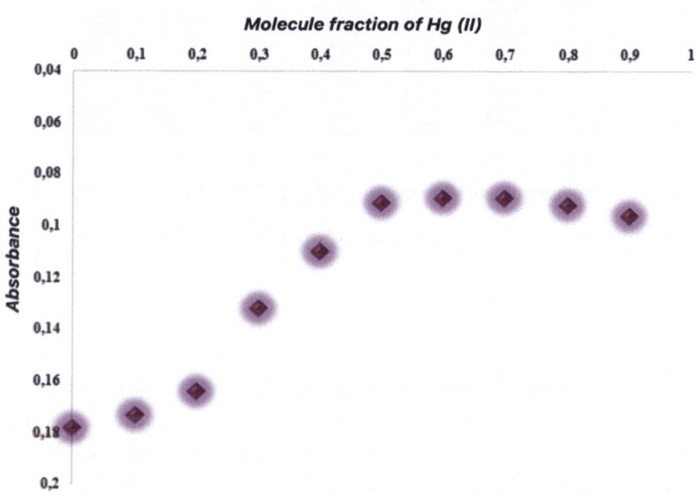

Fig. 19.3 The complex stoichiometry between Bodipy and Hg (II) ion with Job method

and I_0; In the absence of Hg (II) ion, the emission intensity of Bodipy is I; the emission intensity in the presence of Hg (II) ions, Q; Hg (II) ion concentration and K is the binding constant.

In order to understand how fast the interaction of Bodipy with Hg (II) sensitivity with mercury ion takes place, reaction time experiments were performed. For this,

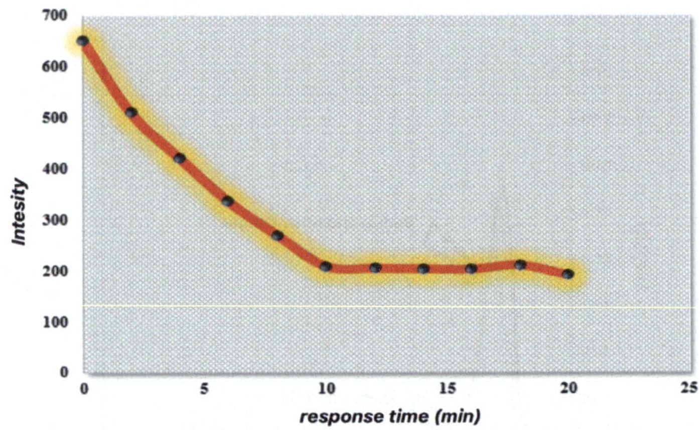

Fig. 19.4 The response time effect of Hg (II) ion on Bodipy

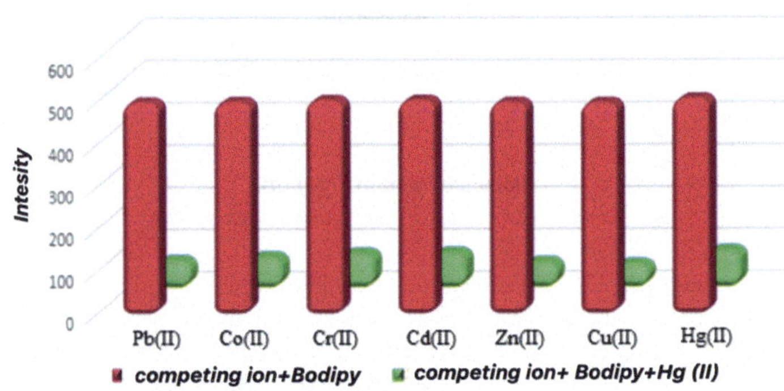

Fig. 19.5 The competing ion effect on selectivity of Bodipy for Hg (II) ion

0, 2, 4, 6, 8, 10, 12, 14, 16, 18 and 20 min were determined. The emission intensity reached a fixed value after 10 min. Thus, the most accurate time for the Hg (II) sensitivity of Bodipy 1 was found to be 10 min (Fig. 19.4).

Competing ion experiments were performed to investigate the selectivity of Bodipy against Hg (II) ions. It only affected mercury ion selectivity. In general, the other metal ions studied did not affect the Hg (II) selectivity of Bodipy. Accordingly, it has clearly demonstrated that Bodipy can be used as an Hg (II) sensor (Fig. 19.5).

19.4 Conclusions

The absorption and emission spectra of Bodipy differed only in the presence of Hg (II) ion when spectroscopic studies of Bodipy were designed to include two-way Bodipy nucleus. So much so that the absorption bands; the emission band was markedly reduced. These data revealed the Hg (II) sensitivity of the ligand. Competitive ion tests for Hg (II) selectivity were performed for Bodipy whose sensitivity study was completed and other metal ions were studied.

Hg (II) did not affect selectivity. This result was supported by other spectroscopic experiments (complex stoichiometry, binding constant, etc.). It was concluded that Bodipy can be used as a sensitive and selective fluorescence sensor for Hg (II) ion.

Acknowledgements This research was carried out within the framework of the joint scientific project of the Selcuk University BAP (project Number: 19201054), 2019. Authors thanks to Selçuk University BAP, Prof. Dr. Zafer Yazicigil and Dr. Tugce Gover.

References

1. C. Baki N., E.U. Akkaya, J. Org. Chem. **66**, 1512–1513 (2001)
2. R. Bandichhor, C. Thivierge, N.S.P. Bhuvanesh, K. Burgess, Acta Cryst. Sect. E Struct. Reports Online **62**(10), 4310–4311 (2006)
3. M. Baruah, W. Qin, C. Flors, J. Hofkens, R.A. Valle, D. Van Beljonne et al., J. Phys Chem. A **110**(18), 5998–6009 (2006)
4. O.A. Bozdemir, F. Sozmen, O. Buyukcakir, R. Guliyev, Y. Cakmak, E.U. Akkaya, Organic. Lett. **12**(7), 1400–1403 (2010)
5. O.A. Bozdemir, O. Buyukcakir, E.U. Akkaya, Chem. A Eur. J. **15**(15), 3830–3838 (2009)
6. T. Brixner, J. Stenger, H.M. Vaswani, M. Cho, R.E. Blankenship, G.R. Fleming, Nature **434**, 625–628 (2005)
7. H. Burghart, M.B. Kim, L. Welch, H. Thoresen, J. Reibenspies et al., J. Org. Chem. **64**, 7813–7819 (1999)
8. H. Chang, O. El-Kabbani, D. Tiede, J.R. Norris, M. Schiffer, Biochemistry **30**, 5352–5360 (1991)

Chapter 20
Resistance of Hall Sensors Based on Graphene to Neutron Radiation

I. A. Bolshakova, Ya. Ya. Kost, M. I. Radishevskyi, F. M. Shurygin, O. V. Vasyliev, Z. Wang, D. Neumaier, M. Otto, M. V. Bulavin, and S. A. Kulikov

Abstract An in-situ study of Hall sensors based on single-layer graphene in neutron fluxes of a nuclear reactor up to the fluence of 1.5×20 n·m^{-2} was conducted. The sensitivity of the sensors to the magnetic field remained stable throughout the experiment, while the resistance changes correlated with the increase in sample temperature due to radiation heating. The experiment confirmed the theoretical expectations regarding the high resistance of graphene sensors to neutron irradiation. Necessary further improvement of sensor technology to optimize their characteristics, as well as irradiation testing to determine the maximum permissible neutron fluence.

20.1 Introduction

Single-layer graphene has unique electrophysical properties that allow it to be used to create high-performance electronic devices [9]. Thus, Hall sensors based on exfoliated graphene have a record high current-related magnetic field sensitivity (~5700 V·A^{-1}·T^{-1}), which is much larger than that of sensors based on traditional semiconductor materials such as Si (~100 V·A^{-1}·T^{-1}), GaAs (~1100 V·A^{-1}·T^{-1}) thin films, InAlSb/InAsSb/InAlSb nanosized heterostructures with two-dimensional electron gas (~2750 V·A^{-1}·T^{-1}) [7]. This high sensitivity is due to the small thickness of the active element of the sensor, which is the smallest possible with the use of graphene (only one layer of atoms), as well as high mobility and low density of charge carriers in graphene [18].

The great advantage of graphene compared with "bulk" materials is high resistance to the impact of corpuscular radiation, observed in experiments with charged-particle

I. A. Bolshakova (✉) · Ya. Ya. Kost · M. I. Radishevskyi · F. M. Shurygin · O. V. Vasyliev
Magnetic Sensor Laboratory, Lviv Polytechnic National University, Lviv, Ukraine
e-mail: inessa.bolshakova@gmail.com

Z. Wang · D. Neumaier · M. Otto
Advanced Microelectronic Center Aachen, AMO GmbH, Aachen, Germany

M. V. Bulavin · S. A. Kulikov
Frank Laboratory of Neutron Physics, Joint Institute for Nuclear Research, Dubna, Russia

A. D. Pogrebnjak et al. (eds.), *Nanomaterials in Biomedical Application and Biosensors (NAP-2019)*, Springer Proceedings in Physics 244,
https://doi.org/10.1007/978-981-15-3996-1_20

beams [20, 13, 14]. This feature is primarily attributed to the lack of a bulk crystal structure in graphene: this reduces the probability of collision of a bombarding particle with the sample, and in the event of such a collision makes it impossible to develop a large-scale cascade of atomic displacements, which minimizes the size of the material's damage [21]. Moreover, it has been proved that graphene is almost "transparent" to beams of light charged particles in certain ranges of their energy [16, 4], which even allows to develop on the basis of graphene the beam windows for the extraction of high-energy proton beams in powerful accelerators [22]. The second reason for the high resistance of graphene to irradiation is the effects of "self-healing" of radiation defects that are absent in "bulk" materials [20]. In graphene, they are realized, first of all, through thermally activated processes—a reordering of displaced atoms, as well as a trapping of adatoms by vacancies and nanoholes [3, 24].

Indicated features make graphene promising for the development of a new generation of irradiation-resistant Hall sensors, which, in particular, will enable to create effective systems for the plasma's magnetic diagnostics in the steady-state thermonuclear fusion reactors [19]. These facilities, in particular, include ITER, a research tokamak, which is being built in Cadarache (France), as well as DEMO, a prototype of the fusion power plant, which should give the first electric power in the middle of the 21st century. In DEMO, Hall sensors will be placed behind the blanket where the high fluxes of fast neutrons are present. It is expected that the total fluence of neutrons that sensors accumulate during the reactor lifetime will reach $F \approx 2\times10^{26}$ $n{\cdot}m^{-2}$ [1]. No device based on traditional semiconductors can operate at such high irradiation loads.

Unfortunately, today very little is known about the impact of neutron irradiation on single-layer graphene. Theoretical estimates predict the high stability of its crystal structure in the neutron fluxes, because due to the neutron's small size and absence of electric charge, the probability of its collision with the graphene atoms is only $\sim10^{-5}$ [6]. These expectations are partly confirmed by experimental studies, which, however, use an indirect analysis of the structure and consist in comparing the Raman spectra of graphene before and after irradiation. This approach has shown that irradiation with the fluence of $F \approx 1.6\times10^{21}$ $n{\cdot}m^{-2}$ forms a very small number of structural defects in graphene, which was manufactured by the Chemical Vapor Deposition (CVD) method and transferred onto the Si/SiO_2 substrates [12]. On the other hand, the authors of this work have shown in their previous studies that CVD-graphene on sapphire substrates maintains a high quality of its structure, at least up to the fluence $F \approx 4.5\times10^{19}$ $n{\cdot}m^{-2}$ [2].

The purpose of this work was an in-situ study of the functional characteristics of Hall sensors based on single-layer graphene during their irradiation with neutron fluxes in a research nuclear reactor.

20.2 Experiment Details

20.2.1 Hall Sensors Samples

Single-layer graphene, grown by the CVD method on copper foil (Graphenea, Spain), was used for manufacturing sensors. As a substrate, sapphire (Al_2O_3) with the thickness of 400 μm was used. Graphene was transferred to the substrate according to the procedure described in [15].

The sensors' active elements were made using photolithography and had a symmetrical four-lead cross-like topology (insert in Fig. 20.1), with intersection area dimensions (100×100) μm^2. In order to ctreate the electrical top-contacts to graphene, the layers of Ni (20 nm) and Au (400 nm) were successive deposed onto the ends of active elements arms. The active elements were protected with a layer of Al_2O_3 (80 nm) deposited by the Atomic Layer Deposition (ALD) method.

After completion of the technological procedures, the sapphire plate was diced into chips with sizes of ~(1×1) mm^2, each with single sensor. The chips were glued to special ceramic holders with contact pads that were connected to the sensor pads by Au-wires (Ø 30 μm) with the help of ultrasonic bonding. The holder with two samples is shown in Fig. 20.1.

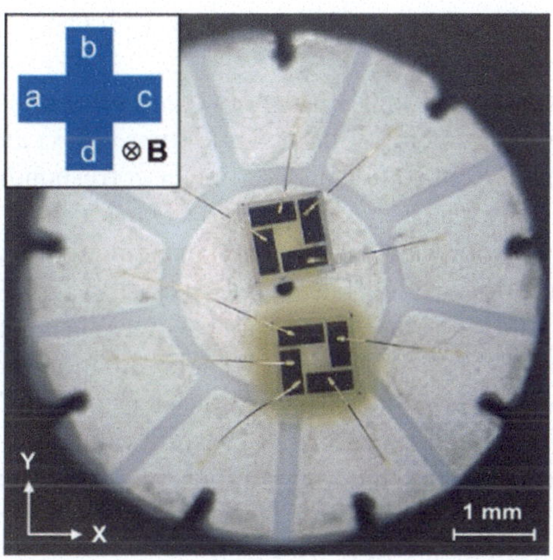

Fig. 20.1 Samples of sensors placed on a ceramic holder. Insert: schematic representation of active element

20.2.2 Sensor Characteristics Measurement Methods

The main investigated parameters in this work were sensors' current-related sensitivity S and graphene's surface resistance R_S. These characteristics were investigated during the selection of samples on laboratory equipment, as well as during the irradiation testing with the help of special control electronics.

When the Hall sensor is biased by a direct current I through one pair of opposite leads, the output signal on another pair S is defined as [17]:

$$S = \left| \frac{V_H(I, B)}{I \cdot B} \right|, \tag{20.1}$$

where I is biasing current; B is normal to the surface component of magnetic induction; $V_H(I, B)$ is Hall voltage.

The problem for S calculating is the presence at the sensor output of the residual voltage (offset) V_0, which is generated, primarily, because of inhomogeneities in the material of the active element, as well as asymmetry in its geometry and location of the leads [17]. The value V_0 does not depend on B, but depends on I and changes the sign when the current is inverted. Due to the offset existence, the output voltage of the sensor V is the sum:

$$V^{bd}(I^{ac}, B) = V_H^{bd}(I^{ac}, B) + V_0^{bd}(I^{ac}), \tag{20.2}$$

where the upper indices near I and V_H denote the leads (see Fig. 20.1) that were attached during the measurement to the positive (first index) and negative (second index) terminals of the current source and voltmeter correspondingly.

For a symmetrical sensor at $\mathbf{B} = const$ and $I = const$ a change in the connection of the arms to the current source and voltmeter in one direction ($I_{ac} \to I_{bd}$ and $V_{bd} \to V_{ca}$) does not change the offset value, but changes its sign ($V_0^{bd} = -V_0^{ca}$), while Hall's voltage preserves both its value and sign ($V_H^{bd} = V_H^{ca}$). When changing connections in different directions ($I_{ac} \to I_{bd}$ and $V_{bd} \to V_{ac}$) the situation is reversed. This is the basis of the spinning-current method, which allows to separate V_H and V_0 [17]. The method consists in sequential measurement of several values of the output voltage V, obtained for different connection's configurations, and their subsequent averaging. In this work, for the effective removal of the offset under the reactor environment, the results of four measurements were averaged:

$$V_H(I, B) = \frac{V^{bd}(I^{ac}, B) + V^{ca}\left(I^{bd}, B\right) + V^{db}(I^{ca}, B) + V^{ac}\left(I^{db}, B\right)}{4}. \tag{20.3}$$

Offset was determined by inversion of the current at $B = 0$:

$$V_0(I) = \frac{V^{bd}(I^{ac}, B = 0) - V^{bd}(I^{ca}, B = 0)}{2}. \tag{20.4}$$

For the implementation of the spinning-current method, special signal switches were used as part of the measuring equipment. It also allowed to determine the surface resistance of graphene R_S by the van der Pauw method, which involves measuring the voltage on a pair of neighboring leads while current supply is through another pair of terminals with subsequent re-switching [17]. For the selected geometry of the active element:

$$Rs = \frac{\pi}{2 \cdot \ln 2} \cdot \left[\frac{V^{bc}(I^{ad}, B = 0)}{I^{ad}} + \frac{V^{dc}(I^{ab}, B = 0)}{I^{ab}} \right]. \quad (20.5)$$

20.2.3 Temperature Testing for Samples Selection

At neutron irradiation the radiation heating of materials is taken place. In particular, preliminary estimates have shown that sensors will heat up to ~50 °C during the in-situ experiment in a nuclear reactor.

Therefore, the preliminary temperature testing of sensors in the laboratory was carried out on a special bench, built on the basis of Keithley and Tektronix high-precision measuring instruments. The bench implements spinning-current and van der Pauw methods, and also allows maintaining a constant temperature of sample T in the range up to 250 °C with accuracy ±0.5 °C.

For each sample, the test was carried out as follows: (i) characterization at $T_1 = 30$ °C; (ii) heating at a fixed rate up to $T_2 = 54$ °C; (iii) time-exposure to achieve thermal equilibrium; (iv) re-characterization. Figure 20.2 shows the distribution of samples on the levels of relative change in their parameters after heating. The area of histogram columns is proportional to the number of samples in them.

As can be seen from Fig. 20.2, increasing of T significantly affects S, V_0 and R_S. At the same time, the changes of these values for each sample are not correlated with each other: for example, a decrease in resistance is not necessarily accompanied by an increase in sensitivity and/or a decrease in the offset. This may indicate that the temperature affects simultaneously several factors that determine the electrophysical properties.

The strongest changes are demonstrated by sensitivity, Fig. 20.2a. Among the studied sensors, there are such ones for that S reduced by almost 90% after heating, but for individual samples, on the contrary, S increased by ~5000% (on Fig. 20.2a this range is not shown for illustrative purposes). Such transformations may be due to changes in the electrostatic interaction of graphene with a substrate and a protective layer, or due to certain uncontrolled aspects of manufacturing technology (resist residuals, nonuniform adhesion, etc.). Asymmetry of histograms on Fig. 20.2b and 2c reflects the fact that after heating the values V_0 and R_S decreased for most samples. For offset, this may be due to the relaxation of mechanical stresses in graphene (although for individual samples the value of V_0 has increased by ~500%). At the

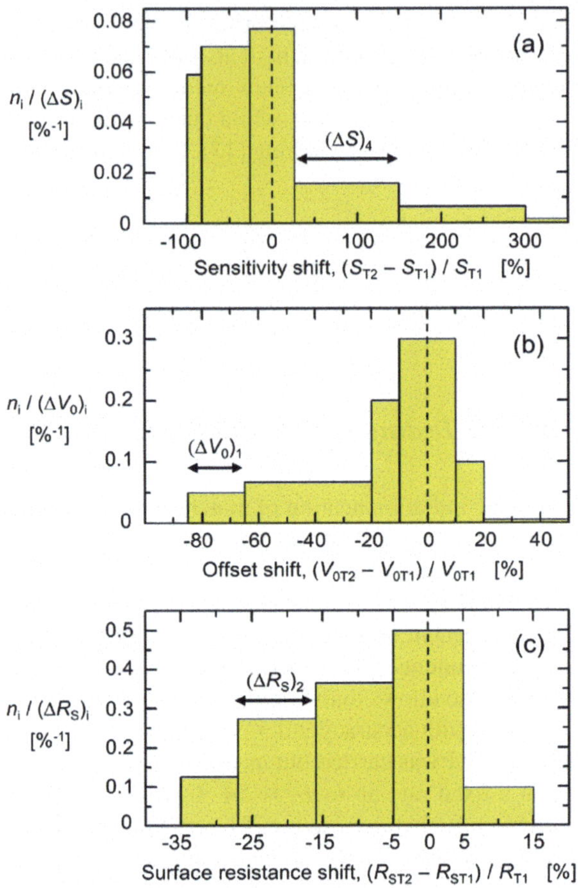

Fig. 20.2 Distribution of samples on the levels of relative change in: **a** the sensitivity S; **b** the offser V_0; **c** the surface resistance R_S during temperature change from $T_1 = 30$ °C to $T_2 = 54$ °C. n_i is the number of samples within the interval $(\Delta\ldots)_i$. Intervals $(\Delta S)_4$, $(\Delta V_0)_1$, $(\Delta R_S)_2$ shown for visualization purposes

same time, the behavior of R_S was not fully understood, as the resistance of graphene should increase from T. The possible reason for the reduction of R_S is some changes in metal/graphene contacts during heating, since the contact resistance may affect the results of measurements by the van der Pauw method [11].

In general, the change in S lies within ±25% for ~29% of the samples; change in V_0 – within ±10% for ~43% of samples; change in R_S – within ±5% for ~36% of samples.

For further investigation in neutron fluxes, it was selected the samples having a high sensitivity S and have demonstrated a satisfactory temperature stability of V_0 and R_S. The average values of their parameters were: $\langle S \rangle \approx 600$ V·A^{-1}·T^{-1}; $\langle V_0 \rangle \approx 150$ mV; $\langle R_S \rangle \approx 3000$ Ω/□.

20.2.4 In-Situ Investigations of Irradiation Stability

In-situ neutron irradiation experiment was conducted in the research nuclear reactor IBR-2 (the Joint Institute for Nuclear Research) in the irradiation channel № 3 [5]. The unique feature of IBR-2 is the pulse modulation of reactivity implemented with the help of movable reflectors, which allows to obtain in the channel № 3 a neutron flux with the high intensity of $\sim 1.5 \times 10^{17}$ n·m^{-2}·s^{-1} (on the surface of the moderator). Moreover, the IBR-2 is a fast-neutron reactor that provides large content (~38%) of fast particles (energy $E \geq 0.1$ MeV) in the neutron flux. These parameters are close to the operating conditions of Hall sensors in DEMO.

The general scheme of the experiment is shown in Fig. 20.3. For the placement of the investigated samples in the neutron flux, it was developed the special fixture based on the irradiation- and thermal-resistant ceramics Macor (Corning, USA). Each fixture contained a platinum resistance thermometer to monitor the samples temperature, as well as a copper solenoid, insert (a) in Fig. 20.3, to create a magnetic field **B**, normal to the Hall sensors surface. Holders with samples, Fig. 20.1, were fixed inside the fixture, which, in turn, was protected by a duralumin housing. Fixtures were installed on an T-beam of the irradiation facility, Fig. 20.3 insert (b), at a distance of 1.2 m from the moderator surface.

For conducting in-situ measurements, it was developed a control electronics that was located outside the irradiation channel at a distance of ~15 m from the samples. To connect samples with electronics, it was used the home-made irradiation-resistant cables like a twisted pair. To manage the electronics operation, data acquisition and processing, it was used PC with special software. The results of the measurements were stored on the remote server.

When conducting an in-situ experiment, the current supply of the sensors was I = 100 μA, and magnetic field was $B = 7$ mT. Neutron flux in the samples location

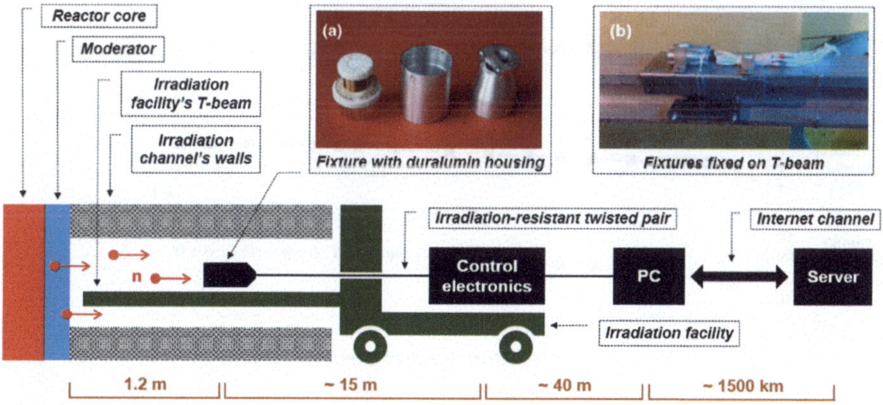

Fig. 20.3 General scheme of the in-situ experiment in the IBR-2 reactor (irradiation channel № 3, side view). Inserts: **a** photo of assembled fixture with protective housing; **b** photo of fixtures with communication lines, placed on a T-beam of the irradiation facility

was ~8.7×10^{14} n·m^{-2}·s^{-1}, which corresponds to the irradiation load level on Hall sensors expected in ITER.

20.3 Results and Discussion

Graphene sensors were irradiated for ~48 h and accumulated fluence $F \approx 1.5\times 10^{20}$ n·m^{-2}. Upon completion of the experiment, 75% of the irradiated samples kept their operability, while the rest stopped conducting the current, and this happened at different values of F. The possible cause of such behavior is the features of metal/graphene top contact. As is known, transferred graphene has weaker adhesion to dielectric layers than to metal ones [23]. At the same time, the thermal contraction of the top contacts, which occurs after the deposition of metals and/or bonding of external leads, may further weaken the adhesion of graphene to sapphire. As a consequence, the probability of graphene detachment from the substrate in the area under the contact pads increases, and, hence, there is a risk of further breaking of the graphene layer continuity under the influence of uncontrolled external factors.

Figure 20.4a, b respectively show measured in-situ dependencies of sensitivity S and surface resistance R_S on fluence F at $F \geq 8.5\times10^{18}$ n·m^{-2} for one of the sensors. For other samples, these dependencies are qualitatively similar. In addition, Fig. 20.4c shows the dependence on the sample temperature $T(F)$, from which it is clear that the sensors during the experiment did not heat up above 55 °C.

The irradiation almost did not change sensors sensitivity: the difference between the initial and final values of S on Fig. 20.4a is ~3%, which is within the measurement error in the noisy environment of the nuclear reactor. However, $S(F)$ has a weakly expressed maximum at $F \approx 7\times10^{19}$ n·m^{-2}. Such a peculiarity cannot be

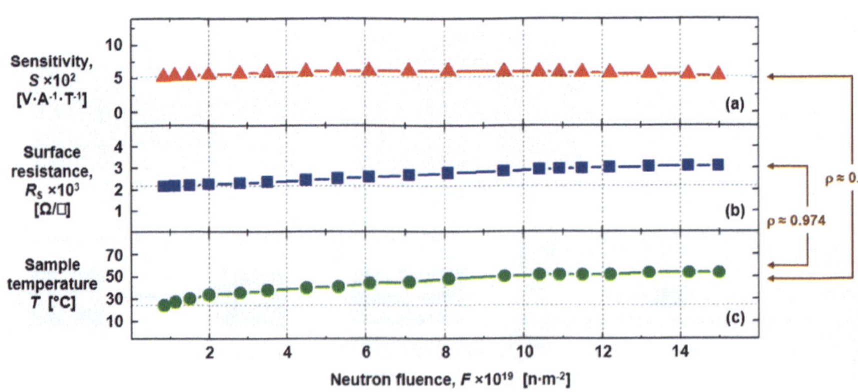

Fig. 20.4 Dependences of the parameters of one of the Hall sensors based on single-layer graphene on the neutron fluence obtained in-situ during the irradiation process: **a** sensitivity $S(F)$; **b** surface resistance $R_S(F)$; **c** temperature $T(F)$. Initial values of sample characteristics ($F = 0$): $S \approx 530$ V·A^{-1}·T^{-1}; $R_S \approx 2150$ Ω/□; $T \approx 25$ °C. ρ—Pearson's correlation coefficient

explained by damage to the crystal structure of graphene by neutrons, since in previous works the authors proved that graphene on a sapphire substrate is defect-free both at lower (4.6×10^{19} n·m^{-2}) and at higher (4.1×10^{20} n·m^{-2}) values of F [2]. On the other hand, $S(F)$ weakly correlates with $T(F)$: Pearson's correlation coefficient $\rho \approx 0.416$. Therefore, the maximum on $S(F)$ also cannot be related exclusively to thermal effects. In addition, during the temperature testing in the laboratory, the dependence $S(T)$ for all samples showed a linear character without extremums in the range 30 °C $\leq T \leq$ 54 °C. The possible cause of the maximum on Fig. 20.4a is that there are specific radiation-stimulated effects in the surrounding materials, resulting in a change in their electrostatic influence on graphene. This should, appropriately, affect the charge carriers density, which is inversely proportional to the sensitivity [17].

In contrast to sensor sensitivity, the graphene surface resistance demonstrated a much greater change after the end of in-situ experiment: the difference between the final and the initial value R_S in Fig. 20.4b is ~44%. Most probably, this result is related to the change in sample temperature T, as evidenced by strong correlation of the $R_S(F)$ and $T(F)$ data, $\rho \approx 0.974$. At the same time, as the literature analysis shows, electrical resistance of graphene is much less dependent on T. For example, the approximation of the results given in [8] for graphene on the SiO_2/Si substrate, shows that when heated from $T = 30$ °C to $T = 55$ °C the resistance of graphene increases only by ~6%. The possible cause of the observed behavior of R_S is the influence of metal/graphene contact resistance on the results of measurements by the van der Pauw method [11]. Neutrons can change the properties of these contacts, in particular, enhance resistance, as it occurs, for example, at the irradiation with low-energy electrons [10].

20.4 Conclusion

For the first time in the world, it was studied in-situ in the neutron fluxes of a research nuclear reactor the characteristics of Hall sensors based on CVD single-layer graphene transferred onto a sapphire substrate. The samples maintained a stable value of sensitivity to the magnetic field throughout the experiment that lasted until the fluence reached 1.5×10^{20} n·m^{-2}. The change in the surface resistance of graphene correlates with a change in temperature, the growth of which is due to radiation heat. The obtained results confirm the predictions regarding the high irradiation resistance of the graphene-based sensors in the neutron fluxes and also indicate their suitability for use in the plasma's magnetic diagnostic systems of fusion reactors. Further tests in the neutron fluxes will allow to determine the limit value of the fluence, at which the graphene sensors still remain operable. However, for conducting such research it is advisable to solve a number of technological problems, in particular, to optimize the selection of surrounding materials to minimize external impact on graphene during the irradiation and heating. This will allow to improve the stability of the samples, to reduce the spread of their parameters, as well as to improve the metal/graphene electrical contacts.

Acknowledgements The work was supported by the Ministry of Education and Science of Ukraine and the Federal Ministry of Education and Research of Germany (Joint R&D Project M74-2018/RaMaG); the European Commission within the Graphene Flagship project (Contract No. 785219); the Joint Institute for Nuclear Research (Project 04-4-1122-2015/2020). The authors express their gratitude to Horst Windgassen for helping in deposition of gold for contact pads of sensors, and Stefan Scholz for helping in the bonding of gold leads.

References

1. W. Biel, R. Albanese, R. Ambrosino, M. Ariola, M.V. Berkel, I. Bolshakova et al.: Diagnostics for plasma control—from ITER to DEMO. Fusion Eng. Des. **146**(A), 465 (2019)
2. I. Bolshakova, D. Dyuzhkov, Ya. Kost, M. Radishevskiy, F. Shurigin, O. Vasyliev et al., in *Abstract Book of Graphene Week 2018 Conference*, San Sebastian, Spain, 10–14 September 2018, We-39
3. T. Botari, R. Paupitz, P. Alves da Silva Autreto, D.S. Galvao: Graphene healing mechanisms: a theoretical investigation. Carbon **99**, 302 (2016)
4. J. Buchheim, R.M. Wyss, I. Shorubalko, H.G. Park.: Understanding the interaction between energetic ions and freestanding graphene towards practical 2D perforation. Nanoscale **8**(15), 8345 (2016)
5. M. Bulavin, A. Cheplakov, V. Kukhtin, E. Kulagin, S. Kulikov et al.: Irradiation facility at the IBR-2 reactor for investigation of material radiation hardness. Nucl. Instrum. Meth. B **343**, 26 (2015)
6. D. Codorniu-Pujals, A. Bermudez-MartΗnez, arXiv (2012). https://arxiv.org/abs/1210.4099
7. J. Dauber, A.A. Sagade, M. Oellers, K. Watanabe, T. Taniguchi, D. Neumaier, C. Stampfer: Ultra-sensitive hall sensors based on graphene encapsulated in hexagonal boron nitride. Appl. Phys. Lett. **106**(19), 193501 (2015)
8. B. Davaji, H.D. Cho, M. Malakoutian, J.-K. Lee, G. Panin et al.: A patterned single layer graphene resistance temperature sensor. Sci. Rep. **7**, 8811 (2017)
9. A.C. Ferrari, F. Bonaccorso, V. Falko, K.S. Novoselov, S. Roche, P. Boggild et al.: Science and technology roadmap for graphene, related two-dimensional crystals, and hybrid systems. Nanoscale **7**(11), 4598 (2015)
10. F. Giubileo, A. Di Bartolomeo, N. Martucciello, F. Romeo, L. Iemmo, P. Romano et al.: Contact resistance and channel conductance of graphene field-effect transistors under low-energy electron irradiation. Nanomater. **6**(11), 206 (2016)
11. G. Gonzalez-Diaz, D. Pastor, E. Garcia-Hemme, D. Montero, R. Garcia-Hernansanz, J. Olea et al.: A robust method to determine the contact resistance using the van der Pauw set up. Measurement **98**, 151 (2017)
12. C.N. Kryworuk, Thesis for the degree of Master of Science, Virginia Polytechnic Institute and State University, 2013
13. S. Kumar, A. Tripathi, S.A. Khan, C. Pannu, D.K. Avasthi: Radiation stability of graphene under extreme conditions. Appl. Phys. Lett. **105**, 133107 (2014)
14. S. Lee, J. Seo, J. Hong, S.H, Park, J.-H. Lee, B.-W. Min et al.: Proton irradiation energy dependence of defect formation in graphene. Appl. Surf. Sci. **344**, 52 (2015)
15. X. Li, Y. Zhu, W. Cai, M. Borysiak, B. Han, D. Chen et al.: Transfer of large-area graphene films for high-performance transparent conductive electrodes. Nano Lett. **9**(12), 4359 (2009)
16. J.B. Park, Y.-J. Kim, S.-M. Kim, J.M. Yoo, Y. Kim, R. Gorbachev et al.: Non-destructive electron microscopy imaging and analysis of biological samples with graphene coating. 2D Mater. **3**(4), 045004 (2016)

17. R.S. Popovic, *Hall Effect Devices*, 2nd edn. (IOP Publishing, Bristol, 2004)
18. G. Song, M. Ranjbar, R.A. Kiehl: Operation of graphene magnetic field sensors near the charge neutrality point. Commun. Phys. **2**, 65 (2019)
19. E.J. Strait, E.D. Fredrickson, J.-M. Moret, M. Takechi: Chapter 2: Magnetic diagnostics. Fusion Sci. Technol. **53**(2), 304 (2008)
20. L. Sun, F. Banhart, J. Warner: Two-dimensional materials under electron irradiation. MRS Bull. **40**(1), 29 (2015)
21. R.C. Walker II, T. Shi, E.C. Silva, I. Jovanovic, J.A. Robinson: Radiation effects on two-dimensional materials. Phys. Status Solidi A **213**(12), 3065 (2016)
22. H. Wang, C. Meng, H. Qu, X. Sun, P. Wang, D. Zhu, in *Proceedings of 9th International Particle Accelerator Conference*, Vancouver, Canada, 29 April–4 May 2018, pp. 47–50
23. T. Yoon, W.C. Shin, T.Y. Kim, J.H. Mun, T.-S. Kim, B.J. Cho: Direct measurement of adhesion energy of monolayer graphene as-grown on copper and its application to renewable transfer process. Nano Lett. **12**(3), 1448 (2012)
24. R. Zan, Q.M. Ramasse, U. Bangert, K.S. Novoselov: Graphene reknits its holes. Nano Lett. **12**(8), 3936 (2012)

Chapter 21
Adhesive and Barrier Sublayers for Metal Nanofilms Active Elements of Hall Sensors

I. A. Bolshakova, Ya. Ya. Kost, M. I. Radishevskyi, F. M. Shurygin, O. V. Vasyliev, I. S. Vasil'evskii, and T. Kuech

Abstract It was investigated the thermal stability of two types of Hall sensors based on gold nanofilms deposited on sapphire: (i) with the titanium adhesive sublayer (Ti/Au) and (ii) with the titanium adhesive sublayer and the platinum barrier sublayer of (Ti/Pt/Au). Vacuum annealing for 3 h at 400 °C significantly changes all investigated electrophysical parameters of samples with the Ti/Au metallization, which is explained by the titanium diffusion into gold. At the same time, the sensitivity and the resistivity of sensors with the Ti/Pt/Au metallization remain unchanged within a few percent, allowing them to be used in the plasma's magnetic diagnostic systems of the ITER and DEMO fusion reactors where operating temperatures will exceed 300 °C.

21.1 Introduction

It is assumed that the plasma's magnetic diagnostic system in the new generation's steady-state fusion reactors ITER and DEMO will include, as the front-end components, the Hall sensors with the high irradiation and thermal resistance, since the high neutron fluences $F \approx (10^{24}–10^{26})$ nm^{-2} and temperatures $T \geq 300$ °C are expected in their locations [1]. Traditional semiconductors, which form the basis of modern sensor technics, rapidly degrade in such environment, so the development of sensors based on other types of functional materials is topical.

Earlier, the authors of this work have shown the promise of the use of polycrystalline gold (Au) nanofilms in the ITER- and DEMO-relevant neutron fluxes [3].

I. A. Bolshakova (✉) · Ya. Ya. Kost · M. I. Radishevskyi · F. M. Shurygin · O. V. Vasyliev
Magnetic Sensor Laboratory, Lviv Polytechnic National University, Lviv, Ukraine
e-mail: inessa.bolshakova@gmail.com

I. S. Vasil'evskii
Institute of Functional Nuclear Electronics, National Research Nuclear University MEPhI, Moscow, Russia

T. Kuech
College of Engineering, University of Wisconsin-Madison, Madison, USA

A. D. Pogrebnjak et al. (eds.), *Nanomaterials in Biomedical Application and Biosensors (NAP-2019)*, Springer Proceedings in Physics 244,
https://doi.org/10.1007/978-981-15-3996-1_21

In-situ measurements in the research fast-neutron fission reactor have shown that the sensitivity of sensors based on Au nanofilms is preserved during their irradiation up to fluence $F \approx 10^{24}$ nm^{-2} without any signs of the material degradation (the given F value is related only to the available time of the reactor operation). At the same time, sensors based on nanoheterostructures of InAs/i-GaAs—the material with highest irradiation resistance among known semiconductors—are stable only up to $F \approx 10^{22}$ nm^{-2} [2]. The additional advantage of gold in comparison with semiconductors is the weak dependence of Hall coefficient on temperature in the DEMO-relevant range of its values [5, 15].

The Hall sensor's general stability is determined not only by its active element, but also by the substrate and the adhesion strength between them. Currently, authors of this work for the investigations of Au-based sensors use the substrates made from sapphire (single-crystal Al_2O_3), which has the high resistance to neutron irradiation and therefore is compatible with the fusion reactors' environment [14]. Modern technologies of the single crystals' growth and polishing allow to make sapphire substrates with a surface, sufficiently smooth (roughness <0.5 nm) to deposit on them the thickness-uniform metal nanofilms, and also with high thermal conductivity (~46 W·mK^{-1}), sufficient to remove heat from the active element when sensor operates in the radiation fluxes.[1]

Films of oxides and metals join with each other through oxygen ions, while the adhesion strength is determined by the activity of metal towards oxygen [7]. Gold is chemically inert and therefore has weak adhesion with Al_2O_3 oxide. The standard approach for its increasing is to deposit between Au and Al_2O_3 the titanium (Ti) nanosized adhesive sublayer which interacts much more strongly with oxygen in sapphire, but also has a high adhesion to gold [12]. The Ti/Au metallization is stable at comparatively low temperatures, but at $T > 175$ °C the Ti and Au mutual diffusion significantly increases [9, 10, 16]. Moreover, at $T \geq 250$ °C the AuTi, Au_4Ti, Au_2Ti and Ti_3Au intermetallic compounds are formed [10]. These processes may change irreversibly the electrophysical parameters of Hall sensors based on Au nanofilms, which authors already observed in previous investigations [4]. To solve the diffusion problem, additional barrier sublayers are used—made from platinum (Pt), palladium (Pd), ruthenium oxide (RuO), etc.—which prevent to the Au and Ti atoms' movement [11, 12]. In particular, it has been shown that on the Al_2O_3 substrates, the Ti/Pt/Au metallization system can preserve value of the gold's surface resistance when heated up to 500 °C [11]. This corresponds to the limit thermal loads for Hall sensors, which are expected in the DEMO reactor [1].

Goal of this work was to study an efficiency of the platinum barrier sublayer's use for the thermal stability improvement of the irradiation-resistant Hall sensors, based on gold nanofilms, deposited on sapphire with titanium adhesive sublayer.

---

[1] http://www.semiwafer.com/sapphire-wafer.html.

21.2 Experiment Details

21.2.1 Samples

Three batches of sensors with the same symmetric cross-like topology with four contact pads were made, Fig. 21.1a. Two batches had the Ti/Au metallization of active element (without a barrier sublayer) and differed only in the Au layer thickness: $h = 70$ nm and $h = 50$ nm, Fig. 21.1b. The third batch had the Ti/Pt/Au metallization (with barrier sublayer), with thicknesses of the Au layer of $h = 35$ nm and of the Pt-sublayer of $p = 5$ nm, Fig. 21.1c. For all samples, the thickness of the adhesive Ti sublayer thickness was the same $t = 4$ nm. Industrial sapphire plates with the thickness $a = 400$ μm were used as substrates.

Metals were sequentially deposited on sapphire by the Vacuum Thermal Deposition (VTD) method—for Ti/Au(50 nm) and Ti/Pt/Au; as well as by the Electronic Beam Evaporation (EBE) method—for Ti/Au(70 nm). To prevent the Ti oxidation and the water adatoms precipitation from the atmosphere on the interlayer interfaces, all active element's metals were deposited under same conditions without the vacuum deterioration.

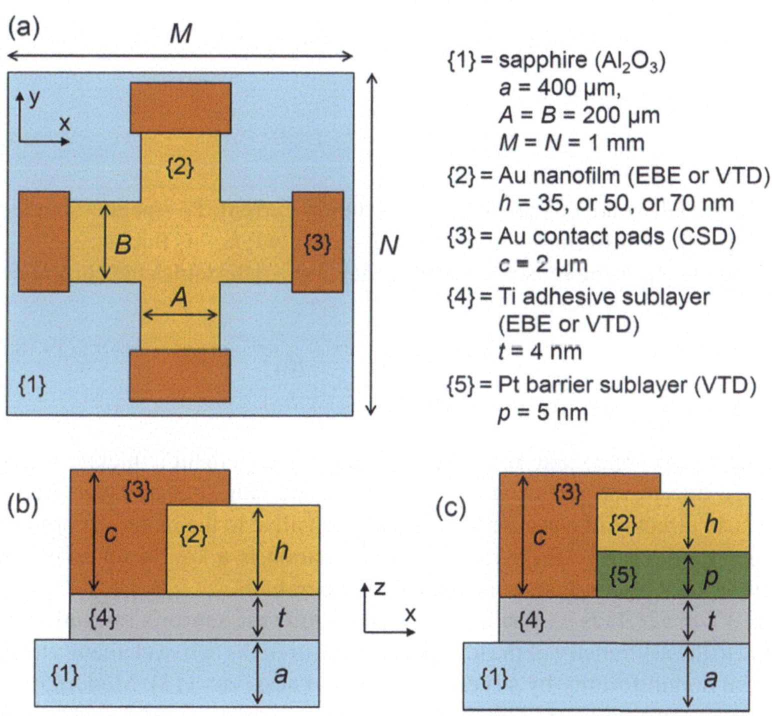

Fig. 21.1 Hall sensors' geometry (not to scale): **a** Chip with active element and contact pads; **b** Structure of layers for samples with the metallization of Ti/Au and **c** of Ti/Pt/Au

The sensor topology was formed by the lithography with lift-off. The active element's cross-hair sizes were $A = B = 200$ μm, Fig. 21.1a. In addition, the Au contact pads with the thickness of $c = 2$ μm, Fig. 21.1, were made at the active elements' ends with the using of lithography with following Chemical Solution Deposition (CSD). After that, the sapphire plates were diced into chips with sizes of $M = N = 1$ mm. Each chip carried single sensor.

In order to connect sensors with measuring devices, external leads (gold wires with the 30-μm diameter) were bonded to thick Au contact pads with the help of thermo-compression bonding machine.

21.2.2 Measurement Technique

The Hall sensors' thermal resistance was studied by comparing their parameters—current-related sensitivity S_I, residual voltage (offset) V_0, resistivity ρ—before and after vacuum annealing.

As it is known, when applying magnetic field **B** perpendicular to the Hall sensor's surface and transmitting current through one pair of its opposite leads, the output voltage is generated on the other pair [13]:

$$V(I, B) = V_H(I, B) + V_0(I), \tag{21.1}$$

$$V_H(I, B) = \frac{R_H}{h} \cdot I \cdot B \tag{21.2}$$

where, V_H is Hall voltage; V_0— offset; I—supply current; B—magnetic induction; R_H—Hall coefficient for material of the active element; h—its thickness.

Hall voltage V_H is the sensor's useful signal and it determines the current-related sensitivity to magnetic field:

$$S_I = \left| \frac{V_H(I, B)}{I \cdot B} \right| = \left| \frac{R_H}{h} \right|. \tag{21.3}$$

It is seen from (21.3) that S_I depends on the active element's thickness h, which can change due to the diffusion of atoms from the adhesive sublayer. Moreover, diffusion can affect Hall constant R_H, which is sensitive to the material composition. Consequently, the diffusion processes' minimization is a key factor to ensure the sensitivity's stability, especially at elevated temperatures.

Offset V_0 in (21.1) is a spurious signal that limits the sensor's resolution. Offset arises due to the asymmetry of the leads placement or of the active element's topology, its material nonuniformity by structure or by thickness, etc. [13]. Modern magnetic measuring systems use special methods of signal processing to eliminate offset and to extract V_H from the output signal (21.1). However, in cases when offset changes

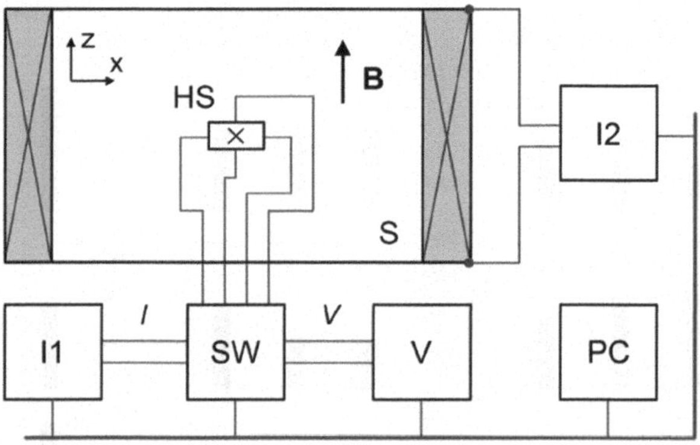

Fig. 21.2 General layout of measuring bench (*see details in text*)

uncontrollably, for example, due to diffusion, the effectiveness of these methods can reduce. It requires to study separately the stability of both sensitivity and offset.

I this work, the spinning-current method was used to determine V_H and V_0 [13]. It lies in the averaging of several output voltage's values V, measured at different configurations of the sensor's opposite leads connection to the current source and voltmeter. Measurements were carried out at the bench which general layout is shown in Fig. 21.2. Leads of Hall sensor (HS) were connected through the 4 × 4 switch (SW) to the Keithley 220 stable current source (I1) and Keithley 2182 nanovoltmeter (V). In order to create a magnetic field **B** it was used a solenoid (S) powered by the Tektronix PWS2323 current source (I2). The computer (PC) with special software carried out the measuring process control, data collecting, processing and storing.

Using the bench, the samples resistivity ρ also was determined by the van der Pauw method [13]. It involves the calculation of the sensor's transverse resistance R_T, based on results of the output voltage measuring the on one pair of its adjacent leads when transmitting a current through another pair and at $B = 0$. Respectively [13]:

$$\rho = \frac{\pi}{\ln 2} \cdot h \cdot \langle R_T \rangle, \tag{21.4}$$

where $<R_T>$ is average value of the transverse resistance's two values, obtained at the current transmitting in mutually perpendicular directions.

21.3 Results and Discussion

Samples was annealed in the vacuum (residual pressure ~10^{-3} Pa) for 3 h at temperature of $T \approx 400$ °C, which corresponds to the average level of thermal loads on Hall sensors expected in DEMO [1].

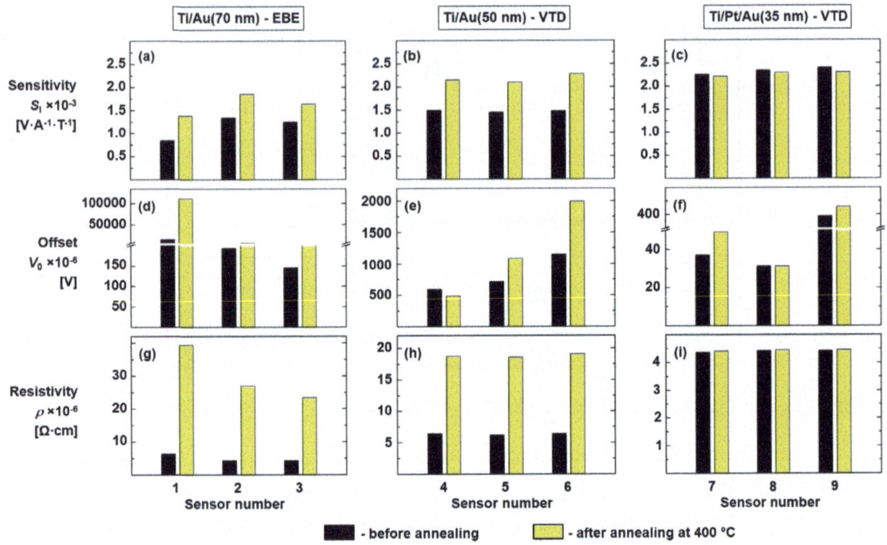

Fig. 21.3 Parameters of Hall sensors with the active element's metallization of Ti/Au(70 nm) (*left*), Ti/Au(50 nm) (*middle*) and Ti/Pt/Au(35 nm) (*right*) before (*black columns*) and after (*yellow columns*) vacuum annealing for 3 h at $T \approx 400$ °C: **a–c** current-related sensitivity S_I; **d–f** residual voltage (offset) V_0; **g–i** resistivity ρ. Sample numbers match the data in Table 21.1

In order to demonstrate the observed effects it was randomly selected three sensors from each samples batch. Figure 21.3 compares for them the sensitivity S_I, offset V_0 and resistivity ρ values measured before (black columns) and after (yellow columns) annealing. Data for S_I are represented on the same vertical axis scale, while for V_0 and ρ the axis scale and breaks are selected for the perception convenience. Relative changes of these parameters (ΔS_I, ΔV_0 and $\Delta\rho$) are given in Table 21.1.

Table 21.1 Relative change in sensors' parameters after annealing (sample numbers coincide with the numbers in Fig. 21.3)

Metallization	No	ΔS_I (%)	ΔV_0 (%)	$\Delta\rho$ (%)
Ti/Au(70 nm) EBE	1	64	702	520
	2	38	3145	533
	3	31	667	447
Ti/Au(50 nm) VTD	4	44	−18	194
	5	45	53	201
	6	54	73	201
Ti/Pt/Au(35 nm) VTD	7	−2	32	1
	8	−2	0	1
	9	−4	9	1

From Fig. 21.3 it is seen, that before annealing, each samples batch had its own average sensitivity value, since it is determined by the gold layer's thickness h, set when manufacturing. Resistivity for all samples was within the range of $\rho \approx (4-6) \times 10^{-6}$ Ω cm, which is consistent with the dimensional effects studies in Au nanofilms [8]. Insignificant spread in S_I and ρ within each batch can be explained by divergence of h for different samples, which occurs because a thickness of the film deposited by the EBE or VTD method, as a rule, tends to slightly decrease in the direction from the substrate's center to its periphery. Regarding offset, then the spread of its value V_0 in each batch was much stronger than the spread of S_I and ρ. This fully corresponds to the nature of origin of this spurious signal, which is sensitive to any inhomogeneities in the material. It should be noted that in sensors with the Ti/Pt/Au metallization, the average offset was smaller. The possible reason for this is the Pt sublayer effect on the Au film growth process, resulting in less inhomogeneities.

As it is seen from Fig. 21.3 and Table 21.1, annealing almost did not affect (within measurement error) the sensitivity S_I and resistivity ρ of sensors with the Ti/Pt/Au metallization, but significantly changed these parameters for sensors with Ti/Au. Such changes can be explained by the diffusion of titanium into gold, which occurs without barrier sublayer.

To do this, consider the Ti/Au interface, Fig. 21.4, in which, after annealing, Ti atoms diffused into Au at certain average depth Δh. Obviously, due to diffusion, the resistivity in the gold layer has distributed in the nonhomogeneous manner along its thickness between the two limit values ρ_{Au} (in Au) and ρ_{Ti} (in Ti). For simplicity, resistivity profile ρ(z) in Fig. 21.4 is shown schematically, whereas ρ_{Au} and ρ_{Ti} values are taken for bulk samples of pure metals: $\rho_{Au} \approx 2 \times 10^{-6}$ Ω·cm and $\rho_{Ti} \approx 42 \times 10^{-6}$ Ω·cm [6]. Since $\rho_{Au} << \rho_{Ti}$, then the current supplied to the sensor flows mainly through the near-surface Au layer with thickness of $(h - \Delta h) < h$ with the minimal resistance, as shown schematically by the current density vector's distribution $\mathbf{j}$(z)

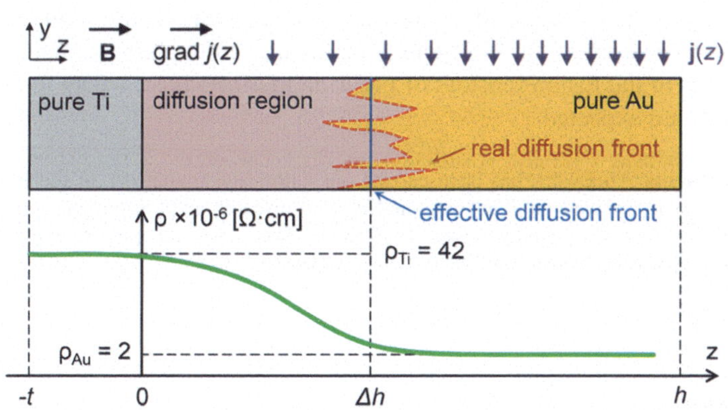

Fig. 21.4 Ti/Au interface with the diffusion region formed after annealing (*top*), and schematic resistivity distribution (*bottom*). $\mathbf{j}$(z) is a supply current density. Values ρ_{Au} and ρ_{Ti} are taken for pure bulk samples of Ti and Au from [6]

in Fig. 21.4. According to (21.3), this should increase the sensor sensitivity S_I, as observed in the experiment, Fig. 21.3a, b. In addition, when measuring with the van der Pauw method, the decrease in thickness of the layer through which the current flows should increase the measured value of the sample's transverse resistance $<R_T>$. Accordingly, the application of (21.4) without correction for Δh will result in an increase in calculated value of ρ, as is it seen in Fig. 21.3g, h.

Offset changes after annealing are noticeable for all sensors studied, although, in general, they have less law-governed character than those observed in the case of S_I and ρ: for most samples, the value of V_0 has increased after annealing, but for some ones, on the contrary, decreased, Fig. 21.3 and Table 21.1. The offset change can be connected with the recrystallization effects in the gold active elements, or with the mechanical stresses' initiation due to the thermal expansion coefficients' mismatch for sapphire and gold, etc.

It should be noted that when using the Pt barrier layer, change in V_0 on average was less, Table 21.1. This may also indicate a certain role of diffusion in the offset change. For example, it is known that the diffusion of titanium into polycrystalline gold occurs both in the individual grains volume and in the intergranuale space [9]. This can form the developed relief of diffusion front, Fig. 21.4, which, in turn, leads to the non-uniform distribution of resistivity in the active element's volume, and the distortion of the electric field's equipotential lines in the material when current transmits through it.

21.4 Conclusions

The use of additional platinum barrier sublayer enhances the thermal stability of Hall sensors based on gold nanofilms, deposited on sapphire with titanium adhesive sublayer, during vacuum annealing at 400 °C for 3 h. This fact, together with the previously proven high resistance of such sensors to the intense neutron fluxes' impact, allows them to be used as the front-end components of the plasma's magnetic diagnostic systems in the fusion reactors of ITER and DEMO type, where temperatures above 300 °C are expected.

Acknowledgements This work has been carried out within the framework of the EUROfusion Consortium and has received funding from the Euratom research and training programme 2014–2018 and 2019–2020 under grant agreement No 633053. The views and opinions expressed herein do not necessarily reflect those of the European Commission.

References

1. W. Biel, *Report Presented at the International School of Fusion Reactors Technology*, Erice, Italy, 28 Apr–4 May 2017

2. I. Bolshakova, S. Belyaev, M. Bulavin, V. Brudnyi, V. Chekanov, V. Coccorese et al., Experimental evaluation of stable long term operation of semiconductor magnetic sensors at ITER relevant environment. Nucl. Fusion **55**(8), 083006 (2015)
3. I. Bolshakova, M. Bulavin, N. Kargin, Ya. Kost, T. Kuech, S. Kulikov et al., Metal hall sensors for the new generation fusion reactors of DEMO scale. Nucl. Fusion **57**(11), 116042 (2017)
4. I. Bolshakova, F. Shurygin, A. Moroz, Y. Kost, Y. Mykhashchuk, M. Radishevskiy et al., in *Proceedings of 14th International Conference on Advanced Trends in Radioelectronics, Telecommunications and Computer Engineering* (Lviv-Slavske, Ukraine, 20–24 Feb 2018), p. 381
5. V. Frank, Hall coefficient of technically pure metals from 80°K to 800°K. II. Results for Zr, W, Mo, Ta, Nb and Al. survey of results for the 4d- and 5d-transition group of metals. Appl. Sci. Res. B **7**(1), 41 (1959)
6. I.S. Grigoriev, E.Z. Meilikhov (eds.), *Handbook of Physical Quantities* (CRC Press, Boca Raton, 1997)
7. L.-H. Lee (ed.), *Fundamentals of Adhesion* (Springer Science + Business Media, New York, 1991)
8. W.G. Ma, H.D. Wang, X. Zhang, W. Wang, Experiment study of the size effects on electron-phonon relaxation and electrical resistivity of polycrystalline thin gold films. J. Appl. Phys. **108**(6), 064308 (2010)
9. W.E. Martinez, G. Gregori, T. Mates, Titanium diffusion in gold thin films. Thin Solid Films **518**(10), 2585 (2010)
10. K. Masahiro, S. Noboru, Effects of temperature, thickness and atmosphere on mixing in Au-Ti bilayer thin films. J. Mater. Sci. **28**(18), 5088 (1993)
11. M. Matsui, Thermal stability of Au thin film deposited on Al_2O_3 substrate with RuO_2 adhesion layer. J. Ceram. Soc. Jpn. **109**(6), 574 (2001)
12. D.M. Mattox, *Handbook of Physical Vapor Deposition (PVD) Processing* (Elsevier Inc., Amsterdam, 2010)
13. R.S. Popovic, *Hall Effect Devices*, 2nd edn. (IOP Publishing, Bristol, 2004)
14. T. Shikama, S.J. Zinkle, Long term degradation of electrical insulation of Al_2O_3 under high flux fission reactor irradiation. J. Nucl. Mater. **258–263**, 1861 (1998)
15. C.R. Tellier, Thin metal film senosors. Active Passive Electron. Comp. **12**(1), 9 (1985)
16. M. Todeschini, A.B. da Silva Fanta, F. Jensen, J.B. Wagner, A. Han, Influence of Ti and Cr adhesion layers on ultrathin Au films. ACS Appl. Mater. Interf. **9**(42), 37374 (2017)

Chapter 22
Morphology and Luminescence Properties of Cellulose-CNT-$BiPO_4$:Pr^{3+} Composites

V. P. Chornii, V. V. Boyko, S. G. Nedilko, M. S. Slobodyanyk, V. P. Scherbatskyi, and K. V. Terebilenko

Abstract The peculiarities of preparation and properties of composites which consist of cellulose matrix, multiwall carbon nanotubes and praseodymium-doped bismuth phosphate nanoparticles are reported. The SEM images show that oxide particles are incorporated into cellulose when their sizes below 100 nm and can pierce out of composite surface in case of bigger particles. The carbon nanotubes are clearly observed only at the regions near oxide particles that can be explained by close adhesion between these components of composites. Chemical element analysis shows the composites of lower content of nanotubes possess better homogeneity from viewpoint of elements distribution. The studied composites reveal intensive luminescence under excitation at 405 and 473 nm. The photoluminescence spectra depend on the PL excitation wavelength and related mostly with cellulose host emission centers. The Pr^{3+}-related emission clearly observed only in the case of composites of higher content of nanotubes at direct PL excitation of $^3H_4 \rightarrow {}^3P_1 + {}^1I_6$ absorption transitions in praseodymium ions.

22.1 Introduction

Cellulose is one of the most common natural polymers with possible application in various branches of technology e.g. biofuels, eco-friendly sorbents and "paper electronics" [1–3]. The variety of its possible application is determined by structure of cellulose (porous material with easy incorporation of micro/nanoparticles), its good mechanical properties and relatively low costs of production due to abundance of cellulose sources in nature.

V. P. Chornii (✉) · V. V. Boyko
National University of Life and Environmental Sciences of Ukraine, Kiev, Ukraine
e-mail: vchornii@gmail.com

V. P. Chornii · S. G. Nedilko · M. S. Slobodyanyk · V. P. Scherbatskyi · K. V. Terebilenko
Taras Shevchenko National University of Kyiv, Kiev, Ukraine

A. D. Pogrebnjak et al. (eds.), *Nanomaterials in Biomedical Application and Biosensors (NAP-2019)*, Springer Proceedings in Physics 244,
https://doi.org/10.1007/978-981-15-3996-1_22

Functionalization of cellulose or design of cellulose-based composites can considerably improve certain properties of this natural polymer according to application goals. In particular, adding of metal salts and oxides improves antibacterial and UV-protecting properties of cellulose [4]. Novel photocatalytic materials can be elaborated on base of TiO_2-doped cellulose [4]. Micro/nanoparticles of some oxide compounds when being added to cellulose, significantly affect photoluminescence (PL) properties and dielectric constants of the latter one [5–7]. The essential advantage of oxide luminescent compounds as modifiers is their stability, thus it is difficult to remove them from the cellulosic fiber even using a special treatment. That is why, cellulose fibers with inorganic luminescent particles are perspective for various applications in the textile industry, as well as for the protection of documents [7]. At the same time adding of inorganic filler to cellulose host influences on its mechanical properties [8]. Addition of zirconia to cellulose fibers decreased tenacity and elongation at break values [7]. From other hand mechanical, thermal and electrical properties of various polymers, in particular cellulose can be greatly improved by adding some amount of carbon nanotubes (CNT) [9–14]. CNTs have good light absorption in all visible spectral range, thus it is expected that absorption edge of composites containing of CNTs should move to lower energies.

In this work the data on preparation of the composites consisted of micro/nanocellulose, $BiPO_4:Pr^{3+}$ oxide and carbon nanotubes as well as results of composites' morphology and luminescent properties study are reported.

22.2 Experiment Details

The pure cellulose tablets (produced by ANCYR-B, Ukraine) were grinded and dispersed using high-speed rotation mill. The amount of starting cellulose in all prepared samples of composites was equal to 2.00 g and amount of $BiPO_4:0.01\%Pr^{3+}$ oxide component was equal to 0.20 g. Carbon nanotubes in weight of 0.01 g and 0.02 g were added to cellulose in composite samples denoted as C1 and C2 respectively. The weighted amount of cellulose, oxide and nanotubes were thoroughly grinded in porcelain crucibles. The used in experiments powders were pressed under pressure 1.8×10^4 kPa/m^2 into discs with average diameter near 10.3 mm.

The Scanning electron microscope (SEM) JAMP-9500F Field Emission Auger Microprobe (JEOL, USA) equipped with X-ray microanalyzer INCA PentaFETx3 (Oxford instruments) was used for SEM measurements. Besides the SEM imaging, microelement analysis of various areas of the samples was also performed using the same microscope.

Photoluminescence (PL) properties of the composites were measured in 400–750 nm spectral range using DFS-12 double diffraction grating monochromator equipped with photomultiplier FEU-79. Diode-pumped lasers (λ_{em} = 405 and 473 nm) were used as excitation light sources. The PL studies were performed at room temperature. All the PL emission spectra were corrected on system responses.

22.3 Results and Discussion

The SEM images for the samples C1 and C2 are shown on Fig. 22.1 at top and bottom of figure, respectively. For both samples the incorporation of oxide nanoparticles (grains sizes below 100 nm) into cellulose host is clearly observed. Larger oxide particles (sizes above 100 nm) pierce out of the composite surface (bottom right part of images). Thus, it is likely to be the nanoparticles of $BiPO_4$ oxide that easily occupy pores in cellulose forming "ceramic-like" samples but incorporation of larger oxide particles leads to local inhomogeneity of samples. This observation reveals that from viewpoint of homogeneity the used method for preparation of composites not as good as that one described in [8]. The CNTs effect on structure is difficultly described from SEM results. Some amount of the carbon nanotubes can be found only at the middle part of SEM image of C2 sample, while no CNTs can be found on SEM image of C1 sample. Interestingly, the CNTs lay only close to oxide particles. This observation can mean that nanotubes have been adsorbed on oxide surfaces.

The chemical elements analysis was performed in three different areas of each of the samples. Spectra 1 and 2 were recorded in the "homogenous" region and spectrum 3—on grains associated with large oxide particles. The results are collected in Table 22.1. It is seen from table, the content of carbon is much higher in the case of sample C2. Such effect can be explained by higher content of CNTs in composite

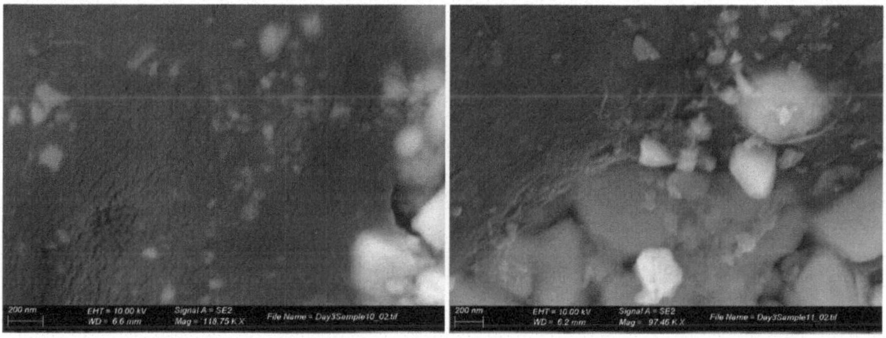

Fig. 22.1 SEM images of C1 (left) and C2 samples (right)

Table 22.1 Content of some chemical elements in different areas of samples C1 and C2 (in at. %)

Sample	Region	C	O	P	Bi
C1	1	23.09	52.53	11.59	12.79
	2	24.27	50.68	11.62	13.43
	3	21.11	53.46	12.09	13.35
C2	1	85.54	13.97	0.04	0.46
	2	83.46	16.32	0.03	0.19
	3	57.05	34.30	3.87	4.78

C2. Interestingly, the sample C1, from viewpoint of chemical element distribution, is more homogenous than C2. The ratios between the elements are very similar to corresponding ratios for regions without pierced out oxide particles even for areas of C1 close to oxide particles. At the same time, chemical elements analysis showed that content of elements strongly depends on the areas of the sample C2 chosen for experiments. This observation can be ascribed to increased interaction between components of studied composites with increasing content of carbon nanotubes.

It is worth noting that excited beam had diameter near 3 mm and therefore studied luminescence properties can be treated as integrated ones in relatively large area of the samples. We suggest the local inhomogeneity of samples does not affect much on the PL properties. In order to verify this suggestion the PL emission was studied at 3 different areas at each sample and it was found that PL spectra at different area of samples are very similar. The PL emission spectra of composites measured under excitation with $\lambda_{ex} = 473$ nm are shown on Fig. 22.2. It is seen the PL emission for both composites consist of broad band with maxima near 550 nm. This band was earlier ascribed to luminescence centers of various origins in the micro/nanocellulose [8]. Increasing of CNTs content in composites leads to significant decreasing of the cellulose host photoluminescence. At the same time, two additional bands appear in the spectrum of C2 sample. These bands have maxima near 608 and 660 nm that matched well with typical positions of $^3P_0 \rightarrow {}^3H_6$ and $^3P_0 \rightarrow {}^3F_2$ radiation transitions in Pr^{3+} ions [15]. The luminescence of Pr^{3+} ions was expected because the excitation light with $\lambda_{ex} = 473$ nm correspond to $^3H_4 \rightarrow {}^3P_1 + {}^1I_6$ absorption transitions in $BiPO_4{:}Pr^{3+}$ compound [16].

It is looks like the PL excitation processes for luminescence centers in cellulose and oxide components of composites compete each other. Due to small amount of oxide component the Pr^{3+}-related centers, obviously, have much smaller concentration than cellulose-related centers. As result, a luminescence of cellulose is dominant in PL spectra of both composites. At the same time, carbon nanotubes actively

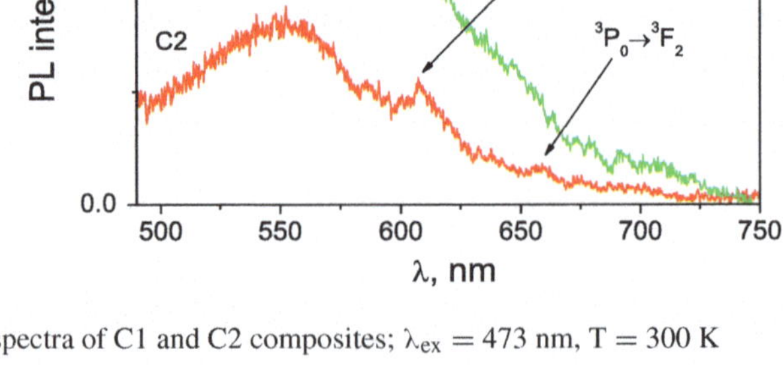

Fig. 22.2 PL spectra of C1 and C2 composites; $\lambda_{ex} = 473$ nm, T = 300 K

participate in luminescence processes, resulting in suppression of cellulose-related luminescence. The effect of CNTs on PL emission of Pr^{3+} ions is unclear and requires further studies.

PL emission was studied also under excitation at $\lambda_{ex} = 405$ nm (this wavelength corresponds to photon energy 3.06 eV). Such photon energy allows excluding direct excitation of Pr^{3+} ions, and also it is much lower than needed for band-to-band absorption of $BiPO_4$ (the band gap of this oxide compound is near 4.8 eV). Thus, only light absorption by cellulose and CNTs take place. As Fig. 22.3 shows, under excitation with $\lambda_{ex} = 405$ nm only broad bands in emission spectra for both C1 and C2 composites are observed. The PL bands have same shape and maxima positions (located near 500 nm). The only difference between spectra on Fig. 22.3 is intensities. Similarly to the case of 473 nm excitation, luminescence intensity of C2 composite is near of 2 times lower than for C1 one.

Decomposition of the photoluminescence spectra on elementary bands allows easily distinguishing spectral changes. Such decompositions were performed for PL spectra using Gaussian curves. The positions of the band maxima ($h\nu_{max}$), full width at half maximum (FWHM), and areas under bands are shown in Table 22.2.

It is seen from table that he main PL emission bands have their maxima at 2.34 eV (530 nm) and 2.24 eV (554 nm) for PL excitation at 405 and 473 nm, respectively. Increasing of CNT content in composites decreases total intensity (area under corresponding band) in almost three times. There are additional weak bands in PL spectra at mentioned PL excitations. In case of $\lambda_{ex} = 405$ nm the additional band have maximum at higher energies and for $\lambda_{ex} = 473$ nm at lower energies in respect to main PL bands. The band at 2.6 eV (477 nm) can be ascribed to cellulose host, because its width is relatively large for the Pr^{3+} emission bands in that region. The band with maximum at 1.94 eV (640 nm) have relatively small FWHM and can be ascribed to unresolved Pr^{3+} emission related with superposition of radiation transitions from

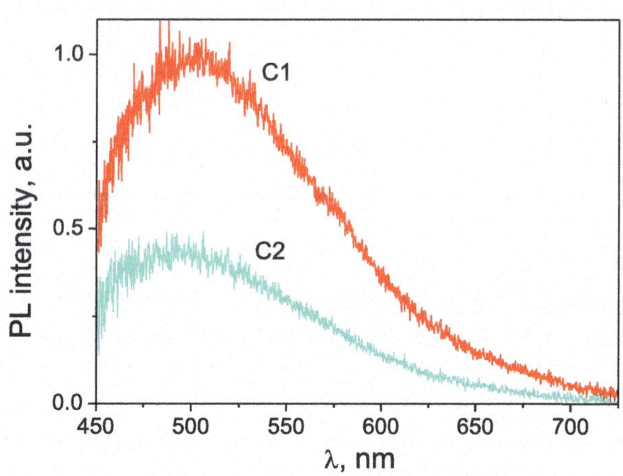

Fig. 22.3 PL spectra of C1 and C2 composites; $\lambda_{ex} = 405$ nm, T = 300 K

Table 22.2 Parameters of PL spectra decompositions

Sample	λ_{ex} (nm)	$h\nu_{max}$ (eV)	FWHM (eV)	Area (a.u.)
C1	405	2.34	0.60	0.097
		2.60	0.35	0.015
	473	1.94	0.15	0.003
		2.24	0.47	0.090
C2	405	2.34	0.60	0.037
		2.63	0.46	0.013
	473	2.24	0.43	0.032

$^3P_0$ on $^3H_6$ and $^3F_2$ levels. The PL emission related with these transitions clearly observed for the case of C2 sample under excitation at 473 nm.

22.4 Conclusions

Two sets of cellulose-CNT-$BiPO_4$:Pr composites have been prepared. Chemical elements analysis and scanning electron microscopy showed that composite samples of lower carbon nanotubes content are more homogeneous. Both composites reveal intensive luminescence under PL excitations at 405 and 473 nm. The most intensive PL bands were ascribed to luminescence centers in a cellulose host. The Pr^{3+}-related emission centers are clearly observed for the case of composites with higher content of carbon nanotubes.

References

1. K. Nelson, T. Retsina, M. Iakovlev, A. van Heiningen, Y. Deng, J.A. Shatkin, A. Mulyadi, in *Materials Research for Manufacturing*, ed. by L.D. Madsen, E.B. Svedberg (Springer, Cham, 2016), p. 267
2. S.G. Nedilko, S.L. Revo, V.P. Chornii, V.P. Scherbatskyi, M.S. Nedielko, J. Sens. Sens. Syst. **4**, 31 (2015)
3. S. Yun, S.D. Jang, G.Y. Yun, J.H. Kim, J. Kim, Appl. Phys. Lett. **95**, 104102 (2009)
4. H.E. Emam, Cellulose **26**, 1431 (2019)
5. S.G. Nedilko, Acta Phys. Pol. A **133**, 829 (2018)
6. L. Zhang, S. Lyu, Z. Chen, S. Wang, Nanomaterials **8**, 352 (2018)
7. P. Kulpinski, A. Erdman, M. Namyslak, J.D. Fidelus, Cellulose **19**, 1259 (2012)
8. M. Nedielko, S. Hamamda, O. Alekseev, V. Chornii, M. Dashevskii, M. Lazarenko et al., Nanoscale Res. Lett. **12**, 98 (2017)
9. S. Zhang, F. Zhang, Y. Pan, L. Jin, B. Liu, Y. Mao et al., RSC Adv. **8**, 5678 (2018)
10. A. Pantano, in *Carbon Nanotube-Reinforced Polymers*, ed. by R. Rafiee (Elsevier, Amsterdam, 2018), p. 201
11. A. Salama, A. Mohamed, N.M. Aboamera, T. Osman, A. Khattab, Adv. Polym. Technol. **37**, 2446 (2018)

12. Z. Pang, X. Sun, X. Wu, Y. Nie, Z. Liu, L. Yue, Vacuum **122**, 135 (2015)
13. M.R. Loos, K. Schulte, in *Carbon Nanotube-Reinforced Polymers*, ed. by M.R. Loos (Elsevier, Amsterdam, 2015), p. 207
14. R. Khare, S. Bose, J. Min. Mater. Character. Eng. **4**, 31 (2005)
15. V.K. Tikhomirov, S.A. Tikhomirova, J. Non-Cryst. Solids **274**, 50 (2000)
16. V.P. Chornii, S.G. Nedilko, K.L. Bychkov, K.V. Terebilenko, M.S. Slobodyanik, V.V. Boyko, Acta Phys. Pol. A **133**, 843 (2018)

Chapter 23
Time Dependence of X-Ray Luminescence from Yttrium Oxide Nanoceramics

S. Kononenko, R. Skiba, I. Mysiura, O. Kalantaryan, V. Zhurenko, V. Chishkala, and M. Azarenkov

Abstract 250–750 nm luminescence of yttrium oxide nanoceramics which are of great interest for several biomedicine applications was measured under the influence of X-ray radiation with up to 60 keV energy. The ceramics was produced from yttrium oxide nanopowder by sintering technique. The time dependence of the luminescent intensity at the 350 nm was investigated. It was shown that the decreasing intensity curve was good approximated by two-exponential function, which indicated the presence of fast and slow processes inducing the luminescence. The cyclic irradiation of the same sample showed that during the absence of irradiation an evident regeneration of the luminescent intensity occurred. The initial value of the intensity registered for each subsequent reirradiation significantly exceeded the final value of the intensity measured for the previous cycle. Wherein, the intensity decreased to a certain saturation value at the end of each cycle.

23.1 Introduction

Yttrium oxide (Y_2O_3) is promising phosphor material in terms of fast response and light yield, in particular, in the field of registration of ionizing radiation [1, 2]. Moreover yttrium oxide is a perspective material for biological imaging applications. It is widely used in medicine, for example, as ceramics for prostheses [3], as an element with upconversion of infrared radiation into visible radiation in cancer cell detection markers [4], and as a functional element of toxic methanol ultrasensitive sensors at room temperature [5]. Considerable practical interest to Y_2O_3 nanoceramics is connected with its possible use as a radiation detector for X-ray quanta with energies of about ten kiloelectronvolts and γ-radiation. Here, knowledge of variability of Y_2O_3 ceramic spectral and kinetic characteristics depending on different irradiation factors is important.

S. Kononenko (✉) · R. Skiba · I. Mysiura · O. Kalantaryan · V. Zhurenko · V. Chishkala · M. Azarenkov
V. N. Karazin Kharkiv National University, Kharkiv, Ukraine
e-mail: sergiy.i.kononenko@gmail.com

A. D. Pogrebnjak et al. (eds.), *Nanomaterials in Biomedical Application and Biosensors (NAP-2019)*, Springer Proceedings in Physics 244,
https://doi.org/10.1007/978-981-15-3996-1_23

Luminescence of Y_2O_3 powder and ceramic samples was studied for different types of irradiation, namely iono- [6], heat stimulated- [7], cathode- [8] and X-ray luminescence [9, 10], as well as upon excitation by a nitrogen laser ($\lambda = 337.2$ nm) [10, 11]. Wide band luminescence with maximum at 350 nm was typical for light spectra of Y_2O_3 exposed to various radiations. The same band we also observed for Y_2O_3 powder and ceramics irradiated by 60 keV X-ray photons earlier [12].

It is well known that Y_2O_3 crystals with 6 eV width of band gap have a cubic syngony with the space group T7h [13]. When wide-gap dielectrics is exposed to ionizing radiation, decay of excitons, especially self-trapped excitons (STE), play an important role both in light generation processes and defect formation processes. A number of authors considered radiation decay of STE to be the main reason for a luminescence light emission at the band with maximum 350 nm (see, e.g., Fukabori [2], Lushchik [14], Guerassimova et al. [15], Novosad et al. [16]). For example, X-ray excitation of Y_2O_3 ceramic material at 85–295 °K temperature induced the intense luminescence, which spectrum was good fitted into the elementary Gaussian shape bands with several maxima [16]. The authors stated that 3.4 eV band of the luminescence was caused by STE of $(YO_6)_9$—complex, when the cation was localized in the field of the trigonal (C3i) and monoclinic (C2) symmetries. However, it is worth noting that exact source of this emission peak is not finally determined yet. So the authors [17] pointed out that in the case of transparent Y_2O_3 ceramics and radio-luminescence induced by 241Am α-ray excitation the origin of 350 nm emission peaks was trapped excitons, i.e. excitons were trapped by point defects, dislocations, grain boundaries (in the case of ceramics), and stacking faults.

Most often, yttrium oxide spectral characteristics study are aimed to identify features associated with the methods of specimen production, doping it with various elements (see, e.g., Novosad et al. [16], Stanton et al. [18], Robindro Singh [19]). Luminescent radiation is the resultant effect of external radiation on defect centres, i.e., it carries information about the structure of a solid, being the dynamics of competing processes of excitation, formation and annealing of defects. At the same time, the spectral characteristics of the samples can undergo significant changes under the influence of intense ionizing radiation. Therefore, it is interesting to study the dynamics of changes of the luminescent radiation intensity as a function of time (or irradiation dose) of X-ray exposure especially for biomedicine applications.

The paper deals with experimental study of changes in the luminescent intensity at wavelength of 350 nm for yttrium oxide nanoceramics, depending on the irradiation time (so-called luminescence kinetics).

23.2 Experiment

The measurements of X-ray luminescence were done at room temperature on the spectrometric complex schematically shown on Fig. 23.1. The irradiation was provided by X-ray unit. The vacuum tube with copper anode and beryllium window was a source of X-ray radiation. Operating voltage of the X-ray tube was up to 60 kV

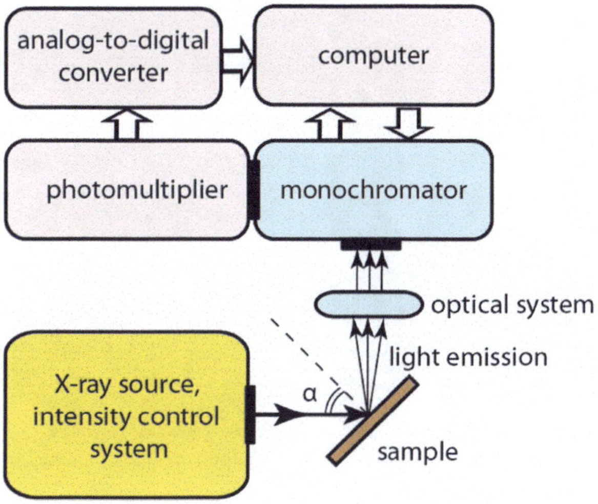

Fig. 23.1 Experimental setup

and current value was 10 mA. The intensity of X-ray radiation was controlled and stabilized by X-ray intensity control system during the experiments. The irradiation duration for each sample under study was 7200 s and the maximum exposure dose was approximately 12.37 C/kg.

The excitation radiation fell on the sample at an angle of 45°. Luminescent light was focused by means of optical system on the entrance slit of grating monochromator (wavelength range 250–750 nm). The resulting signal was detected by photomultiplier connected to the entrance amplifier unit of the analog-digital converter and further transmitted to computer. The spectrometric complex was calibrated by spectrophotometric filament tungsten lamp. The luminescent spectra were corrected to the spectral sensitivity of the apparatus.

The experiments were carried out with samples made from Y_2O_3 powder with purity of 99.99%. The powder consisted of both nano- and micro-sized particles (Fig. 23.2). The ceramics samples were made by sintering technique. At the beginning the compression was performed by the compaction of the powder in a cylindrical tablet 10 mm in diameter and 5 mm thickness (approximately 4×10^3 kg/cm^2). In the second stage the sintering took place in vacuum electric furnace (residual gas pressure 5×10^{-4} Pa) in an alumina crucible at temperature of 1500 °C. The sintering duration was 1 h. At the last stage the samples were cleaned and prepared for installation into the spectrometric complex. The structural state of the samples was examined using X-ray diffractometer DRON-4-07. We used Cu-Kα radiation applying Ni selectively absorbing filter.

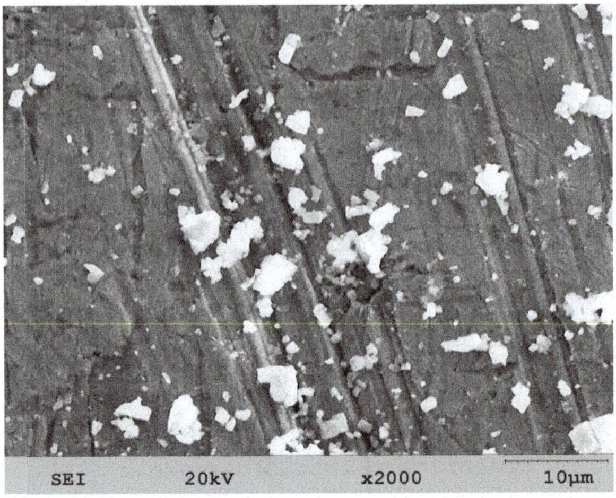

Fig. 23.2 Scanning electron microscope image of Y_2O_3 powder

23.3 Results and Discussions

Figure 23.3 shows the typical XRD pattern for Y_2O_3 sample. The structure is single phase; it contains cubic yttrium oxide peaks (space group #206). The lattice parameters of yttrium oxide are slightly higher than handbook values, but for all samples the measurements are the same within the measurement error. The intensity distribution of the diffraction lines corresponds to the polycrystalline non-textured state; the lines have a small half-width, that is, the sample is in a large-crystalline state.

We measured luminescent spectra for Y_2O_3 ceramics (Fig. 23.4). The spectra were in good agreement with our previous results [12]. For all the samples under

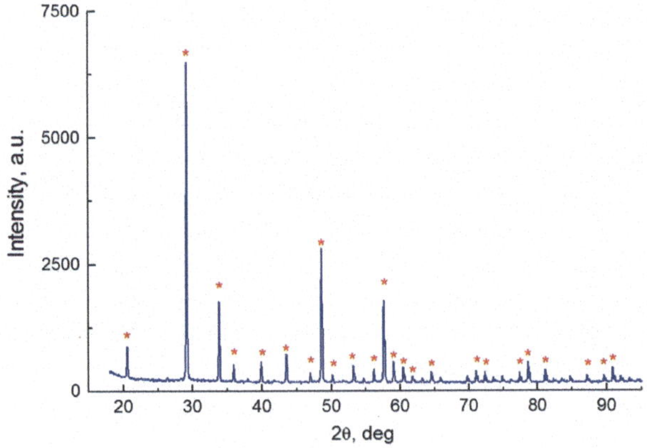

Fig. 23.3 XRD pattern of Y_2O_3 ceramics

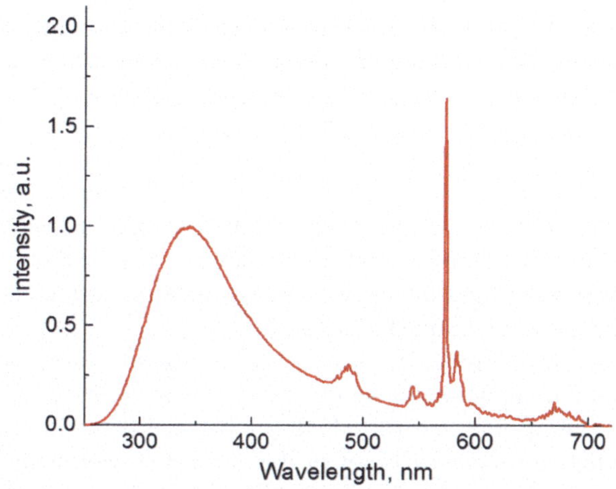

Fig. 23.4 Typical luminescence spectrum of Y_2O_3 ceramics

study the shapes of the spectra were similar. The most pronounced wide band was observed in the UV range with maximum at 350 nm. The maximum amount of light emission is integrally concentrated at this wavelength band. As we mentioned above, intense luminescence at a given wavelength range is often associated with emission of STE decay [2, 14–16].

We found that prolonged X-ray exposure caused a change of 350 nm peak intensity. Figure 23.5 shows the typical time evolution of the peak intensity (dose dependence) for Y_2O_3 ceramics.

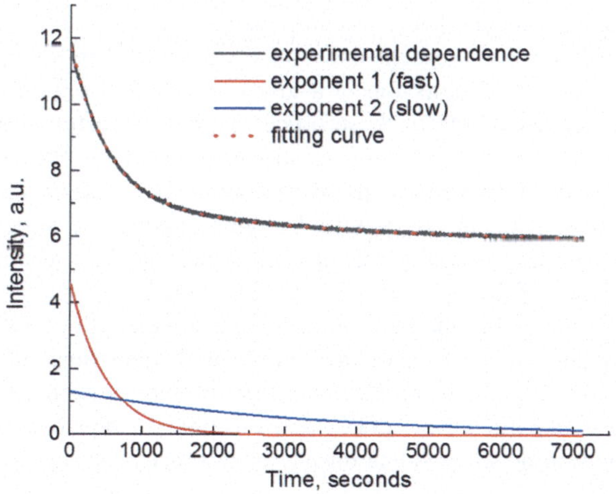

Fig. 23.5 The experimental time evolution of Y_2O_3 luminescent intensity at 350 nm and the result of exponential fitting (we also show the fast and slow exponents of the fitting curve for visualization)

It is well known that the process of relaxation of STE in a substance occurs in three ways. The first one is the excitation of a luminescent center, which is subsequently de-excited by light emission. The second way is de-excitation of the defect, which leads to its annealing. The third one is subthreshold formation of a defect due to the breaking of the bond in the lattice between neighboring atoms and the subsequent displacement of atoms by a distance, which does not allow restoring this bond. It should be noted that STE can migrate along the lattice at many atomic distances, and the probability of its relaxation is significantly higher when the STE encounters the defect [20]. Luminescent light intensity is the result of the instantaneous dynamic equilibrium of the above mentioned processes.

As it can be seen from Fig. 23.5, the decrease of intensity curve means that with increasing exposure time, nonradiative processes prevail over the radiative exciton decay. However, during long-term irradiation a balance between competing non-radiative and radiative processes is set in the substance. Herewith, the radiation intensity becomes almost two times less than its initial value. The fitting of experimental curves was carried out. It turned out that the most accurate result was obtained for experimental curve fitting by the sum of two exponents (fast and slow) in the following form (see Fig. 23.5):

$$I = I_0 + A_1 e^{-\frac{t}{t_1}} + A_2 e^{-\frac{t}{t_2}}.$$

The curve had two intervals of the rate of intensity change. The first interval is a high rate of intensity change; the second is a large exposure time and a slow decrease. As can be seen, the time dependence is determined by two processes characterized by significantly different times. The first one is fast process having characteristic time of several hundred seconds, namely $t_1 \approx 500$ s, while the second slow process has a characteristic parameter 5–6 times longer than the first, approximately $t_2 \approx 3300$ s.

As far as time dependence had two components, it is reasonable to assume that a decrease in the X-ray luminescent intensity is determined by two processes that contribute to the total light emission in a given wavelength region. As the X-ray irradiation dose increases, both of these processes, having different dynamic characteristics of luminescence, affect the overall balance of the emission processes in different ways. At the initial moments of irradiation ($t < 500$ s), the fast process prevails, manifesting itself in the rapidly decaying exponential dependence of the light emission. Beginning from the time of 1500 s, the role of the second mechanism in the overall balance of the light emission greatly increases.

The time evolution of luminescent intensity at a wavelength of 350 nm was also studied upon repeated irradiation of the sample with long time interval between exposures (Fig. 23.6). Cyclic irradiation (total irradiation for each sample consisted of 3 cycles, the cycle time was 7200 s and the interval between exposures was 24 h) of the same sample showed that during the absence of irradiation a significant regeneration of the luminescent intensity occurred. It was observed that after pause of 24 h from the previous irradiation, the initial value of the luminescent intensity was significantly higher than the value measured at the end of the previous cycle. At

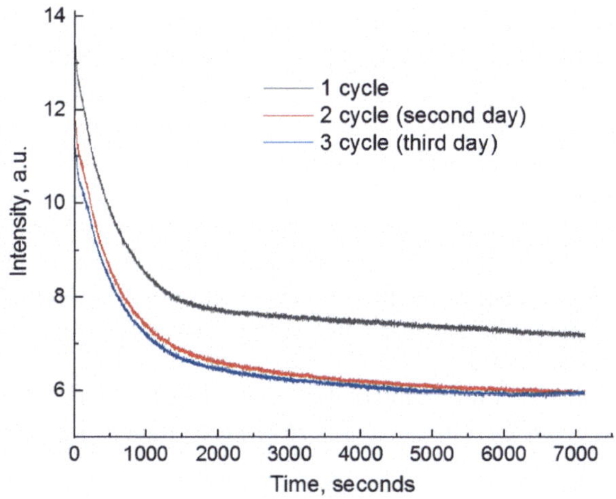

Fig. 23.6 The total time evolution of Y_2O_3 luminescent intensity at 350 nm for cyclic irradiation

the same time, the initial intensity of luminescence for each next irradiation cycle was somewhat less than one for the previous cycle. In addition, the intensity of the luminescence tended to a certain saturated value at the end of each irradiation cycle. As can be seen from Fig. 23.6, for the second and third irradiation cycles the saturated values of intensities at the end of irradiation are very close.

It should be noted that the tendency of two-exponential decrease in the time dependence has always been fulfilled for all irradiation cycles. The time constant of the fast process (first exponent) increased slightly with each subsequent irradiation cycle. The second time constant of the slow process (second exponent) remained unchanged.

An analysis of the time dependence of the luminescent intensity during cyclic irradiation allows us to conclude that there are several types of defects and their balance in the sample. As we mentioned above light generation is associated with STE decay in this wavelength region, so the following scenario of defect balance can be considered. Three types of defects are involved in the dynamics of luminescence.

The first type is a defect that is not annealed during one cycle of exposure to X-ray radiation. Figure 23.6 shows that such defects can be at least half of the total amount of defects that contribute to light emission. But on the other hand it is possible that they can be modified into a defect of other type, which will be annealed during the next irradiation cycle.

The second type is a defect, which is annealed under the influence of X-ray due to a change in the position of the atom and bond reduction with the neighboring atom or a change in the charge state. This bond is maintained during exposure to X-rays. However, after the irradiation termination the influence of thermal vibrations of the lattice leads to breakage of the bonds and finally the defects will be restored (the number of the defects will be the same as before exposure). During subsequent irradiation (next cycle) these defect will again be a source of light radiation for some time until its annealing is repeated. It is precisely this fact that explains the decrease in the intensity of light emission during prolonged X-ray irradiation.

We attribute the third type of defects with one that is finally annealed after the influence of X-ray radiation. As a result of the annealing process, stable bonds are formed and subsequently these defects do not participate in luminescence processes. The portion of these defects can be defined as the difference between the intensities of light radiation in different irradiation cycles at the initial moments of time or after prolonged exposure, when the intensity of light radiation has stabilized (see the difference between intensities for the first cycle and both the second and the third cycles at the beginning of irradiation $t = 0$ on Fig. 23.6; also see the difference at the end of irradiation $t = 7200$ s).

For a more detailed determination of the nature of various types of defects, as well as their evolution under the influence of X-ray radiation and temperature, additional studies are necessary.

23.4 Conclusions

As a result of long-term X-ray irradiation of Y_2O_3 ceramics sintered from nano-sized powder, the evident decrease (almost double) in the intensity of luminescent radiation at 350 nm occurred. This band is often associated with light emission through STE decay. It was found that the time dependence of the luminescence intensity was good fitted by exponential functions and therefore can be characterized by two processes: fast with a time constant of about 500 s and slow with a time constant of about 3000 s. It was observed that if the sample was exposed to X-ray reirradiation after 24 h pause, then the initial luminescence intensity would considerably exceed the value measured at the end of the previous irradiation. That is, it can be argued that a regeneration of the luminescence centers responsible for light emission at this band occurred during the interval between exposures. As a result of long-time cyclic irradiation, the luminescence intensity approached a saturation value at the end of each cycle, and finally the saturation values for the second and third cycles became practically the same. An analysis of the time evolution of Y_2O_3 luminescence at the wavelength of 350 nm during cyclic irradiation allows us to conclude that several types of defects are involved in the dynamics of light generation.

Acknowledgements We thank to the staff of protecting coating and ceramic laboratory for preparing and sintering the ceramics. We are also grateful S. Bogatyrenko for electron microscope images of the samples.

References

1. A. Fukabori et al., J. Cryst. Gr. **318**, 823 (2011). https://doi.org/10.1016/j.jcrysgro.2010.10.211
2. A. Fukabori et al., J. of Appl. Phys. **107**, 073501 (2010). https://doi.org/10.1063/1.3330407
3. G.V. Lyamina et al., Nanotechnol. Russ. **13**(5–6), 337 (2018). https://doi.org/10.1134/S1995078018030102
4. D.H. Chavez, K. Juarez-Moreno, G.A. Hirata, Nanobiomedicine **3**, 1 (2016). https://doi.org/10.5772/62252
5. J. Shruthia, N. Jayababua, P. Ghosalb, M. V. Ramana Reddy, Ceram. Int. **45**(17) (2019). https://doi.org/10.1016/j.ceramint.2019.07.141
6. N.J. Shivaramu, B.N. Lakshminarasappa, K.R. Nagabhushana, F. Singh, Rad. Meas. **71**, 518 (2014). https://doi.org/10.1016/j.radmeas.2014.03.027
7. O.M. Bordun, I.M. Bordun, J. Appl. Spectr. **64**, 361 (1997). https://doi.org/10.1007/BF02675099
8. V.V. Osipov, A.V. Rasuleva, V.I. Solomonov, Opt. Spectr. **105**, 524 (2008). https://doi.org/10.1134/S0030400X08100068
9. O.M. Bordun, I.M. Bordun, S.S. Novosad, J. Appl. Spectr. **62**, 1060 (1995) https://doi.org/10.1007/BF02606760
10. O.M. Bordun, J. Appl. Spectr. **69**, 430 (2002). https://doi.org/10.1023/A:1019763518857
11. O.M. Bordun, J. Appl. Spectr. **68**, 304 (2001). https://doi.org/10.1023/A:1019228505728
12. S. Kononenko et al., in *Paper presented at the 8th IEEE International Conference on "Nanomaterials: Applications & Properties"*, Odessa, September 2018
13. G. Blasse, L.H. Brixner, Eur. J. Solid State Inorg. Chem. **28**, 767 (1991). https://doi.org/10.1002/chin.199136016
14. A. Lushchik et al., J. Lumin. **232**, 87–89 (2000). https://doi.org/10.1016/S0022-2313(99)00271-9
15. N. V. Guerassimova et al., HASYLAB Annu. Rep. **603** (2005)
16. S.S. Novosad et al., Acta Phys. Polonica A **133**, 806 (2018) https://doi.org/10.12693/APhysPolA.133.806
17. A. Fukabori et al., Ceram. Int. **38**, 2119 (2012). https://doi.org/10.1016/j.ceramint.2011.10.052
18. I.N. Stanton et al., Nanoscale **6**, 5284 (2014). https://doi.org/10.1039/c4nr00497c
19. L. Robindro Singh, Nanotechnology **19**, 1 (2008). https://doi.org/10.1088/0957-4484/19/05/055201
20. L.H. Abu-Hassan, P.D. Townsend, R.P. Webb, Nucl. Instr. Meth. Phys. Res. B **19–20**, 927 (1987). https://doi.org/10.1016/S0168-583X(87)80186-6

Chapter 24
Electrochemical Formation of 'Synthetic Receptors' Based on Conducting Polymers

A. Ramanavicius, A. Tereshchenko, I. Plikusiene, V. Ratautaite, M. A. Deshmukh, V. Smyntyna, Ya. Oztekin, U. Bubniene, and A. Ramanaviciene

Abstract Analysis of pharmaceuticals and biologically active materials recently can be performed by biosensors, which can be based on conducting polymers. Therefore, in this Mini Review some achievements in the synthesis and application of conducting polymer—polypyrrole (Ppy), which is often used in the design of sensors and biosensors, are overviewed. Some perspective methods of conducting polymer synthesis are outlined. Significant attention has been paid to electrochemical, chemical and biochemical synthesis of conducting polymers (CPs) , which were developed by authors. The applicability of polypyrrole based functional layers in the design of electrochemical biosensors is overviewed. The adaptability of enzyme—glucose oxidase (GOx), which can be applied as (i) biological recognition element—in the design of glucose biosensors, (ii) a biocatalyst—in the synthesis of some above mentioned conducting polymers, is discussed. Part of biocompatibility related

A. Ramanavicius (✉) · A. Tereshchenko · I. Plikusiene · V. Ratautaite · M. A. Deshmukh · Ya. Oztekin · U. Bubniene
Department of Physical Chemistry, Faculty of Chemistry and Geosciences, Vilnius University, Vilnius, Lithuania
e-mail: arunas.ramanavicius@chf.vu.lt

A. Ramanavicius · I. Plikusiene · V. Ratautaite · U. Bubniene
Laboratory of Nanotechnology, State Research Institute Center for Physical Sciences and Technology, Vilnius, Lithuania

A. Tereshchenko · M. A. Deshmukh · Ya. Oztekin · A. Ramanaviciene
Faculty of Chemistry and Geosciences, NanoTechnas - Centre of Nanotechnology and Material Science, Vilnius University, Vilnius, Lithuania

A. Tereshchenko · V. Smyntyna
Department of Experimental Physics, Faculty of Mathematics, Physics and Information Technologies, Odesa I.I. Mechnikov National University, Odesa, Ukraine

M. A. Deshmukh
Department of Physics, RUSA - Center for Advanced Sensor Technology, Dr. Babasaheb Ambedkar Marathwada University, Aurangabad, Maharashtra 431 004, India

Ya. Oztekin
Faculty of Science, Department of Chemistry, Selcuk University, Konya, Turkey

A. D. Pogrebnjak et al. (eds.), *Nanomaterials in Biomedical Application and Biosensors (NAP-2019)*, Springer Proceedings in Physics 244,
https://doi.org/10.1007/978-981-15-3996-1_24

aspects of some conducting polymers are also discussed and some insights in the application of polypyrrole-based coatings for implantable sensors are outlined.

24.1 Introduction

Demands for advanced pharmaceutical and biomedical analysis are evolving very rapidly. Therefore, various new and advanced technologies based on sensors and biosensors [1] are finding the application in the solution of challenging problems of pharmaceutical and biomedical analysis [2]. Various polymer-based nanomaterials recently are applied in the design of sensors and biosensors in order to advance their sensitivity and selectivity and in this way to extend their bioanalytical applicability [3, 4]. It should be noted that conducting polymers are among the most promising materials, which can extend analytical characteristics of sensors [5–7], including their sensitivity and selectivity [8]. These achievements are possible due to electrical conductivity, electrochemical activity, mechanical elasticity and environmental stability of conducting polymers [9–11]. In addition, unique capabilities of transferring electric charge from biological objects towards electrodes are also very attractive property for sensorics related application of conducting polymers [12]. Simple doping and dedoping of conducting polymers enables to manipulate the properties of material makes conducting polymer ideal material for sensing applications [13, 14]. Among a variety of conducting polymers polypyrrole (Ppy), polyaniline (PANI) and polythiophene (PTH) are of the greatest interest due to their high technological potential, which has been exploited in the design of super capacitors [15], rechargeable batteries [16], corrosion prevention [17], sensors [18, 19] solar cells [20] and biosensors [21]. Technologically it is advantageous that various conducting polymer based nanocomposites with entrapped enzymes [22], receptor-like proteins [13], antibodies [23] and DNA [24] can be formed by different methods overviewed and discussed in this review. Therefore, such conducting polymer based composite materials can show unique sensing properties, which are determined by entrapped materials and dopants.

The aim of this 'Mini Review' is to evaluate the methods of synthesis and bioanalytical application of some conjugated polymers including polypyrrole, polythiophene and polyaniline.

24.2 Synthesis of Conducting Polymers

24.2.1 Synthesis of Conducting Polymers

It was demonstrated that conducting polymers might be produced by chemical [25], electrochemical [13, 26] and even by biotechnological approaches [3].

Therefore, these conducting polymers synthesis methods are taken for consideration in this review.

Chemical synthesis is one of the most popular methods to form conducting polymers. This method is relatively cheap and simple based on usage of strong oxidators such as $FeCl_3$ or H_2O_2 [17, 27]. Spherical particles of polypyrrole can be formed using H_2O_2, which has oxidation potential sufficient to initiate the polymerization of pyrrole and other monomers that are forming conducting polymer [28], this method is very attractive because excess of H_2O_2 can be easily degraded into H_2O and O_2, therefore pure particles of the conducting polymer can be formed, which possesses a good biocompatibility with living stem cells [29, 30] and even when injected in mice peritoneum [31].

The advantage of chemical synthesis is that using this method can be formed large quantities of nanoparticles, based on the conducting polymer, which are suspended in a solution. During the next technological steps these nanoparticles eventually can be modified and applied for biomedical purposes. However, this method is not always well suited for the formation of films, because chemically formed conducting polymer layers mostly are sparse and not stable enough for further technological applications.

Some authors presented in their work where polypyrrole nanostructures such as nanowires and nanotubes were chemically synthesized inside aluminium template [32]. The response surface methodology based on central composite design was used to determine the relationship between the morphology of the nanostructures (nanowires or nanotubes) and synthesis conditions [32]. Other authors showed that polypyrrole nanostructures can be gained chemically synthesizing Ppy within the pores of microporous and nanoporous particle track-etched membranes [33]. Chemical synthesis of polypyrrole nanoparticles is possible using facile one-step chemical oxidative polymerization method at room temperature. Using this method doping of conjugated-polypyrrole nanoparticles into electrically conductive adhesives (ECAs) can be obtained. The purpose of such polymerization procedure is to prepare low-electrical resistivity interconnecting materials [34].

Biochemical synthesis of conducting polymers can be performed in several ways: (i) enzymatic—using redox enzymes, e.g. glucose oxidase or (ii) microbiological—using whole microorganisms [35, 36].

Enzymatic synthesis of conducing polymers. Catalytic reactions of Glucose oxidase (GOx) E.C. 1.1.3.4. from *Penicillium vitale* can be applied for the synthesis of various conducting polymers such as: polypyrrole [3, 13, 28, 29], polyaniline [37] and polytiophene [38], polyphenanthroline [39] and some other conducting polymer based layers and nanoparticles. This is a unique approach, which is based on catalytic action of GOx. Actually polymerization of monomers, which are forming conducting polymer, is initiated by hydrogen peroxide that is formed during the catalytic action of GOx. Immobilized and in water dissolved enzymes have been successfully applied in the enzymatic synthesis of conducting polymer based layers or particles with entrapped enzymes, which are producing hydrogen peroxide. Due to remaining activity of entrapped enzyme, such particles and layers are well suitable for the creation of amperometric glucose biosensors and biofuel cells.

Microbiological synthesis of polypyrrole. Some redox processes, which are part of metabolism occurring in living cells can be adapted for the synthesis of Ppy within cell wall of living cells: yeast cells [22].

Electrochemical synthesis is also very attractive method used for the deposition of conducting polymer based layers. The adjustment of various electrochemistry-related parameters such as polymerization-inducing potential, current, scan rate and polymerization duration enables to create polymeric layers of different characteristics [13, 40, 41]. This method enables to manipulate properties of conducting polymers for particular application by controlling the process parameters [42]. In addition, electrical conductivity and some electrochemical properties of conducting polymers can be tailored and controlled by variation of polymerizable monomer concentrations, pH of polymerization bulk solution and different dopant concentrations [43–45]. Electrochemical formation of conducting polymer based layers is attractive because both morphology and thickness of formed layer can be controlled by the adjustment of (i) potential-profile and the duration of potential pulses or (ii) potential sweep rate [13], pH of the electrolyte [46].

24.2.2 Some Chemical and Physical Properties of Conducting Polymers

In our researches we have determined that conducting polymer—polypyrrole—exhibit unique electrochemical, affinity [32] and/or optical [33] properties. Therefore, changes of one or more of these physicochemical properties (e.g. resistance of CP-based biological recognition layer, variation of electrical capacitance, changes of optical properties etc.) can be determined by particular signal transducer and can be precisely monitored by registration device. It should be noted that among number of CPs—polypyrrole mostly has been used in the design of enzymatic biosensors as enzyme immobilization matrixes [3].

24.3 Conducting Polymers for Biosensor Design

Conducting polymers in the design of amperometric biosensors. Due to relatively low permeability of Ppy layer for substrates and reaction products, apparent Michaelis constant (K_{Mapp}) of immobilized enzymes increases significantly. This effect enables to extend linear range of such enzymatic sensors [14], which can be used for biosensors based on enzymes, which have low K_{Mapp} and therefore are not suitable for the investigation of samples with high substrate concentration, e.g. glucose concentration in blood is significantly higher in comparison with K_{Mapp} of the most popular glucose oxidases [34]. This methodology enables to tune analytical characteristic such as lower and upper limits of detection and linear range of biosensor [3, 23, 24]. In addition to the above mentioned properties, some conducting polymers are capable to transfer charge directly from redox enzymes towards electrodes [39]. Therefore, electrochemically generated conducting polymers are very promising as advanced immobilization matrixes of amperometric catalytic biosensors [14] biofuel cells and other bioelectronic devices [3, 47–49]. However, most of enzymes have redox centers deeply buried within protein 'shell'. Therefore, charge transfer from these enzymes is complicated even if they are entrapped within conducting polymer. In order to solve this problem in some of our researches we have demonstrated that charge transfer capabilities of electrochemically formed conducting polymers can be improved by grafting to the surface of carbon-based electrodes [39]. Glucose biosensors, which are mostly based on GOx, are important tools, which are used not only for biomedical purposes, but also as a model system in the development of enzymatic sensors. Therefore, very different conducting polymer and GOx based structures have been formed and applied in glucose sensor design [14, 50]. Single-step procedure has been established for the modification of electrode by composite material consisting of polypyrrole, Prussian blue (PB) and glucose oxidase (GOx) [31].

Conducting polymers in the design of affinity biosensors can serve as (i) immobilization matrixes [13, 31], signal transduction systems [13, 39] and even analyte recognizing components [28]. Electrochemically deposited layers of conducting polymers modified with entrapped proteins (antibodies or antigens) have been applied in the design of various types of ***immunosensors*** [13], in such sensors conducting polymer layer is enhancing potentiodynamically generated electrochemical signal [13] or reducing the influence of interfering materials if photoluminescence based analytical signal is registered [51]. The most promising among affinity sensors are ***molecularly imprinted polymers (MIPs) based sensors***, because they do not need very expensive biological recognition materials [52–54]. Overoxidized polypyrrole seems to be one of the most promising among all MIPs because it can be prepared in very simple electrochemical way using single polymerizable monomer. Polypyrrole can be easily imprinted by large [28, 29, 55–57] and low [7, 10, 58, 59] molecular weight molecules (e.g. by Caffeine [10], theophylline [58], L-aspartic acid [59] and histamine [7]) and even by large molecular weight biomolecules (e.g. by DNA and proteins [28, 29, 55–57]).

24.4 Conclusions

Conducting polymers have found application in various sensors and biosensors, which can be used in the analysis of pharmaceuticals and biologically active materials. Electrochemical synthesis of conducting polymers seems promising, because it can be controlled by adjustment of electrical current profile. Characteristics of conducting polymer based sensors are determined by the thickness of polypyrrole layer. Enzymatic and microbial synthesis of polypyrrole demonstrated that in such way formed polypyrrole/bio-composite materials are also suitable for the development of biosensors and biofuel cells. Also chemical synthesis of polypyrole reveal possibilities to form different nanostructures that can be used as biosensing platforms. The applicability of similar methods for the synthesis of other conducting polymers such as polyaniline, polythiophene, and some others conducting or π-π conjugated polymers is predicted.

Acknowledgements This research was supported by Ukrainian-Lithuanian Research project Lithuanian Research Council project No. P-LU-18-53.

References

1. N.K. Bakirhan, G. Ozcelikay, S.A. Ozkan, J. Pharm. Biomed. Anal. **159**, 406 (2018)
2. A. Sankiewicz, L. Romanowicz, M. Pyc, A. Hermanowicz, E. Gorodkiewicz, J. Pharm. Biomed. Anal. **150**, 1 (2018)
3. N. Aydemir, J. Malmström, J.T. Sejdic, Phys. Chem. Chem. Phys. **18**, 8264–8277 (2016)
4. J.-M. Moon, N. Thapliyal, K.K. Hussain, R.N. Goyal, Y.-B. Shim, Biosens. Biolect. **102**, 540–552 (2018)
5. A. Ramanavicius, A. Kausaite, A. Ramanaviciene, Analyst **133**, 1083 (2008)
6. K.C. Persaud, Mater. Today **8**, 38–44 (2005)
7. V. Ratautaite, M. Nesladek, A. Ramanaviciene, I. Baleviciute, A. Ramanavicius, Electroanalysis **26**, 2458 (2014)
8. I. Baleviciute, V. Ratautaite, A. Ramanaviciene, Z. Balevicius, J. Broeders, D. Croux, M. McDonald, F. Vahidpour, R. Thoelen, W. De Ceuninck, K. Haenen, M. Nesladek, A. Reza, A. Ramanavicius, Synth. Met. **209**, 206–211 (2015)
9. V. Ratautaite, A. Ramanaviciene, Y. Oztekin, J. Voronovic, Z. Balevicius, L. Mikoliunaite, A. Ramanavicius, Colloids Surf. A **418**, 16 (2013)
10. M. Tomczykowa, M.E.P. Brzezinska, Polymers **11**, 350 (2019)
11. H. Yoon, Nanomaterials **3**, 524–549 (2013)
12. Y. Oztekin, A. Ramanaviciene, Z. Yazicigil, A.O. Solak, A. Ramanavicius, Biosens. Bioelectron. **26**, 2541 (2011)
13. Z. Ma, W. Shi, K. Yan, L. Pan, G. Yu, Chem. Science **25**, 6232–6244 (2019)
14. N. Karaoğlan, C. Bindal, Eng. Sci. Technol. Int. J. **21**, 1152–1158 (2018)
15. Y. Wang, Y. Chen, Y. Liu, W. Liu, P. Zhao, Y. Li, Y. Dong, H. Wang, J. Yang, Electrochim. Acta **295**, 989 (2019)
16. Z. Zhao, T. Yu, Y. Miao, X. Zhao, Electrochim. Acta **270**, 30 (2018)
17. J.O. Iroh, W. Su, Electrochim. Acta **46**, 15 (2000)
18. V. Ratautaite, D. Plausinaitis, I. Baleviciute, L. Mikoliunaite, A. Ramanaviciene, A. Ramanavicius, Sens. Actuators B: Chem. **212**, 63 (2015)
19. M. Holguín, O.E.R. Álvarez, C.A. Arizabaleta, W. Torres, Comput. Theor. Chem. **1147**, 29 (2019)
20. N. Sangiorgi, A. Sangiorgi, F. Tarterini, A. Sanson, Electrochim. Acta **305**, 322 (2019)
21. A. Ramanavicius, Y. Oztekin, A. Ramanaviciene, Sens. Actuators B **197**, 237 (2014)
22. N. German, A. Ramanavicius, J. Voronovic, A. Ramanaviciene, Colloids Surf. A **413**, 224 (2012)
23. F.S. Felix, L. Angnes, Biosens. Bioelectron. **102**, 470 (2018)
24. A. Ramanaviciene, A. Ramanavicius, Anal. Bioanal. Chem. **379**, 287 (2004)
25. K. Leonavicius, A. Ramanaviciene, A. Ramanavicius, Polymerization model for hydrogen peroxide initiated synthesis of polypyrrole nanoparticles. Langmuir **17**, 10970 (2011)
26. V. Syritski, J. Reut, A. Öpik, K. Idla, Synth. Metals **102**, 1326 (1999)

27. N.Y. Abu-Thabit, J. Chem. Educ. **939**, 1606–1611 (2016)
28. E. Falletta, C. Della Pina, M. Rossi, J. Adv. Cata. Sci. Technol. **1**, 6–14 (2014)
29. A. Vaitkuviene, V. Kaseta, J. Voronovic, G. Ramanauskaite, G. Biziuleviciene, A. Ramanaviciene, A. Ramanavicius, J. Hazard. Mater. **250–251**, 167 (2013)
30. A. Vaitkuviene, V. Ratautaite, L. Mikoliunaite, V. Kaseta, G. Ramanauskaite, G. Biziuleviciene, A. Ramanaviciene, A. Ramanavicius, Coll. Surf. A **442**, 152 (2014)
31. A. Ramanaviciene, A. Kausaite, S. Tautkus, A. Ramanavicius, J. Pharm. Pharmacol. **59**, 311 (2007)
32. V. Ratautaite, S.N. Topkaya, L. Mikoliunaite, M. Ozsoz, Y. Oztekin, A. Ramanaviciene, A. Ramanavicius, Electroanalysis **25**, 1169 (2013)
33. A. Ramanavicius, N. Ryskevic, A. Kausaite-Minkstimiene, U. Bubniene, I. Baleviciute, Y. Oztekin, A. Ramanaviciene, Sens. Actuators B: Chem. **171–172**, 753 (2012)
34. N. German, A. Kausaite-Minkstimiene, A. Ramanavicius, T. Semashko, R. Mikhailova, A. Ramanaviciene, Electrochim. Acta **169**, 326 (2015)
35. A. Kisieliute, A. Popov, R.M. Apetrei, G. Cârâc, I. Morkvenaite-Vilkonciene, A. Ramanaviciene, A. Ramanavicius, Eng. J. **356**, 1014 (2019)
36. A. Ramanavicius, E. Andriukonis, A. Stirke, L. Mikoliunaite, Z. Balevicius, A. Ramanaviciene, Enzyme Microb. Technol. **83**, 40 (2016)
37. A. German, A. Popov, A. Ramanaviciene, A. Ramanavicius, Polymer **115**, 211 (2017)
38. V. Krikstolaityte, J. Kuliesius, A. Ramanaviciene, L. Mikoliunaite, A. Kausaite-Minkstimiene, Y. Oztekin, A. Ramanavicius, Polymer **55**, 1613 (2014)
39. S.J. Park, C.S. Park, H. Yoon, Polymers **9**, 155 (2017)
40. R. Xiao, S.I. Cho, R. Liu, S.B. Lee, J. Am. Chem. Soc. **129**, 4483–4489 (2007)
41. D.E. Labaye, C. Jérôme, V.M. Geskin, P. Louette, R. Lazzaroni, L. Martinot, R. Jérôme, Langmuir **18**, 5222–5230 (2002)
42. Y. Long, M. Mengli, C. Gu, M. Wan, J.L. Duvail, Z. Liu, Z. Fan, Prog. Polym. Sci. **36**, 1415 (2011)
43. M.A. Rahman, P. Kumar, D. Park, Y. Shim, Sensors **8**, 118 (2008)
44. B. Luc, G. Street, Acc. Chem. Res. **18**, 309 (1985)
45. S. Srilalitha, K. Jayaveera, S. Madhvendhra, Int. J. Innov. Res. Sci. Eng. Technol. **2**, 2694 (2013)
46. T.H. Le, Y. Kim, H. Yoon, Polymers **9**, 150 (2017)
47. S. Ghosh, T. Maiyalagan, R.N. Basu, Nanoscale **8**, 6921–6947 (2016)
48. Beenish, Inamuddin, M.I. Ahamed, A.M. Asiri, K.A. Al Amry, Mater. Sci. Energy Technol. **1**, 63–69 (2018)
49. I. Inamuddin, K.M. Shin, S.I. Kim, I. So, Electrochim. Acta **54**, 3979–3983 (2009)
50. A. Ramanavicius, A.I. Rekertaitė, R. Valiūnas, A. Valiūnienė, Sens. Actuators B: Chem. **240**, 220 (2017)
51. A. Ramanavicius, N. Ryskevic, Y. Oztekin, A. Kausaite-Minkstimiene, S. Jursenas, J. Baniukevic, J. Kirlyte, U. Bubniene, A. Ramanaviciene, Anal. Bioanal. Chem. **398**, 3105 (2010)
52. G. Zhang, Y. Yu, M. Guo, B. Lin, L. Zhang, Sens. Actuators B: Chem. **288**, 564 (2019)
53. A. Tretjakov, V. Syritski, J. Reut, R. Boroznjak, O. Volobujeva, A. Öpik, Microchim. Acta **180**, 1433 (2013)
54. A. Menaker, V. Syritski, J. Reut, A. Öpik, V. Horváth, R.E. Gyurcsányi, Adv. Mater. **21**, 2271 (2009)
55. A. Ramanaviciene, A. Ramanavicius, Biosens. Bioelectron. **20**, 1076 (2004)
56. Z.D. Kojabad, S.A. Shojaosadati, Mater. Des. **96**, 378 (2016)
57. J. Duchet, R. Legras, S. Demoustier-Champagne, Synth. Met. **2**, 113 (1998)
58. V. Ratautaite, S.D. Janssens, K. Haenen, M. Nesládek, A. Ramanaviciene, I. Baleviciute, A. Ramanavicius, Electrochim. Acta **130**, 361 (2014)
59. V. Syritski, J. Reut, A. Menaker, R.E. Gyurcsányi, A. Öpik, Electrochim. Acta **53**, 2729 (2008)

Chapter 25
Optical Immunosensor Based on Photoluminescent TiO_2 Nanostructures for Determination of Bovine Leucosis Proteins. Model of Interaction Mechanism

A. Tereshchenko, V. Smyntyna, U. Bubniene, and A. Ramanavicius

Abstract The main aspects of the interaction mechanism between nanostructured TiO_2 layer and BLV proteins *gp*51 have been evaluated during the formation of photoluminescence-based immunosensor. *Bovine leucosis* protein *gp*51 was adsorbed on the surface of a nanostructured TiO_2 thin film, formed on glass substrates. A photoluminescence (PL) peak shift from 517 to 499 nm was observed after modification of TiO_2 surface by adsorbed *gp*51 (i.e. formation of the biosensitive layer *gp*51/TiO_2). An incubation of *gp*51/TiO_2 in a solution containing anti-*gp*51 antibodies resulted in the formation a of a new structure (anti-*gp*51/*gp*51/TiO_2) and the backward PL peak shift from 499 nm to 516 nm. The PL shifts are attributed to the variations in the self-trapped exciton energy level, which were induced by the changes of electrostatic interaction between positively charged atoms and groups, provided by the adsorbed *gp*51 protein and negatively charged surface of TiO_2. The charge–charge-based interaction in the double charged layers *gp*51/TiO_2 can also be interpreted as a model based on 'imaginary capacitor', formed as a result of the electrostatic interaction between oppositely charged protein *gp*51 layer and the TiO_2 surface.

25.1 Introduction

Bovine leucosis is a lethal cancerous disease caused by *Bovine leukemia* virus (BLV) that belongs to one of the most monitored family of retro-viruses in the world. There is a significant risk that BLV can infect other mammalians like it was in the case

A. Tereshchenko (✉) · V. Smyntyna
Department of Experimental Physics, Faculty of Mathematics, Physics and Information Technologies, Odesa I.I. Mechnikov National University, Odesa, Ukraine
e-mail: alla_teresc@onu.edu.ua

A. Tereshchenko · U. Bubniene · A. Ramanavicius
Department of Physical Chemistry, Faculty of Chemistry and Geosciences, Vilnius University, Vilnius, Lithuania

A. D. Pogrebnjak et al. (eds.), *Nanomaterials in Biomedical Application and Biosensors (NAP-2019)*, Springer Proceedings in Physics 244,
https://doi.org/10.1007/978-981-15-3996-1_25

of Human immunodeficiency virus (HIV). Due to lethal epidemical nature, the outbreaks of *Bovine leucosis* pose a great threat to the environment and eco-systems. As the food quality control is an inseparable part of human safety and wellbeing thus an advanced determination of *Bovine leucosis* can decrease the risks of human diseases caused by this and other dangerous viruses. For this reason, timely diagnosis of virus-induced diseases is an important direction in environmental monitoring. For such monitoring very efficient bio-analytical systems are required but old bio-analytical methods are cumbersome and slow. Therefore, the development of advanced bioanalytical systems based on photoluminescence immunosensors seems to be the most promising direction [1, 2]. However, despite of many reports on photoluminescence-based immunosensors, the mechanism of interaction of biomolecules with semiconducting materials, which are used as analytical signal transducers in the most promising PL-based immunosensors, is still very poorly evaluated [3].

Nanostructured Titanium dioxide (TiO_2) is known as a material of intense photoluminescence at room temperature [4, 5]. The application of TiO_2 photoluminescence properties in optical biosensors and immunosensors have been reported in the range of works [1–3]. TiO_2 is widely studied material as a wide-band gap semiconductor with a great combination of physical and chemical properties [6–8]. A good biocompatibility of TiO_2 nanostructures, their applicability at physiological pHs in the range of 5.5–7.0, non-toxicity and excellent chemical stability have resulted in the extensive application of TiO_2 in various biosensors [1, 2, 9]. Optical biosensors are increasingly studied class of biosensors because optical detection methods have a number of advantages. Optical methods allows to evaluate some inter-molecular interactions contactless, i.e., without contamination and/or deterioration of the aliquot, which contains biological compounds [2, 3] and it allows to avoid application of special tags or markers and enables to determinate the analyte concentration directly without any chemical/physical labels [1–3]. Among the large variety of biosensors, a special attention is paid to the development of immunosensors, based on the specific interaction between antibody and antigen that can be applied for the determination of wide variety of analytes in complex biological samples [1, 2, 9].

The changes in the photoluminescence spectra (shift of photoluminescence maximum and the variation of photoluminescence signal intensity) were exploited as analytical signals for the determination of target analyte [2, 3, 9, 10]. However, the interaction mechanism of proteins with TiO_2 and the origin of the changes in the photoluminescence spectra were not discussed. Although the mechanism of the interaction between semiconductor TiO_2 and proteins is the key in solving many of problems, which are still arising during the development of TiO_2-based immunosensors, such as an improvement of sensitivity and selectivity.

This work is aiming to highlight the origin of the changes in the photoluminescence spectra of TiO_2 resulted after the protein adsorption on its surface during the formation of biosensitive layer, and after the interaction of biosensitive layer with the analyte. The proposed interaction mechanism provides the general understanding of the interaction between TiO_2 and proteins, what is a key in the development of new

PL-immunosensors and solving of many issues related to an improvement of performance of PL-based immunosensors, first of all, related to the sensors' sensitivity and selectivity.

25.2 Experiment Details

Nanostructured TiO_2 layers containing TiO_2 nanoparticles were formed by deposition of colloidal suspension of TiO_2 nanoparticles (Sigma Altrich, 99.7%, particle size of 32 nm) dissolved in ethanol. The concentration TiO_2 nanoparticles was about 0.01 mg/ml. TiO_2 layers were dryed at room temperature and annealed at 350 °C. Structural and surface characterization of the obtained samples showed that TiO_2 layers kept the anatase structure and TiO_2 nanoparticles formed a high surface area porous structure suitable for the formation of biosensitive layer. More detailed information on the deposition and characterization procedures applied for characterization of nanostructured TiO_2 layers is reported in earlier researches [3, 10].

Optical characterization of nanostructured TiO_2 layers was performed by photoluminescence measurements using 355 nm solid state laser as the excitation source. Optical setup is described previous our researches [3, 9]. The spectra were recorded in the range of wavelength from 360 to 800 nm.

The immobilization of biological molecules was carried out by incubation in a solution containing Bovine leukemia virus proteins—*gp*51, similar to the immobilization procedure described in earlier works [3, 9, 10]. In brief: a solution of PBS containing *gp*51 antigens at a high concentration was directly immobilized on the TiO_2 surface. Then the sample was placed into a Petri cup for the incubation in a medium saturated with water vapor at 25 °C. After 10 min of incubation, the surface of the sample was washed with PBS solution in order to remove non-immobilized antigens on the TiO_2 surface. As a result, an adsorption-sensitive layer TiO_2/*gp*51 was formed, selective to the one type of bio-molecules—anti-*gp*51 antibodies against leukemia proteins *gp*51. To prevent a nonspecific interaction (i.e., binding of anti-*gp*51 antibodies directly to unmodified TiO_2 surface), the surface of TiO_2 was further treated with a solution of bovine serum albumin (BSA), which filled possible adsorption sites that remained free after the modification of TiO_2 surface with *gp*51. Thus it is expected that the selectivity of the structure of TiO_2/*gp*51 was improved by this procedure based on treatment with BSA, which is frequently applied during the development of immunosensors devoted for the determination of proteins.

25.3 Results and Discussion

Analysis of the interaction between TiO_2 and *gp*51 proteins was based on the evaluation of photoluminescence of TiO_2 nanoparticles. The protein *gp*51 is specifically binding with antibodies against *gp*51 (anti-*gp*51) that were interpreted as analyte in this research. The photoluminescence properties of the TiO_2 nanoparticles have been previously investigated by authors in the research papers [3, 9, 10] as well as the influence of *gp*51 protein adsorption on the optical properties of TiO_2 and the development of photoluminescence based immunosensor for the determination of *gp*51 antibodies [3, 9, 10].

The process of immobilization of *gp*51 and formation of TiO_2/*gp*51 structure was similar to that reported in earlier our works [3, 9, 10]. The immobilization of *gp*51 antigens leads to an increase in the intensity of the photoluminescence signal of TiO_2 nanostructures and a UV-shift in the position of the maximum of the photoluminescence spectra (Fig. 25.1). The application of BSA, which were used to block the free adsorption centers on the TiO_2 surface, also leads to a slight increase in the photoluminescence intensity, but the spectrum shift in this case was not observed.

Interaction between TiO_2/*gp*51 and anti-*gp*51 led to the inverse changes in the photoluminescence spectrum (Fig. 25.1), i.e. a decrease in the integral intensity of the photoluminescence and the IR-shift of spectra. Therefore, the response of the immunosensor TiO_2/*gp*51 to anti-*gp*51 can be estimated by two parameters: (i) the photoluminescence intensity and (ii) the position of the PL-maximum. The sensitivity of TiO_2/*gp*51 based immunosensor towards anti-*gp*51 was in the range of 2–8 mkg/ml [9, 10].

TiO_2 (anatase) is known as a semiconductor of n-type conductivity, usually with an 'upward' band bending of the energy levels when closing the surface of TiO_2 (Fig. 25.2) [11], which indicates the accumulation of a negative charge (bound at surface levels) on its surface. The adsorption of the most of molecules is known to

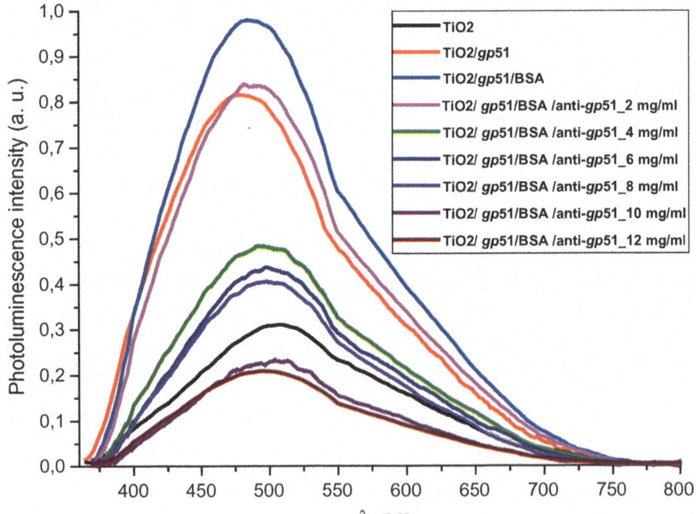

Fig. 25.1 Photoluminescence spectra of TiO_2 nanoparticles before and after the immobilization of *gp*51 antigens on the TiO_2 surface, subsequent BSA deposition and after the interaction of TiO_2/*gp*51 based immunosensor after with analyte (anti-*gp*51 antibodies) of different concentrations

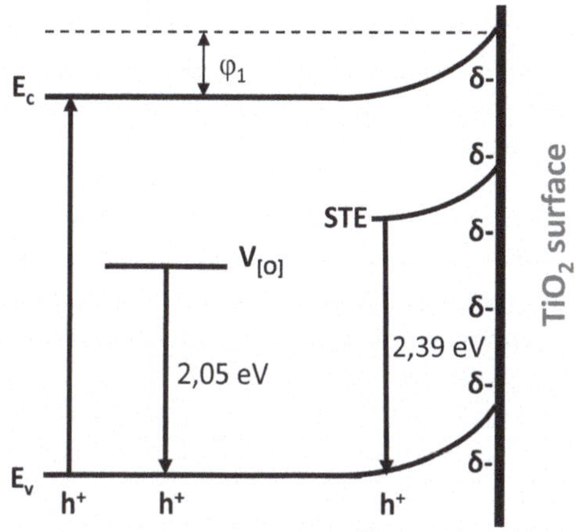

Fig. 25.2 Energetic levels of TiO_2: E_c, E_v—conductive and valence bands respectively, STE—self-trapped exiton level, $V_{[o]}$ —oxygen vacancies level

introduce an additional charge on the solid state surface and it can change the existing surface energy levels or form the additional ones that are involved in the exchange of charges with the volume of a solid material [12].

The proteins consist of amino acids that might contain positively and/or negatively charged radicals that are determining the charge of the different protein domains [13]. A large quantity of negatively charged groups such as aldehyde (–CHO), hydroxyl (–OH), carboxyl (–COOH) and positively charged primary amine (–NH_2) and some other groups, which are involved into the structure of amino acids, are responsible for the partial (δ+ and δ−) charges of particular protein domains (Fig. 25.3) [14]. Therefore the proteins are characterized by electrostatic properties, and sometimes even significant electrostatic 'asymmetry of protein molecule' because the atoms

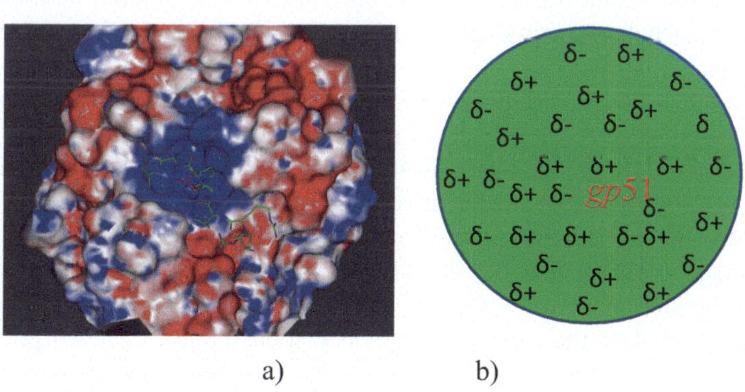

Fig. 25.3 **a** Charge distribution in calciumneurin protein: positive charge—blue color, negative charge—red color, neutral charge—white color [15]; **b** schematic image of charge distribution in *gp*51 antigen protein: δ+ and δ− are partial positive and negative charges respectively

and functional groups forming the protein molecules are charged differently both in their sign and in absolute charge value. Naturally, the charges at least partly are compensating each other, but since the ternary structure of proteins is relatively rigid and the charged groups have only limited degree of freedom to move within the protein globule, therefore in some parts of the protein some uncompensated charge on the surface and inside of the protein still remains [14]. The distribution of charged groups on the surface of the protein depends on the sequence of amino acids, which is pre-determined by the genome that was developed during billions of years lasting evolution and selected genes promoting the synthesis of proteins whose structure the most efficiently matches their function.

It should be taken into account that even if the structure of the most proteins is at some extent 'rigid' there is some degree of flexibility because both secondary and tertiary structures of the protein are supported by a large number of hydrogen bonds but many of them are not very strong [13, 16]. The electrostatic bonds, which are based on Coulomb forces, between the opposite charges, van der Waals forces and disulfide bonds also play an important role in the formation of both secondary and tertiary structures of protein.

25.3.1 Mechanism of Interaction Between TiO_2 and Proteins

A *gp*51 protein molecule has a molecular mass of 51 KDa. The characteristic geometric size of the *gp*51 molecules adsorbed on the TiO_2 surface is about 6 nm in diameter [17]. The authors, which have published a research on the formation of *gp*51 virus based capsid of BLV, have constructed an image of *gp*51 virus structure from the X-ray crystallography data and they have reported that this protein is extra-flexible, which provides very high functionality and the ability to associate and/or dissociate of BLV capsid from the membrane of BLV infected cell [18]. Therefore, it is expected that on the surface of TiO_2 *gp*51 forms well-ordered monolayer. The formation of such layer was confirmed in other our researches by spectroscopic ellipsometry [10, 19, 20].

*Gp*51 protein is not a redox-protein therefore the charge transfer between *gp*51 and TiO_2 is not possible [18, 14]. However the *gp*51 protein like many others, contains a number of partially charged groups and domains, represented as partial charges "δ−" and "δ+" in Fig. 25.3b, which mostly are lower in value than the total electron charge (1.6×10^{-19} coulombs) per charged atom or group. The presence of these partial charges suggests that the electrostatic influence on the surface charge of TiO_2 from the side of partially uncompensated charges in those parts of the *gp*51 protein that located on the surface of TiO_2 is responsible for the adsorption of this protein on the TiO_2 surface. The Coulomb interaction takes place between charged groups in the *gp*51 protein and the negatively charged surface of the TiO_2 because such electrostatic interactions are very strong at a distances ranging from several Angstroms to few nanometers. Therefore, among the others interactions such as hydrogen bonds, disulfide bonds, Van der Waals interaction, etc., which also have significant role during the adsorption of proteins, the electrostatic interaction plays one of the most important role during the adsorption of proteins to electrically charged surfaces, such as TiO_2. In addition, the local electric fields of charged domains of adsorbed proteins are affecting the photoluminescence centers of TiO_2 and it causes the shift in the photoluminescence spectra of TiO_2 nanoparticles. Therefore, the photoluminescence maximum caused by STE shifts from 517 to 499 nm (i.e., to 18 nm), which corresponds to ~0.086 eV that is less than 0.1 eV, and it is one of the proofs of electrostatic interaction based physical adsorption of *gp*51 [12].

The splitting of the photoluminescence spectra into Gaussian curves at each stage of the experiment shows that after the adsorption of *gp*51 protein molecules on the TiO_2 surface the energy value of excitation levels, which are responsible for the luminescence and are associated with oxygen vacancies $I_{V[O]}$, almost does not change remaining at a value of 605 ± 2 nm. At the same time, the photoluminescence maximum caused by recombination of self-trapped excitons (STE) [14, 21] shifts to short wavelengths, changing its position from 517 ($STE_1 = 2.39$ eV) nm to 499 ($STE_2 = 2.48$ eV) nm. Since the involvement of the STE level in the process of radiative recombination is regulated by the surface, this indicates that STE level is located either on the surface plane or not very deeply within the surface layer of TiO_2. The displacement of the light emitting recombination peak indicates that the energy level of STE is complex and has its 'basic' and 'excited' states [18]. The appearance of luminescence in the region of 499 nm indicates a radiative transition from the excited STE level. This indicates that the charge at the TiO_2/*gp*51 boundary controls the energy level of the STE and shows that the electronic demarcation level practically coincides with the position of the STE level, i.e. is approximately 2.39 eV above the valence band. Therefore, the appearance of proteins on the TiO_2 surface leads to a shift in the energy levels, including the light emitting centers, relatively to the electron demarcation level E_{dn}.

The blue-shift of the photoluminescence maximum by 18 nm as a result of adsorption of the *gp*51 protein, which corresponds to $\Delta E_{STE} = STE_2 - STE_1 = 0.086$ eV, also indicates that the initial value of the potential barrier on the TiO_2 surface has decreased by a value of 0.086 eV. Variation of the potential barrier means that the value of negative charge localized on the TiO_2 surface has also changed, due to the charge-charge-based interaction with adsorbed protein *gp*51. Positively charged atoms and groups, which are provided by the *gp*51 protein, partially compensates the surface charge of TiO_2 and reduces the energy of electrons localized at the surface levels, which are the most responsible for the generation of photoluminescence signal. Taking into account the fact that the total negative charge predominates on the TiO_2 surface, the positively charged parts of the *gp*51 protein electrostatically interact with the negatively charged TiO_2 surface. As a result, a partial decrease of the surface charge reduces the electric field in the TiO_2 surface region. Further interaction of TiO_2/*gp*51 structure with anti-*gp*51, which is also a protein, leads to the inverse changes in the photoluminescence spectra, i.e., to UV-shift the spectrum (Fig. 25.1) and decrease the photoluminescence intensity to the value that corresponds to the pure TiO_2. The latter effect is based on the formation an immune complex between immobilized antigens *gp*51 and anti-*gp*51 antibodies, which were present in aliquot.

Formation of this immune complex besides the van der Waals interaction and other interactions at a very high extent is based on the interaction between oppositely charged domains, functional groups and atoms in *gp*51 and anti-*gp*51 molecules (including the formation of number of hydrogen bounds, which can be estimated as specific kind of electrostatic interaction). It can be assumed that uncompensated charges ($\delta+$ and $\delta-$) of both proteins are involved in electrostatic interactions during the formation of immune complex. As a result, some of the charged groups that were originally involved in the interaction between *gp*51 and TiO_2 are at least partially compensated by the opposite charge of the anti-*gp*51 protein groups, thereby reducing the direct electrostatic effect from immobilized *gp*51 proteins to the charged surface of TiO_2 and to light emitting centers.

The effects described above cause the shift of photoluminescence maximum and decrease in the potential barrier on TiO_2/*gp*51 interface due to the charge-charge interaction between TiO_2 and *gp*51. The potential barrier at the interface between TiO_2 and *gp*51 has greater value in TiO_2/*gp*51 structure in comparison with that in TiO_2/*gp*51/anti-*gp*51 due to partial compensation (decrease in value) and/or delocalization of charges, which were initially involved into interaction between TiO_2 and *gp*51 after formation of TiO_2/*gp*51 structure.

The distribution of charges in TiO_2/*gp*51 structure can also be interpreted as a model based on an 'imaginary flat capacitor' (Fig. 25.4), formed as a result of the electrostatic interaction between oppositely charged protein *gp*51 layer and the TiO_2 surface. The capacitor is formed as a result of protein *gp*51 adsorption on TiO_2 surface, after which the charges are distributed in energetically most favorable way, partially compensating each other. Consequently, the positive 'imaginary capacitor plate' is based on the positive charges, which are predominant in the protein *gp*51 area that after adsorption appears in close proximity to TiO_2/*gp*51 interface and/or due to the negative electrostatic effect of TiO_2 are induced/attracted closer to negatively charged surface.

These charged atoms/groups/domains of *gp*51 that are localized in the close proximity to the TiO_2 surface and they electrostatically affect the TiO_2 emission centers and the energy value of the surface potential barrier. Hence, the position of the

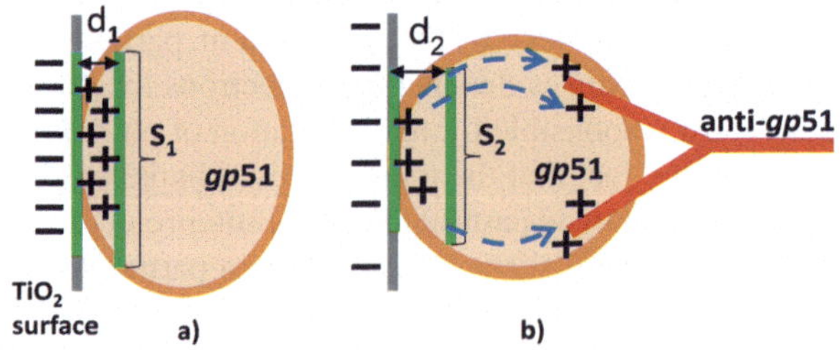

Fig. 25.4 Flat capacitor based model of the charges interaction between TiO_2 surface and *gp*51 proteins: **a** electrostatic interaction of uncompensated charges of immobilized protein *gp*51 with charges located on the surface of TiO_2; **b** model of interaction that takes into account the electrostatic interaction of charges within *gp*51 antigens and anti-*gp*51 antibodies

energy levels of the TiO_2 emission maximum depends on TiO_2 surface modification stage (TiO_2 or TiO_2/*gp*51) shifts from/backwards the initial position of the demarcation level. Figure 25.4a represents an imaginary flat capacitor consisting of a negatively charged plate on the surface of TiO_2 and an 'imaginary positively charged plate' formed in *gp*51 protein in close proximity to TiO_2/*gp*51 interphase. Hence, the interaction of TiO_2/*gp*51 with anti-*gp*51 antibodies and the formation of *gp*51/anti-*gp*51-based immune complex leads to a 'deformation' and the reduction of charge 'stored' on 'the positive imaginary capacitor plate' (Fig. 25.4b). This is mainly due to the redistribution and partial compensation of charges during the formation of the *gp*51/anti-*gp*51 immune complex, which in turn reduces the charge of 'the imaginary capacitor plate' based on *gp*51 ($q_2 < q_1$). Due to this reduced charge it can be interpreted as the reduction of the area of the same plate (S_2) and/or the increase of the distance (d_2) between the two imaginary capacitor plates based on *gp*51 and TiO_2 which leads to the decrease of capacitance according to (25.1).

$$C = \frac{\varepsilon \varepsilon_0 S}{d} \tag{25.1}$$

This effect is observed because some of the *gp*51 protein charges move from the TiO_2/*gp*51 interface towards interacting anti-*gp*51 protein and are partially compensated by charge present in anti-*gp*51, whereby the imaginary positive *gp*51-based capacitor plate of the capacitor is reduced in imaginary surface area and/or correspondingly moving apart from the negative TiO_2 plate. This effect leads to a decrease in the capacitance of this imaginary capacitor and the electric field induced by *gp*51 becomes reduced. Therefore, after the interaction of TiO_2/*gp*51 with anti-*gp*51 antibodies and the formation of *gp*51/anti-*gp*51 complex, which is involved into TiO_2/*gp*51/anti-*gp*51 structure, the electrostatic effect of *gp*51 initially adsorbed on TiO_2 towards the TiO_2 surface significantly decreases.

25.4 Conclusions

The main aspects of the interaction mechanism between nanostructured TiO_2 layer and bovine leukemia virus proteins *gp*51, during the formation of photoluminescence based immunosensor, have been developed. Bovine leukemia virus protein *gp*51, adsorbed on the surface of nanostructured TiO_2 thin film, formed the biosensitive layer (glass/TiO_2/*gp*51) that resulted in the TiO_2 photoluminescence peak shift from 517 to 499 nm. The interaction glass/TiO_2/*gp*51 structure with specific antibodies against *gp*51 (anti-*gp*51) has shifted the photoluminescence peak backwards from 499 nm to 516 nm. These photoluminescence shifts are attributed to the variation of STE energy level, which was induced by changes of electrostatic interaction between adsorbed *gp*51 and negatively charged TiO_2 surface. The displacement of the light emitting recombination peak confirms that the energy of STE level is complex and has its ground and excited states. The blue-shift of the photoluminescence maximum by 18 nm as a result of adsorption of the *gp*51 protein, which corresponds to $\Delta E_{STE} = I_{STE2} - I_{STE1} = 0.086$ eV, indicates that the initial value of the potential barrier on the TiO_2 surface has decreased by a value of 0.086 eV. Variation of the potential barrier means that the value of negative charge localized on the TiO_2 surface has changed due to the charge-charge-based interaction with adsorbed protein *gp*51. Positively charged atoms and groups, provided by the *gp*51 protein, partially compensate the surface charge of TiO_2 and reduce the energy of electrons localized at the surface levels, which are the most responsible for the generation of photoluminescence signal.

The charge-charge-based interaction in the double charged layers TiO_2/*gp*51 can also be interpreted as a model of 'flat capacitor', formed as a result of the electrostatic interaction between oppositely charged protein *gp*51 layer and the TiO_2 surface. The capacitor is formed as a result of *gp*51 protein adsorption on TiO_2 surface, after which the charges are distributed in energetically most favorable way, partially compensating each other. Consequently, the positive 'imaginary capacitor plate' appears in the close proximity to TiO_2/*gp*51 interface, which is based on the positive charges of protein *gp*51, predominant after its adsorption on the TiO_2 surface. The positive charges are attracted closer to negatively charged surface due to the negative electrostatic effect of TiO_2. The interaction of TiO_2/*gp*51 with anti-*gp*51 antibodies and formation of *gp*51/anti-*gp*51-based immune complex leads to a deformation and reduction of charge in 'the positive imaginary capacitor plate', caused by redistribution and partial compensation of charges during the formation of the *gp*51/anti-*gp*51 immune complex, which in turn reduces the charge of 'the imaginary capacitor plate' based on *gp*51 adsorbed on the TiO_2 surface.

The highlighted origin of the changes in the photoluminescence spectra of TiO_2 as a result of the formation of biosensitive layer and after its interaction with the analyte, bring us closer to an understanding of the interaction mechanism between TiO_2 and proteins, that is the key in the solving of many issues related to an improvement of biosensor performance.

Acknowledgements This research was supported by Joint Ukrainian-Lithuanian research project "Application of hybrid nanostructures which are based on TiO_2 or ZnO and modified by biomolecules, in optoelectronic sensors", supported by Ministry of Education and Science of Ukraine (agreement No. M/20-2019) and Lithuanian Research Council (project No. P-LU-18-53).

References

1. A. Tereshchenko, M. Bechelany, R. Viter, V. Khranovskyy, V. Smyntyna, N. Starodub, R. Yakimova, Sensors Actuat B Chem **229**, 664 (2016)
2. A. Tereshchenko, V. Fedorenko, V. Smyntyna, I. Konup, A. Konup, M. Eriksson, R. Yakimova, A. Ramanavicius, S. Balme, M. Bechelany, Biosen Bioelectron **92**, 763 (2017)
3. R. Viter, A. Tereshchenko, V. Smyntynab, J. Ogorodniichuk, N. Starodub, R. Yakimova, V. Khranovskyy, A. Ramanavicius, Sensors Actuat B Chemical **252**, 95 (2017)
4. R. Plugaru, A Cremades and J Piqueras, J. Phys.: Condens. Matter 16 (2004)

5. J. Preclíková, P. Galář, F. Trojánek, S. Daniš, B. Rezek, I. Gregora, Y. Němcová, P. Malý, J. Appl. Phys. **108**, 113502 (2010)
6. X. Chen, S.S. Mao, Chem. Rev. **107**(7), 2891–2959 (2007)
7. A. Kumar Tripathi, M.K. Singh, M.C. Mathpal, S.K. Mishra, A. Agarwal, J Alloy Compd. **549**, 114 (2013)
8. X. Li, C. Gao, J. Wang, B. Lu, W. Chen, J. Song, S. Zhang, Z. Zhang, X. Pan, E. Xie, J. Power Sources **214**, 244–250 (2012)
9. A. Tereshchenko, V. Smyntyna, A. Ramanavicius, RSC Adv. **8**, 37740–37748 (2018)
10. R. Viter, V. Smyntyna, N. Starodub, A. Tereshchenko, A. Kusevitch, I. Doycho, S. Geveluk, N. Slishik, J. Buk, J. Duchoslav, J. Lubchuk, I. Konup, A. Ubelis, J. Spigulis, Procedia Eng. **47**, 338 (2012)
11. S. M. Gupta, M. Tripathi, Chin. Sci. Bull. **56**, 1639 (2011)
12. V. Smyntyna, *Electron and Molecular Phenomena on the Surface of Semiconductors* (Nova Publishers, New York, 2013)
13. D.L. Nelson, M.M. Cox, A.L. Lehninger, *Principles of Biochemistry* (Worth Publishtrs Inc., New York, 2000)
14. I. Sildos, A. Suisalu, V. Kiisk, M. Schuisky, H. Mändar, T. Uustare, J. Aarik, Proc. SPIE **4086**, 427
15. A. Tempczyk, C Kissinger, E. Villafranca, Catalytic *Mechanism of Calcineurin Serine* (Threonine Protein Phosphatase, Inc. 2010)
16. V. Shewale, P. Joshi, S. Mukhopadhyay, M. Deshpande, R. Pandey, S. Hussain, S. Karna, J. Phys. Chem. **115**, 10426 (2011)
17. G. Obal, F. Trajtenberg, F. Carrión, L. Tomé, N. Larrieux, X. Zhang, O. Pritsch, A. Buschiazzo, Science **5182**, 1–7 (2015)
18. T. Ogawa, *Biochemistry, Genetics and Molecular Biology, Volume 'Protein Engineering—Technology and Application'* (ISBN 978-953-51-1138-2)
19. Z. Balevicius, I. Baleviciute, S. Tumenas, L. Tamosaitis, A. Stirke, A. Makaraviciute, A. Ramanaviciene, A. Ramanavicius, Thin Solid Films **571**, 744 (2014)
20. I. Baleviciute, Z. Balevicius, A. Makaraviciute, A. Ramanaviciene, A. Ramanavicius, Biosen Bioelectron **39**, 170 (2013)
21. K. Wakabayashi, Y. Yamaguchi, T. Sekiya, S. Kurita, J. Lumin. **112**, 50 (2005)
22. Z. Balevicius, A. Makaraviciute, G.J. Babonas, S. Tumenas, V. Bukauskas, A. Ramanaviciene, A. Ramanavicius, Sensors Actuat B Chemical **181**, 119–124 (2013)

Chapter 26
Electrical and Photoelectric Properties of Iron/Chromium Oxide Nanolayers Composite Structures

I. Demchenko, S. Mulenko, A. Smirnov, R. Savkina, M. Walczak, and N. Voloshin

Abstract We report here results of our investigation of oxide nanometric films integrated with silicon substrates for new multifunctional applications. Ultraviolet photons of KrF-laser (248 nm) was used for the synthesis of nanometeric films based on iron and chromium oxides ($Fe_2O_{3-X}(0 \leq x \leq 1)$ and $Cr_{3-x}O_{3-y}(0 \leq x \leq 2; 0 \leq y \leq 2)$) with variable thickness, stoichiometry and electrical properties. X-ray photoelectron spectroscopy was used for nanometric films characterization. The roughness value of the samples investigated shows a good quality of the finishing characteristics of films synthesized by the reactive pulsed laser deposition technique. Investigation of the charge carriers' transport kinetics was carried out by surface photovoltage (SPV) and impedance spectroscopy technique. It was found that effect of the iron and chromium oxide layers' combination results in the photovoltage enhancement of 500 times in the spectral region between 600 and 1150 nm. The effect of the long-term relaxation of the photovoltage was revealed also.

I. Demchenko
Faculty of Chemistry, University of Warsaw, Warsaw, Poland

S. Mulenko
G. Kurdyumov Institute of Metal Physics NAS of Ukraine, Kiev, Ukraine

A. Smirnov · R. Savkina
V. Lashkaryov Institute of Semiconductor Physics NAS of Ukraine, Kiev, Ukraine

M. Walczak
School of Materials, University of Manchester, Manchester, UK

N. Voloshin (✉)
National Technical University of Ukraine "Igor Sikorsky Kyiv Polytechnic Institute", Kiev, Ukraine
e-mail: nick.voloshka@gmail.com

A. D. Pogrebnjak et al. (eds.), *Nanomaterials in Biomedical Application and Biosensors (NAP-2019)*, Springer Proceedings in Physics 244,
https://doi.org/10.1007/978-981-15-3996-1_26

26.1 Introduction

Transition-metal oxides are considered to be very fascinating functional materials [1]. Their unique features involve high dielectric constants, efficient charge separation, enhanced surface reactivity as well as magnetization and polarization properties. Applications of composite structures including two or more transition-metal oxides allow even wider diversity in their electronic properties and chemical behavior. For example, in single-phase multiferroic metal-oxide based materials, the magnetoelectric coupling is very weak and the ordering temperature is too low [2]. In contrast, multiferroic composites incorporated ferroelectric and ferromagnetic phases are characterized with giant magnetoelectric coupling response above room temperature. Besides, it is well-known that a variety of changes in both chemical and physical properties of materials under consideration occur in the nanoscale area. Therefore, transition metal-oxides based nanocomposite structures research is still relevant and important in terms of the practical application of the ones.

Isostructural and isovalent iron and chromium oxides are the subject of active experimental and theoretical investigations as members of the transition metal oxide family [3–8]. The polycrystalline α-Fe_2O_3 and Cr_2O_3 under normal conditions crystallize in the lattice with R3c space group symmetry. Thermodynamically, hematite (α-Fe_2O_3) is the most stable in the family of iron (III) oxides. It is a charge transfer insulator with a band gap of $\sim$2.1 eV [6], absorbs $\sim$40% of the solar spectrum [5], and conducts when doped with Ti [4]. The Cr_2O_3 compound has an eskolaite-like structure. It is one of the most important wide band gap (Eg $\approx$ 3.3 eV) p-type semiconductor transition metal-oxide material with efficient dielectric properties. Up to now, Cr_2O_3 has been most promising material for realistic applications close to room temperature in magnetoelectric-controlled spintronic elements like MERAM.

Multilayer structures based on these materials exhibit multiferroic behaviors. We have previously demonstrated magnetic hysteresis and magnetoresistive memory effect for hybrid systems of the alternate layers $Fe_2O_{3-x}(0 \leq x \leq 1)/Cr_{3-X}O_{3-Y}(0 \leq x \leq 2; 0 \leq y \leq 2)$ [9]. Besides, interest in understanding such materials is being renewed because of their distinctive properties including enhanced surface reactivity and efficient charge separation.

Important properties of iron oxide are its low-toxicity, Earth abundant and suitable redox potential for photocatalytic water dissociation as a source of H_2 fuel [7]. But, one of the major limiting factors affecting the utilization of hematite as a photocatalysts include fast electron–hole recombination. This can be improved by the development of the built-in electric fields and charge carriers' separation by exploitation the band offset properties of Fe_2O_3/Cr_2O_3 heterojunctions. Theoretical justification and experimental studies can be found in the following [3, 8] respectively. The Fe_2O_3/Cr_2O_3 heterostructure is known to have a type-II band alignment, where valence band maximum in Fe_2O_3 is lower than that of Cr_2O_3 [8]. This feature results in an occurrence of the potential gradient over several periods of the multilayer structure that may be useful for enhanced effectiveness in spatially separating electrons and holes.

In the present work we report on the studies of the electrical and photoelectric properties of nanometric films of iron and chromium oxides ($Fe_2O_{3-x}(0 \leq x \leq 1)$, $Cr_{3-X}O_{3-Y}(0 \leq x \leq 2; 0 \leq y \leq 2)$) as well as nanocomposite structures of the alternate layers $Fe_2O_{3-x}(0 \leq x \leq 1)/Cr_{3-X}O_{3-Y}(0 \leq x \leq 2; 0 \leq y \leq 2)$ synthesized on silicon substrates by reactive pulsed laser deposition (RPLD) method. Here, we use surface photovoltage and impedance spectroscopy technique for investigation of charge carriers' transport in this material.

26.2 Experiment Details

26.2.1 Materials

We have carried out a study of nanometric films based on oxides of the transitional metals with variable thickness, stoichiometry and electrical properties. Chromium $Cr_{3-X}O_{3-Y}$ $(0 \leq x \leq 2; 0 \leq y \leq 2)$ and iron ($Fe_2O_{3-X}(0 \leq x \leq 1)$ oxide layers as well as their combination ($Fe_2O_{3-Y}/Cr_{3-X}O_{3-Y}/Fe_2O_{3-Y}/Cr_{3-X}O_{3-Y}$) were grown by RPLD techniques in the vacuum stainless steel reactor on the boron-doped (100)-oriented p-type silicon wafers at the substrate's temperature $T_S = 293$ and 800 K. The development and improvement of RPLD technology for the production of nanometric films of iron and chromium oxides is shown in [10, 11]. The thickness of the films (d) was controlled by "Tensor Instruments" model "Alpha-step 100" profilometer with an accuracy of 5%.

26.2.2 Methods

The samples were studied by X-ray photoelectron spectroscopy (XPS) using Kratos Ultra Axis spectrometer with monochromatic Al Kα radiation ($h\upsilon = 1.4866$ keV) from an X-ray source with a spot of $700 \times 300\ \mu m^2$ (due to the analyzer settings in hybrid mode) while operating at 150 W and 15 kV. The high-resolution (HR) XPS spectra were collected with the hemispherical analyzer at the pass energy of 20 eV and the energy step size of 0.1 eV. The photoelectron take-off angle was 0° with respect to the normal to the sample's surface plane. No charge compensation was applied. Samples were mounted on the grounded holder. Binding energies (BEs) of the photoelectrons were calibrated using the gold $4f_{7/2}$ photoelectron peak at 84 eV. The Casa XPS software (version 2.3.17) [12] was used to evaluate the XPS data. Taking into account that sputtering procedure can easily modify electronic structure of studied samples due to intermixing of atoms (see, e.g., [13]), it was decided to abandon sputtering in this study.

All samples were characterized by measuring their surface photovoltage (SPV) spectra. These measurements were performed at 300 K on a spectrometer SPM-2 with a G-60 prism using a lock-in detection scheme with modulation at 200 Hz at low level of homogeneous excitation by a monochromatic light in a wavelength range of 500 ÷ 1200 nm. The light source was a quartz halogen lamp JCD (950 lm; 50 W). A photovoltage arising was recorded without electrical bias by lock-in nanovoltmeter. Contact metallization was photolithographically patterned on the structure top. Impedance characteristics of the samples were studied using the precision impedance meter Z-3000X within the frequency range 1 Hz…3 $\times$ 10^6 Hz with the amplitude of a sinusoidal signal 120 mV.

26.3 Results and Discussion

Our previous X-ray diffraction (XRD) investigation evidenced the polycrystalline structure of the metal oxide films deposited on Si substrate in oxygen atmosphere [14]. Moreover, XRD analysis revealed strong correlation between oxide structural properties and technological parameters. The increase of oxygen pressure in the chamber caused a kinetic energy loss of iron atoms. Correspondingly, the cooling time of films on substrate decreases and the crystallization status is worsening. This accompanied by the decrease of the line intensity in XRD iron oxides spectra. On the other hand, in case of substrates heated at 800 K, the iron oxide lines in XRD spectra are enhanced, the kinetic energy of the ions is increased, resulting in the growth of a new iron oxide phase. Chromium oxide films composition depends strongly on oxygen content in the gas phase and mainly composed of Cr_2O_3 which is the most stable phase in the Cr–O phase diagram [11].

26.3.1 Surface Electronic Structure Characterization

XPS was used to measure of Fe 3p/3s, Cr 3p/3s, and O 2s core levels. Figure 26.1a shows electron spectra partially covering the valence band region up to Fe 3s line at 93.6 $\pm$ 0.2 eV for standard and investigated samples. Spin exchange splitting of the Fe 3s and Cr 3s states is clearly seen in the experimental data. The exchange interaction between cores electrons and those in incomplete shells, here 3d electrons, gives rise to multiplet splitting of the s-electrons themselves [15]. The estimated Cr 3s multiplet splitting for examined samples is the same and equals to 4 eV (BE of the main resonance is 75.4 $\pm$ 0.2 and of the second one is 79.4 $\pm$ 0.2 eV). Obtained values agree well with data shown by Battistoni et al. [16] representing Cr_2O_3 phase. The Cr 3p BE is 43.6 $\pm$ 0.2 eV for both examined samples, and according to [17], it is associated with Cr_2O_3 phase as well. It is clearly seen (see "Mult.lay.#2" data in Fig. 26.1a) that the loss feature of Cr 3p (marked by blue asterisk) superimposes with the Fe 3p states. Nevertheless, the Fe 3p/3s lines development (i.e., intensity of signal dependence for multilayer samples) makes it clearthat the thickness of the top layer of chromium oxide is greater for the sample "Mult.lay.#1". Opposite is true for the

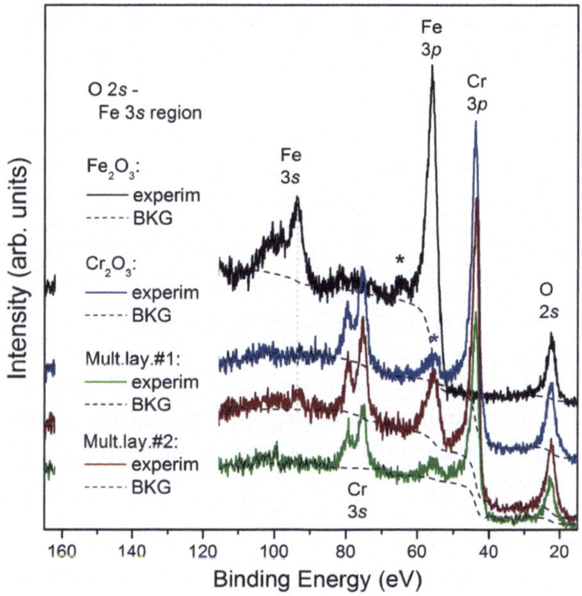

Fig. 26.1 XPS spectra of the Fe3*s*, Cr3*s*, Fe3*p*, Cr3*p*, and O2*s* states measured for Fe_2O_3 (PO_2 = 0.5 Pa, d = 26 HM, N = 5000), Cr_2O_3 (PO_2 = 0.5 Pa, d = 55 HM, N = 5000), and both "Cr_2O_3/Fe_2O_3" multilayer structures: "Mult.lay.#1" (PO_2 = 0.5 Pa, d = 100 HM, N = 2500/2000/2500/2000), "Mult.lay.#2" (PO_2 = 0.5 Pa, d = 10 HM, N = 500/500/500/500). Here and further N is number of the laser impulses. Dashed lines show background signal that was removed before the analysis

sample "Mult.lay.#2", where contribution of Fe 3p/3s lines is difficult to distinguish. Deconvolution of both lines give the flowing BE values: 93.8 ± 0.2 eV for Fe 3s (whereas multiplet splitting is estimated to be 7.3 eV) and 55.6 ± 0.2 eV for Fe 3p. To the same precision, both of them correspond to Fe_2O_3 [17, 18]. Unfortunately, it was found that BEs of the 3d/s lines for Fe_2O_3 are the same or close to Fe_3O_4 phase (there are no differences in BE of the Fe 3s line and only 0.2 eV for the Fe 3p line [19]). Consequently, as a next step, we would like to check the Fe 2p line for the mentioned above compounds noting that the difference of the Fe 2p3/2 resonances in standard (Fe_2O_3 vs. Fe_3O_4) samples is 0.8 eV. The results of this study will be published elsewhere. In any case, the results obtained by XPS at present stage confirm the conclusion of the XRD studies [11, 14] on these samples, namely, existence of Cr_2O_3 and Fe_2O_3 phases.

26.3.2 *Kinetics and Spectral Distribution of the Surface Photovoltage*

All samples were characterized by measuring their surface photovoltage (SPV) spectra. Surface photovoltage is defined as the difference between the surface potential of a sample in the dark and under illumination. A nonzero SPV indicates redistribution of photogenerated free charges. As a modulated illumination was used in our experiment, the surface photovoltage value can be expressed as [20]:

$$V_{\mathrm{PV}}(\omega) = J_{\mathrm{PC}}(\omega) \cdot Z_{\mathrm{eff}}(\omega),$$

where the photocurrent density—J_{PC} is typically extracted from a current-balance calculation and the effective impedance—Z_{eff} is composed of parallel resistance (R_j) and capacitance (C_j) components, corresponding to the contributions of the various photovoltage influencing agents such as minority and majority carrier transport, depletion and inversion layers, surface states, etc.

Figure 26.2 shows the SPV spectra of Si-based multi-component composite systems investigated. SPV spectrum of Si substrate in the original state (not shown) exhibits a selective peak with the spectral position of the ‘red’ boundary, which corresponds with the silicon band gap Eg = 1.158 eV (300 K). The SPV spectra of $Cr_{3-X}O_{3-Y}$ ($0 \leq x \leq 2$; $0 \leq y \leq 2$) nanometric film on Si substrate exhibits a selective peak with maximum corresponding to the silicon band gap Eg at 300 K. It is obvious that chromium oxide layer is transparent in this spectral region. The measured value of the SPV_{max} of “iron oxide on silicon” structure at light power density ~1 W cm^{-2} is about ~4 μV (see Fig. 26.2b). At the same time, it was found a surprising improvement in the photoelectric properties of multilayer systems. A broad band of the photosensitivity between 600 and 1150 nm with a high value of the photovoltage ~2×10^3 μV is typical for $Cr_{3-x}O_{3-y}/Fe_2O_{3-y}/Cr_{3-x}O_{3-y}/Fe_2O_{3-y}$ on Si structures (Fig. 26.3c).

It should be noted, that an increase in the photosensitivity in the visible range showed samples grown on a cold substrate. At the same time, samples grown on a hot substrate (800 K) have exhibited photoluminescence phenomenon in the range 500–600 nm.

The photovoltage kinetics was studied by the example of this structure. The typical time dependence of the photovoltage signal for $Cr_{3-x}O_{3-y}/Fe_2O_{3-y}/Cr_{3-x}O_{3-y}/Fe_2O_{3-y}$ on Si structure excited by radiation with quantum energy 2.53 eV is demonstrated in Fig. 26.2d. The effect of the long-time photorelaxation was found. Photovoltage increases and reaches a maximum within 12 min. The total time of the photorelaxation is about 120 min. The photovoltage signal V_{PV} kinetics can be described by the dependence:

$$V_{\mathrm{PV}}(t) = A_{1,2}\left(1 - b_{1,3}\exp\left(-t/\tau_{1,3}\right) - b_{2,4}\exp\left(-t/\tau_{2,4}\right)\right)$$

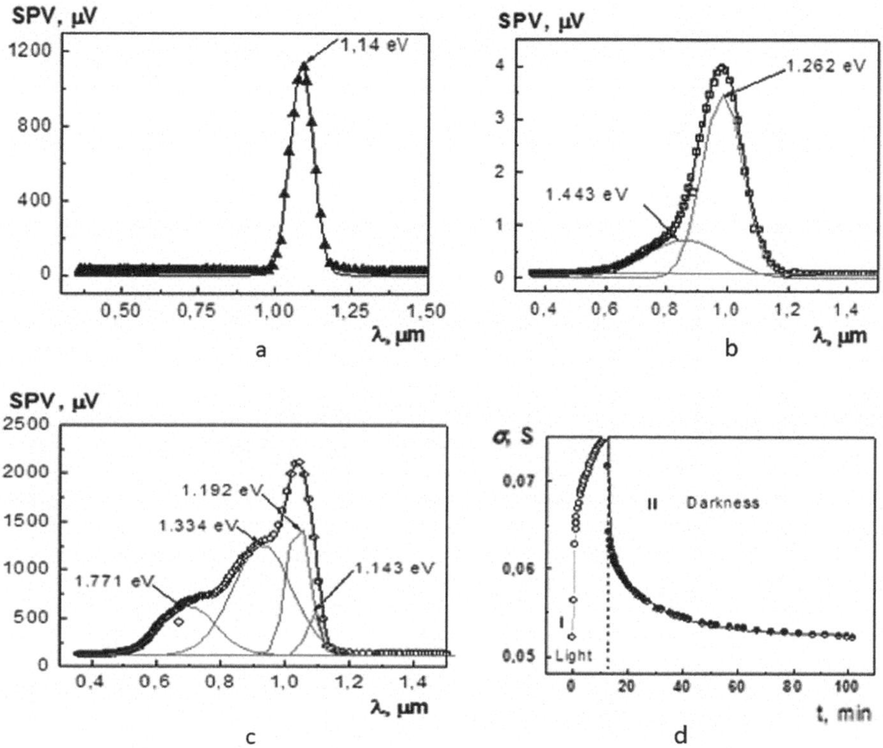

Fig. 26.2 Surface photovoltage spectra of nanometric layers structure deposited on Si substrate corrected on spectral response of setup: **a** $Cr_{3-x}O_{3-y}$/Si, **b** Fe_2O_{3-x}/Si, **c** Mult.lay.#2 ($Cr_{3-x}O_{3-y}/Fe_2O_{3-y}$) × 2 on Si. All SPV spectra have complicated shape that could be described by a sum of number components. In order to determine positions of spectral components, we have used a deconvolution procedure with fitting to Gauss functions. The peak positions are presented in figures. The dots indicate the experimental data and the solid lines present the results of the fitting procedure. The thin-dash lines show approximation. **d** kinetics of photovoltage of multilayer systems excited by blue LED ($\lambda_D = 475$ nm, the electrical power 1 W) in uninterrupted regime: I—region of the photoexcitation; II—region of the relaxation

where $\tau_1 = 0.51$ min, $\tau_2 = 9.03$ min are the characteristic times of the photoexitation (region I in Fig. 26.2d) and $\tau_3 = 0.92$ min, $\tau_4 = 19.05$ min are the characteristic times of the relaxation (region II in Fig. 26.2d). It should be noted that long-term relaxation is typical for wide-gap oxides such as ZnO or SnO_2 [21] and is associated with stoichiometric oxygen deficiency in these compounds. Another view at the long-term relaxation of the photoexitation in multiferroic materials ($BiFeO_3$) is presented in [22], where a photo-induced depolarization with excited localized states was proposed and first-principles calculations of double-potential well were provided. Anyway, to determine the mechanism of long-term relaxation in our case, additional studies are needed.

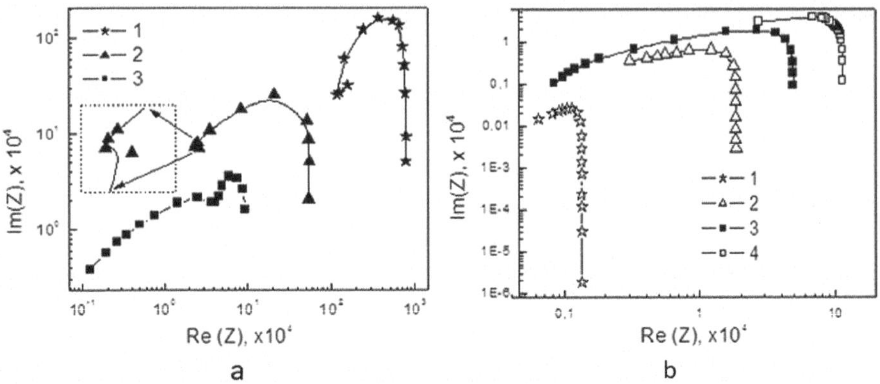

Fig. 26.3 Nyquist plots for **a** $Cr_{3-X}O_{3-Y}$ ($0 \leq x \leq 2$; $0 \leq y \leq 2$)) nanometric films on Si substrate: 1—$PO_2 = 1$ Pa, d = 160 HM, N = 5000; 2—$PO_2 = 0.05$ Pa, d = 40 HM, N = 4000; 3—$PO_2 = 0.5$ Pa, d = 55 HM, N = 5000 and **b** multilayer structures ($Cr_{3-x}O_{3-y}/Fe_2O_{3-y}$)/Si and ($Cr_{3-x}O_{3-y}/Fe_2O_{3-y}$) × 2/Si: 1—$PO_2 = 0.5$ Pa, d = 26 HM, N = 5000; 2—$PO_2 = 0.5$ Pa, d = 100 HM, N = 2500/2000/2500/2000; 3—$PO_2 = 0.5$ Pa, N = 3000/3000, d = 80 HM, $T_{sub} = 293$ K; 4—$PO_2 = 0.5$ Pa, N = 3000/3000, d = 80 HM, $T_{sub} = 800$ K. The impedance measurements were performed at the G-Cp (parallel conductance and capacitance) configuration using Au plates as blocking electrodes. The dots in the figure are experimental data and the solid line is fitting curves obtained by EIS spectrum analyzer

26.3.3 Impedance Spectroscopy

It is known that a semiconductor structure with a potential barrier, in the simplest case, can be considered within the context of an equivalent circuit consisting of the resistance of the depletion region and the barrier capacitance connected in parallel and the resistance connected in series to them. Electrical Impedance Spectroscopy (EIS) is a very sensitive method for research of non-stationary charge transport governed by charge-carrier relaxation in semiconductors structure and development an equivalent circuit. EIS technique has been used for characterization the electron transport kinetics of $Cr_{3-X}O_{3-Y}$($0 \leq x \leq 2$; $0 \leq y \leq 2$), Fe_2O_{3-X}($0 \leq x \leq 1$), ($Cr_{3-x}O_{3-y}/Fe_2O_{3-y}$) and 2 × ($Cr_{3-x}O_{3}$-y/$Fe_2O_{3-y}$) structures on Si substrate prepared by RPLD.

Figure 26.3 Nyquist plots for (a) $Cr_{3-X}O_{3-Y}$ ($0 \leq x \leq 2$; $0 \leq y \leq 2$) nanometric films and (b) multilayer $Cr_{3-x}O_{3-y}/Fe_2O_{3-y}$ structures on Si substrate. The impedance plane plots for multilayer $Cr_{3-x}O_{3-y}/Fe_2O_{3-y}$ structures on Si substrate (Fig. 26.3b, curve 2, 3, 4) as well as single-layer samples of Fe_2O_{3-y} (Fig. 26.3b, curve 1) and $Cr_{3-x}O_{3-y}$ (Fig. 26.3a, curve 3) are in the shape of one or two semicircles, which can be well interpreted by an equivalent circuit composed of resistors R and constant phase element CPE in parallel. One semicircle corresponds to one chain R-CPE, two semicircles—sequences of two chains R-CPE in series. CPE element accounts for a disorder in the charge diffusion dynamics by material inhomogeneity, which can be typically expected in a polycrystalline material.

At the same time, Nyquist plots for $Cr_{3-X}O_{3-Y}$ ($0 \leq x \leq 2; 0 \leq y \leq 2$) nanometric films obtained under $PO_2 = 1$ Pa (Fig. 26.3a, curve 1) and $PO_2 = 0.05$ Pa (Fig. 26.3a, curve 2) are characterized by the presence of a loop at the higher frequency side graphics that corresponds to the inductive-type impedance of these material (see Inset in Fig. 26.3a).

26.4 Conclusions

In summary, we have investigated the light-induced phenomena in nanometric films based on iron and chromium oxides ($Fe_2O_{3-X}(0 \leq x \leq 1)$ and $Cr_{3-X}O_{3-Y}(0 \leq x \leq 2; 0 \leq y \leq 2)$) as well as composite structures of the alternate layers $Fe_2O_{3-X}(0 \leq x \leq 1)/Cr_{3-X}O_{3-Y}(0 \leq x \leq 2; 0 \leq y \leq 2)$ synthesized on silicon substrates by reactive pulsed laser deposition. It was found that "chromium oxide on silicon" structure is transparent and "iron oxide on silicon" structure exhibits weak photosensitivity in the wavelength range of $500 \div 1200$ nm. At the same time, it was found a surprising improvement in the photoelectric properties of multilayer systems. Effect of the nanometric oxide (Cr, Fe) layers combination results in the photovoltage enhancement of 500 times. Moreover, the effect of long-term relaxation with a total time of the photorelaxation up to 2 h characterizes such multilayer structure. The equivalent electrical circuit model was obtained by EIS technique for structures investigated.

We think that integration silicon as an industrial key material and oxide-based composite nanostructures characterized by a close coupling of magnetization and polarization properties (such as transitional metal oxides Fe_2O_3, Cr_2O_3) will help in the generation of the new class of sensing devices which multifunctionality can be controlled not only by applied magnetics or electric fields but also by light.

Acknowledgements This work is done within Ukrainian–Polish joint research project "Characterization of the hybrid system of the nanometric layers based on transition metals oxides useful for spintronic" under the agreement on scientific cooperation between the Polish Academy of Sciences and the National Academy of Sciences of Ukraine.

References

1. S. Maekawa, T. Tohyama, S.E. Barnes, S. Ishihara, W. Koshibae, G. Khaliullin, *Physics of transition metal oxides*, vol. 144 (Springer, Verlag Berlin Heidelberg, 2004)
2. N.A. Hill, J. Phys. Chem. B **104**, 6694–6709 (2000)
3. T.C. Kaspar, D.K. Schreiber, S.R. Spurgeon, M.E. Mc Briarty, G.M. Carroll, D.R. Gamelin, et al., Adv. Mater. **28**(8), 1616–1622 (2016)
4. B. Zhao, T.C. Kaspar, T.C. Droubay, J. McCloy, M.E. Bowden, V. Shutthanandan et al., Phys. Rev. B **84**, 245325 (2011)
5. M.N. Huda, A. Walsh, Y. Yan, S. H. Wei, M.M. Al-Jassim, J. Appl. Phys. **107**(12), 123712 (2010)

6. H. Mashiko, T. Oshima, A. Ohtomo, Appl. Phys. Lett. **99**(24), 241904 (2011)
7. T.W. Hamann, Dalton Trans. **41**(26), 7830–7834 (2012)
8. S.A. Chambers, Y. Liang, Y. Gao, Phys. Rev. B **61**(19), 13223 (2000)
9. A.B. Smirnov, S.B. Kryvyi, S.A. Mulenko, M.L. Sadovnikova et al., Nanoscale Res. Lett. **11**(1), 467 (2016)
10. S.A. Mulenko, YuN Petrov, N.T. Gorbachuk, Appl. Surf. Sci. **258**(23), 9186–9191 (2012)
11. A.P. Caricato, A. Luches, M. Martino, D. Valerini, Y.V. Kudryavtsev et al., J. Optoelecron. Adv. Mater. **12**(3), 427–431 (2010)
12. N. Fairley, *Software Package for the Analysis of XPS Results, Casa XPS Version 2.3.17 dev6.6o* (Casa Software Ltd.). http://www.casaxps.com
13. I.N. Demchenko, W. Lisowski, Y. Syryanyy, Y. Melikhov, I. Zaytseva et al., Appl. Surf. Sci. **399**, 32–40 (2017)
14. N. Serban, C. Ristoscu, G. Socol, N. Stefan et al., Mater. Res. Bull. **50**, 148–154 (2014)
15. G.K. Wertheim, S. Hufner, H.J. Guggenheim, Compounds. Phys. Rev. B **7**(1), 556 (1973)
16. C. Battistoni, J.L. Dormann, D. Fiorani, E. Paparazzo, S. Viticoli, Solid State Commun. **39**(4), 581–585 (1981)
17. G.C. Allen, S.J. Harris, J.A. Jutson, J.M. Dyke, Appl. Surf. Sci. **37**(1), 111–134 (1989)
18. https://srdata.nist.gov/xps/acknowledgment.aspx
19. P. Mills, J.L. Sullivan, J. Phys. D Appl. Phys. **16**(5), 723 (1983)
20. L.W. Martin, Y.H. Chu, R. Ramesh, Mater. Sci. Eng. Rep. **68**(4–6), 89–133 (2010)
21. Tianyou Zhai, Xiaosheng Fang, Meiyong Liao et al., Sensors **9**, 6504–6529 (2009)
22. Gu Jun-xing, Kui-juan Jin, Le Wang et al., J. App. Phys. **118**(20), 204103 (2015)

Subject Index

A. D. Pogrebnjak et al. (eds.), *Nanomaterials in Biomedical Application and Biosensors (NAP-2019)*, Springer Proceedings in Physics 244,
https://doi.org/10.1007/978-981-15-3996-1

The manufacturer's authorised representative in the EU is Springer Nature Customer Service Centre GmbH, Europaplatz 3, 69115 Heidelberg, Germany. If you have any concerns regarding our products, please contact ProductSafety@springernature.com

Printed and bound by CPI Group (UK) Ltd, Croydon, CR0 4YY

30/09/2026

02225507-0011